Biomass Sugars for Non-Fuel Applications

RSC Green Chemistry

Editor-in-Chief:
Professor James Clark, *Department of Chemistry, University of York, UK*

Series Editors:
Professor George A. Kraus, *Department of Chemistry, Iowa State University, Ames, Iowa, USA*
Professor Andrzej Stankiewicz, *Delft University of Technology, The Netherlands*
Professor Peter Siedl, *Federal University of Rio de Janeiro, Brazil*

Titles in the Series:

How to obtain future titles on publication:
A standing order plan is available for this series. A standing order will bring delivery of each new volume immediately on publication.

For further information please contact:
Book Sales Department, Royal Society of Chemistry, Thomas Graham House, Science Park, Milton Road, Cambridge, CB4 0WF, UK
Telephone: +44 (0)1223 420066, Fax: +44 (0)1223 420247
Email: booksales@rsc.org
Visit our website at www.rsc.org/books

Biomass Sugars for Non-Fuel Applications

Edited by

Dmitry Murzin
Abo Akademi University, Turku, Finland
Email: dmurzin@abo.fi

Olga Simakova
Georgia Institute of Technology, Atlanta, GA, USA
Current address: Amgen Manufacturing, Juncos, PR, USA
Email: simakova@amgen.com

THE QUEEN'S AWARDS
FOR ENTERPRISE:
INTERNATIONAL TRADE
2013

RSC Green Chemistry No. 44

Print ISBN: 978-1-78262-113-3
PDF eISBN: 978-1-78262-207-9
ISSN: 1757-7039

A catalogue record for this book is available from the British Library

Published by The Royal Society of Chemistry,
Thomas Graham House, Science Park, Milton Road,
Cambridge CB4 0WF, UK

Registered Charity Number 207890

For further information see our web site at www.rsc.org

Printed in the United Kingdom by CPI Group (UK) Ltd, Croydon, CR0 4YY, UK

Preface

The search for a sustainable source of chemicals as alternatives to petroleum has led to the discovery of multiple routes for renewable biomass conversion into biofuels and value-added platform chemicals. Extensive research in this area has been focused primarily on the diversification of energy resources.

In fact, the amount of biomass available currently is too limited to address all the demands with respect to fuels. Thus *ca.* 30% of the global arable land is needed to cover only 10% of the global fuel demand by 2030. It should be noted that currently the majority of oil is used for the production of fuels, while only 5–8% (5.8 in 2012) of a crude oil barrel is used in the manufacture of chemicals, while the turnover in monetary value is almost the same for fuels and chemicals.

Since the price of fuels is much lower than for chemicals, in the future, limited resources should be used mainly for chemicals, while the growing energy demand must be compensated for by alternative energy sources (solar, hydropower, nuclear, *etc.*).

The major components of lignocellulosic biomass are cellulose (40–50%), lignin (16–33%) and hemicelluloses (15–30%), where the last two are practically built up of sugar units. Therefore, biomass is a rich feedstock for the various sugars and their derivatives.

Sugar derivatives were considered as the top twelve value-added chemicals from biomass by the U.S. Department of Energy (2004). Extensive research was performed to advance the technology of cellulosic biomass transformation into biofuels. However, sugar-based chemicals are also a green and feasible source for the sustainable manufacturing of a variety of valuable products including polymers, surfactants, pharmaceuticals, and others.

The objective of this book is to represent the scope of green sugar-based technologies beyond fuels. The book gives an overview of the current status of sugar-based technology describing the challenges and opportunities with

RSC Green Chemistry No. 44
Biomass Sugars for Non-Fuel Applications
Edited by Dmitry Murzin and Olga Simakova
© The Royal Society of Chemistry 2016
Published by the Royal Society of Chemistry, www.rsc.org

the synthesis of valuable chemical commodities. Such synthesis starts with the so-called biorefinery concept, which would allow the production of chemicals (and fuels) in the same way as it is done nowadays in classical refineries and large (petro)chemical complexes.

One of the challenges in the valorization of lignocellulosic cellulose is related to the primary fractionation. This is addressed in the chapter of Seidl and co-authors. The same authors also describe the Brazilian sugar cane industry, which is of high importance for making fuels and has a strong potential for the synthesis of chemicals. The history and current status of Brazilian sugar cane transformation into chemicals are presented as a real life case study.

The work of Fardim and co-workers focuses on hemicelluloses, in particular xylans, which are now the most studied hemicelluloses. The most effective extraction methods, as well as potential applications of xylan-based materials for fibre-surface engineering, are presented and discussed.

Leino and co-workers describe the application of oligosaccharides as pharmaceuticals, while recent advances in the synthesis of sugar-based surfactants are considered by Kovensky and Grand.

The chapter of Bhaumik and Dhepe gives a comprehensive overview of multiple strategies to design and develop various catalysts and catalytic processes to hydrolyse saccharides (cellulose, hemicelluloses) into sugars.

Further catalytic transformations of sugars, namely hydrogenation to sugar alcohols and aqueous-phase reforming of sugar derivatives, are overviewed respectively by Murzin *et al.* and Seshan and co-workers.

The editors hope that this book will be helpful to scientists working for academia and industry, who are primarily focused on the development of non-fuel applications of sugars.

Contents

RSC Green Chemistry No. 44
Biomass Sugars for Non-Fuel Applications
Edited by Dmitry Murzin and Olga Simakova
© The Royal Society of Chemistry 2016
Published by the Royal Society of Chemistry, www.rsc.org

Chapter 3   Catalytic Hydrogenation of Sugars   89
Dmitry Yu Murzin, Angela Duque, Kalle Arve, Victor Sifontes,
Atte Aho, Kari Eränen and Tapio Salmi

CHAPTER 1

Conversion of Biomass into Sugars

PRASENJIT BHAUMIK AND PARESH LAXMIKANT DHEPE*

Catalysis & Inorganic Chemistry Division, CSIR-National Chemical Laboratory, Dr Homi Bhabha Road, Pune 411008, India
*Email: pl.dhepe@ncl.res.in

1.1 Introduction

In the current circumstances, fossil feedstocks (crude oil, coal and natural gas) are utilized for the synthesis of a range of chemicals and fuels. Yet, their sustainability is at stake due to finite reserves, sporadic prices, volatile geopolitical scenarios and unfavourable effects on the environment (global warming) because of the discharge of a major contributor to the greenhouse gas effect, carbon dioxide (CO_2) into the atmosphere.[1] During World Wars I and II, due to a shortage of crude oil, Germany and a few other countries started extensive research on the production of chemicals and fuels (particularly ethanol and diesel) from alternate sources such as coal and biomass.[2] The world's first ethanol production plant (Skutskär sulfite ethanol plant), based on the sulfite process, was started in 1909 in Sweden.[3] Although a total of 33 plants were started using the same concept in Sweden, since 1983, just one plant has remained operational.[3] After the development of efficient ways throughout the 20th century to explore, extract and process crude oil, research on biomass was decreased. But, following the recent crisis in oil production and for geo-political reasons, there has been a renewed interest in looking for alternative sources for the synthesis of chemicals and fuels. Though, for a long time, Brazil has successfully shown

RSC Green Chemistry No. 44
Biomass Sugars for Non-Fuel Applications
Edited by Dmitry Murzin and Olga Simakova
© The Royal Society of Chemistry 2016
Published by the Royal Society of Chemistry, www.rsc.org

that due to the highest world production of sugarcane (Brazil: 3.3×10^8–7.7×10^8 ton per year in 2000–2013, World: 1.3×10^9–1.9×10^9 ton per year in 2000–2013),[4] it can produce bio-ethanol from bagasse (sugarcane waste after extracting sugar juice) in large quantities for public distribution to run vehicles.[5] Conversely, in the rest of the world, after numerous deliberations and considering history, recently, it has been suggested that the only alternative and sustainable resource, biomass should be leveraged for the synthesis of chemicals and fuels by developing environmentally benign pathways. Since biomass is renewable, carbon neutral, abundant, locally accessible in most countries and has a lower impact on the environment, it becomes a natural choice as an alternate resource.[1,6] In recent times, several countries and industries have disclosed their interests in developing methods for the conversion of biomass into known and new chemicals and fuels.[1] Biomass is a non-fossil and is made up of complex molecules present in plants and animals. It is considered as a rich source of organic products, which have a characteristic chemical composition of C, H, O, N.[7] However, until now, much of the work has reported on the conversion of plant-derived biomass into chemicals. Naturally, plant biomass is produced during the photosynthesis pathway using water, carbon dioxide and sunlight and is classified into two categories, namely edible and non-edible, solely based on human consumption ability. For example wheat, rice, corn, potato *etc.* are made up of a polysaccharide, starch and are considered as edible biomass or a first generation raw material (for the synthesis of fuels and chemicals). Starch is composed of a mixture of linear polysaccharide, amylose (homopolymer of D-glucose linked *via* a α-1,4 glycosidic bond) and branched polysaccharide, amylopectin (homopolymer of D-glucose linked *via* a linear α-1,4 glycosidic bond and branched α-1,6 glycosidic bond). Non-edible biomass, for example crop waste or wood, is called lignocellulosic biomass or lignocelluloses and is considered as a second generation raw material. Lignocelluloses have a composition of *ca.* 45% cellulose (homopolymer or homopolysaccharide of D-glucose linked *via* a β-1,4 glycosidic bond), *ca.* 25% hemicellulose (heteropolymer or heteropolysaccharide of several C_5 and C_6 sugars linked *via* various bonds), *ca.* 20% lignin (amorphous 3D network polymer of several aromatic monomers), some macro and micro nutrients (nitrogen, phosphorus, potassium, calcium, magnesium, sulfur, iron, manganese, copper, boron, zinc, chloride and molybdenum) and extractives (fats, fatty acids, resins, tannins, volatile oils, proteins *etc.*).[7,8] Typically, saccharides or carbohydrates (hydrates of carbon) have a molecular formula of $C_m(H_2O)_n$, where m and n are almost same. For instance, a simple monosaccharide, glucose has a molecular formula of $C_6H_{12}O_6$ while deoxyribose has a molecular formula of $C_5H_{10}O_4$. This makes saccharides rich in oxygen content with an O/C ratio of *ca.* 1 and a H/C ratio of 2. Usually, during the formation of disaccharides or polysaccharides for example cellobiose (glucose dimer or disaccharide) with a molecular weight of 342 and cellulose (glucose polysaccharide) with per unit of glucose molecular weight of 162, loss of one mole of water (H_2O) with a molecular formula of 18 per two moles of monosaccharides is essential. Hence, the O/C ratio in cellulose and hemicellulose (lignocelluloses) has a slightly lower value (*ca.* 0.8).[8] Nevertheless, for a chemical to be used as a fuel or fuel additive, its O/C ratio should be low (biodiesel: *ca.* 0.1, ethanol: 0.5).[8] Consequently, conversions of saccharides into fuels or fuel additive necessitates extra processing for the reduction in O/C ratio. At the same time, conversions of saccharides into chemicals (sugars and its derivatives) for non-fuel applications exempt the extra process of decreasing the O/C ratio. Hence, it is apparent that lignocelluloses should be used for chemical production. Moreover, economic analysis suggests that while lignocelluloses are obtainable at a price of $50 per ton, glucose has a market price of $450–650 per ton and xylose has a market price of $1000–2500 per ton.[9] Further conversion of these monosaccharides (sugars) into various chemicals such as 5-hydroxymethylfurfural (HMF) ($300 000–350 000 per ton), furfural ($2500–3000 per ton), sorbitol ($500–700 per ton) and xylitol ($1000–3000 per ton) adds value to these sugars.[9] Hence, it is understandable that suitable transformations of starch, cellulose and hemicellulose to various sugars (C_6 and C_5) *via* hydrolysis of glycosidic bonds present in polysaccharides are economical. Nevertheless, use of a first generation biomass (polysaccharide), starch for obtaining sugars as platform chemicals to produce a variety of other essential chemicals is a debatable issue since it is principally used as a food. Hence, use of a second generation biomass, lignocelluloses (non-edible biomass)—with a high energy content (*ca.* 2×10^{10} J per ton of dry biomass)[10] is desirable for the synthesis of sugars. Additionally, huge worldwide availability (1.8×10^{12} tones) of plant-derived lignocelluloses including crop (agricultural) wastes and forest residues (90–95% with respect to total plant biomass production)[8] might permit those to be used as a feedstock for better rural economy. On the other hand, in an ideal scenario, it can be considered that from non-edible feedstock, one can produce edible products (sugars). The conversion of di/polysaccharides into chemicals can be done by either thermal (combustion, pyrolysis, gasification, supercritical water), thermo-chemical (acid, alkali) or biological (enzyme) methods. Under thermal conditions, substrates are heated at high temperatures (pyrolysis: >350 °C; gasification: >550 °C, supercritical water: ~300–400 °C) essentially without a catalyst (however, in a few cases such as gasification and treatment in supercritical water, catalysts are added to drive the reaction in a particular direction) to yield sugars, tar, char, gases *etc.* In most of these studies, gases (CO, CO_2, H_2, CH_4 *etc.*) are formed as the main products with a minor quantity of sugars formed (<20–30%). On the contrary, under thermo-chemical conditions at lower temperatures (<250 °C), catalysts are used to obtain sugars in higher quantities by subjecting substrates to hydrolysis. Considering this, in this chapter, discussions are focused on the conversion of di/polysaccharides into sugars by hydrolysis reactions. In the conversion of lignocelluloses to chemicals, multiple steps are involved and these are depicted in Figure 1.1.

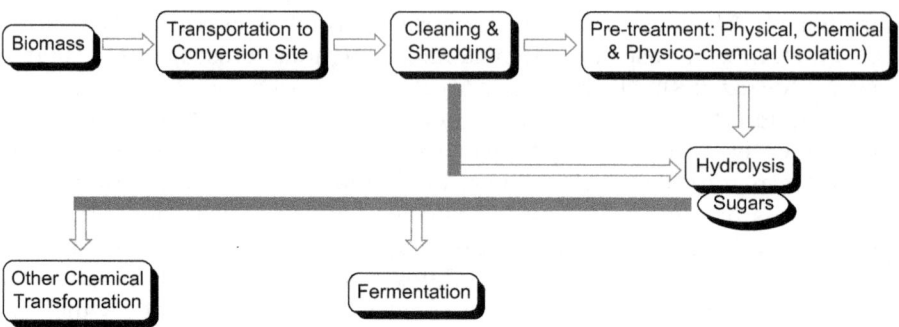

Figure 1.1 Illustration of the multiple steps involved in biomass processing to chemicals.

1.1.1 Potential Source of Sugars

Monosaccharides, or else we call them sugars, are named in two ways: (1) a monosaccharide containing an aldehyde group is called aldose and (2) a monosaccharide containing a ketone group is called ketose. In total, eight C_6 aldo-sugars (glucose, mannose, galactose, allose, altrose, gulose, idose and talose) and four C_5 aldo-sugars (xylose, arabinose, ribose and lyxose) are structurally possible. Besides these aldo-sugars, two more keto-sugars *viz.* fructose and xylulose are also well-known in nature. But, among them, idose and talose are not found in nature. Moreover, the presence of allose, altrose, gulose, ribose and lyxose is very rare in nature and hence discussions on those are not made here. The rest of the sugars are generally present in fruits, edible plants, living bodies, bacteria, proteins *etc.*

In Figure 1.2, likely sources of main C_6 sugars (glucose, fructose, mannose, galactose) and C_5 sugars (xylose, arabinose, xylulose) are illustrated. In general, these monosaccharides (sugars) can be obtained by the hydrolysis (addition of one mole of water per 2 moles of sugars) of their respective disaccharides [maltose: α-1,4-D-glucose disaccharide (found in potatoes, cereal, beverages *etc.*), cellobiose: β-1,4-D-glucose disaccharide, sucrose: disaccharide of α-D-glucose and β-D-fructose linked *via* a 1,2 glycoside bond (found in sugarcane, beet, grains *etc.*), xylobiose: β-1,4-D-xylose disaccharide *etc.*]. Further, several polysaccharides such as starch (α-1,4-D-glucose polysaccharide), cellulose (β-1,4-D-glucose polysaccharide), inulin (fructose polysaccharide), hemicellulose (polysaccharide of several C_5 and C_6 sugars) *etc.* derived from edible and non-edible parts of plant biomass can yield sugars on hydrolysis. Moreover, lignocelluloses are made up of *ca.* 75% of polysaccharides (cellulose, hemicellulose, starch and saccharose)[11] and hence, it will be beneficial to use plant biomass (lignocelluloses) directly for the synthesis of various sugars. Since cellulose is present as a major component (*ca.* 45%) in lignocelluloses, its conversion into chemicals (mainly sugars) is considered as foremost in the bio-refinery concept. In recent times, municipal solid wastes (kitchen waste containing cellulose) have also been increasing and their effective utilization to generate chemicals and

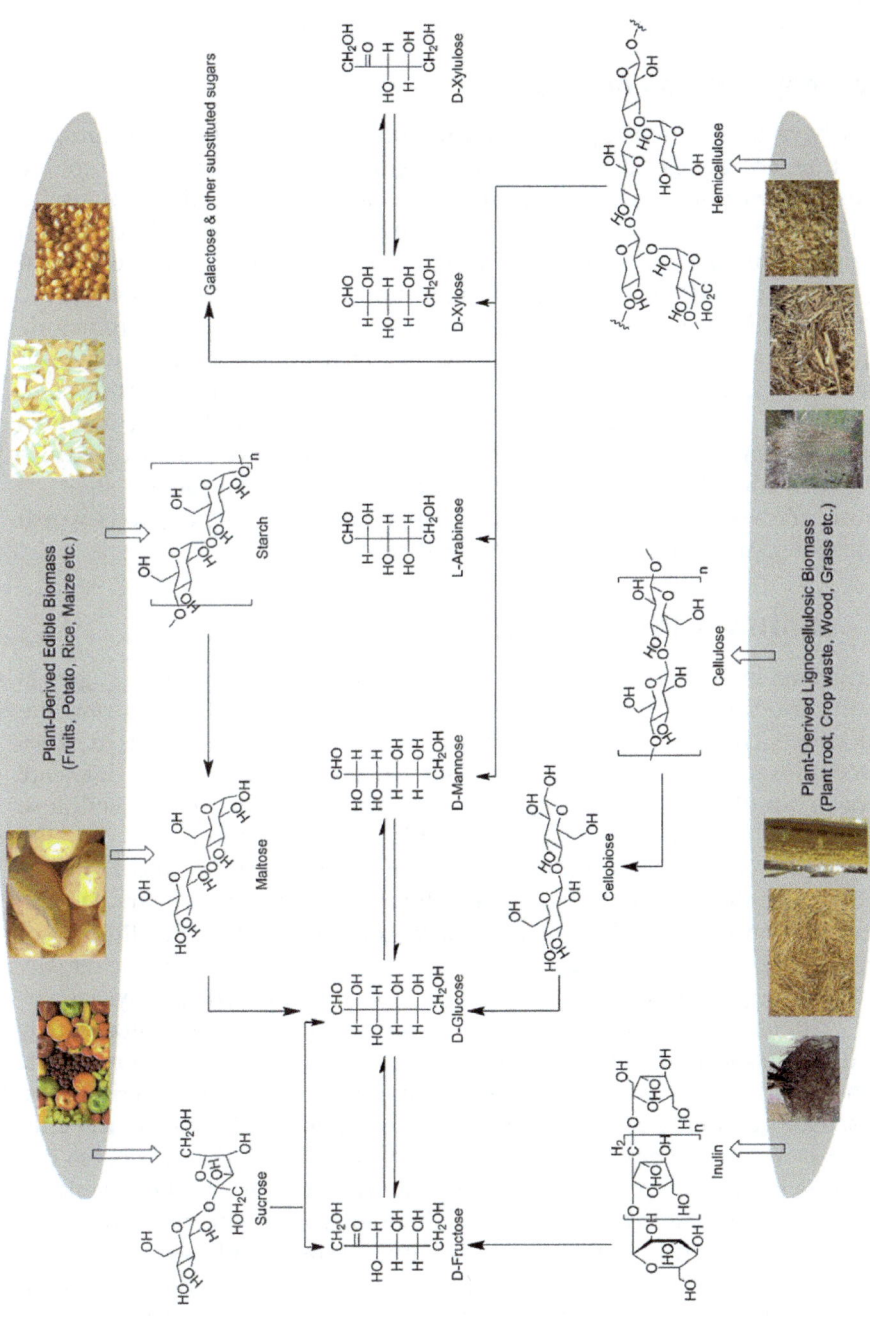

Figure 1.2   Schematic diagram of possible sources of sugars.

fuels may prove to be vital in curbing the problems of landfill and inciner-
ation, which gives rise to pollution by liberating hazardous chemicals and
gases.

1.1.2   Applications of Sugars

Sugars have a variety of applications in fine chemicals, pharmaceuticals,
agriculture, cosmetics *etc.* In most cases, sugars are used as an energy source
(glucose), low-calorie sweetener (xylose) and for the synthesis of many industrially important chemicals such as furans (5-hydroxymethylfurufral and
furfural; precursors for fuel, resin, plastic, nylon, polyester, fine chemical
etc.), sugar alcohols (sorbitol, mannitol, xylitol, arabitol; used as low-calorie
sweetener, adhesive, cosmetics, energy source *etc.*), sugar acids (gluconic
acid, xylonic acid, arabinonic acid; used as chelating agent, cement retard-
ant, cosmetics, medicine *etc.*), acids (succinic acid, itanoic acid, formic acid,
glycolic acid; used in the food industry and polymer industry), alcohols
(ethanol, butanol; used as fuels, solvents *etc.*) and alkyl ethers of sugars
(alkyl glucoside, alkyl xyloside; used as biomass-derived surfactant *etc.*)[12–14]
Because of these extremely significant applications of sugars, it is worth
synthesizing sugars from biomass-derived resources.

1.2   Biomass Pre-treatment

As suggested in Section 1.1.1, the most favourable way to synthesize sugars is
the utilization of lignocellulosic (non-edible) biomass directly as a raw material which consists of polysaccharides (*ca.* 45% cellulose and *ca.* 25%
hemicelluloses) in large quantities.[8] But, in reality, due to very intricate
hydrogen bonding such as intra-, inter-molecular and inter-sheet in cellulose
(Figure 1.3),[15,16] the existence of lignin (aromatic polymer) in lignocelluloses, and multiple bonding between polysaccharides and lignin, it
becomes complicated to process lignocelluloses directly into sugars. During
the conversion of the polysaccharide part (cellulose and hemicellulose) of
lignocelluloses into sugars, lignin remains unconverted because it usually
requires high processing temperatures (>250 °C) compared to polysaccharides (<230 °C).[17] And if conversions of lignin are also tried simultaneously
then degradation reactions of sugars become predominant. In most cases,
unconverted lignin is also capable of poisoning the catalytically active sites.
As discussed earlier, due to the occurrence of multiple H-bonding in cellulose, its structure becomes very rigid and crystalline and thus becomes difficult to degrade. A representative XRD pattern for commercially available
microcrystalline cellulose is presented in Figure 1.4. The pattern shows
peaks due to the amorphous ($2\theta = 15.8°$) and crystalline phases ($2\theta = 22.5°$,
$34.7°$) of cellulose. Additionally, due to the very strong H-bonding in cellulose, it remains insoluble in many common solvents but is soluble in ionic
liquids (ILs), concentrated aqueous $ZnCl_2$ solution and ammoniacal
$Cu(OH)_2$ solution.[18,19] Cellulose also possesses a very high degree of

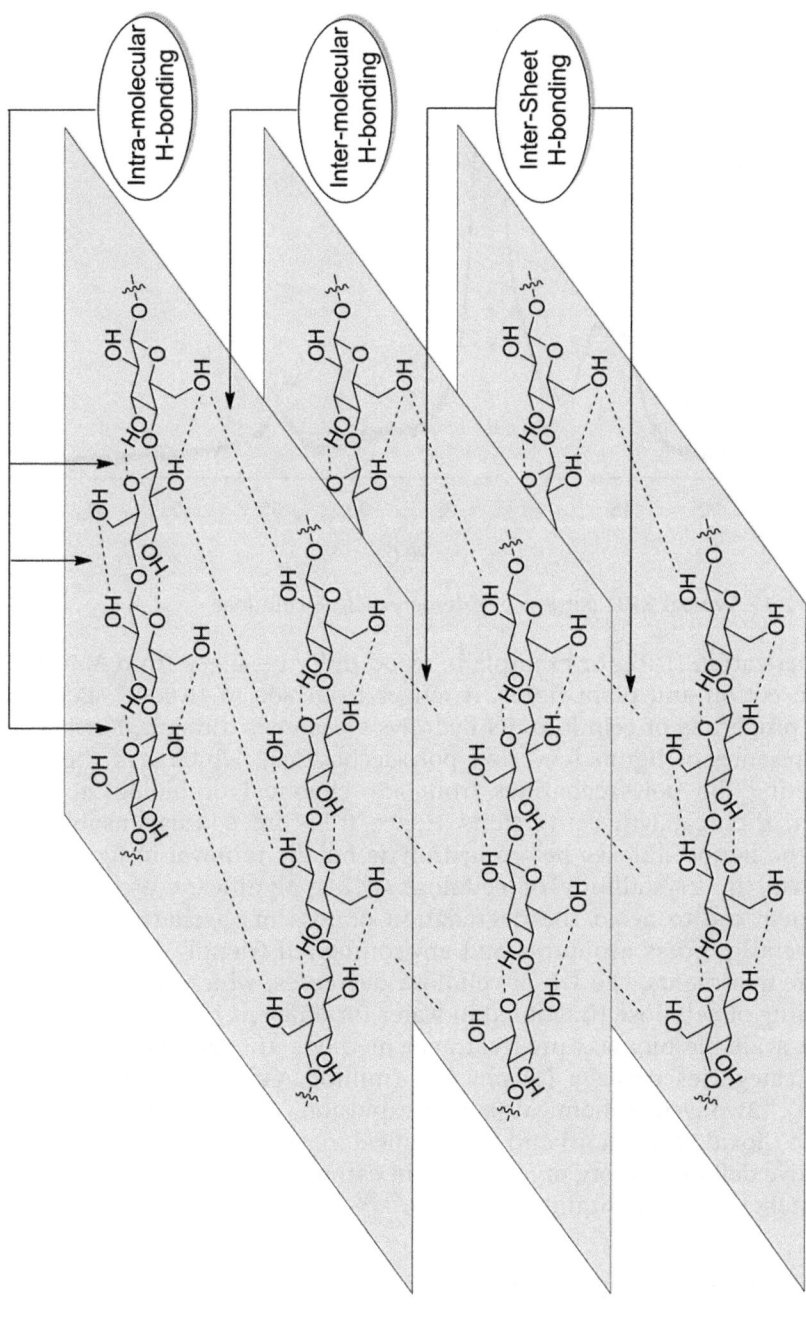

Figure 1.3 Illustration of the H-bonding present in cellulose.

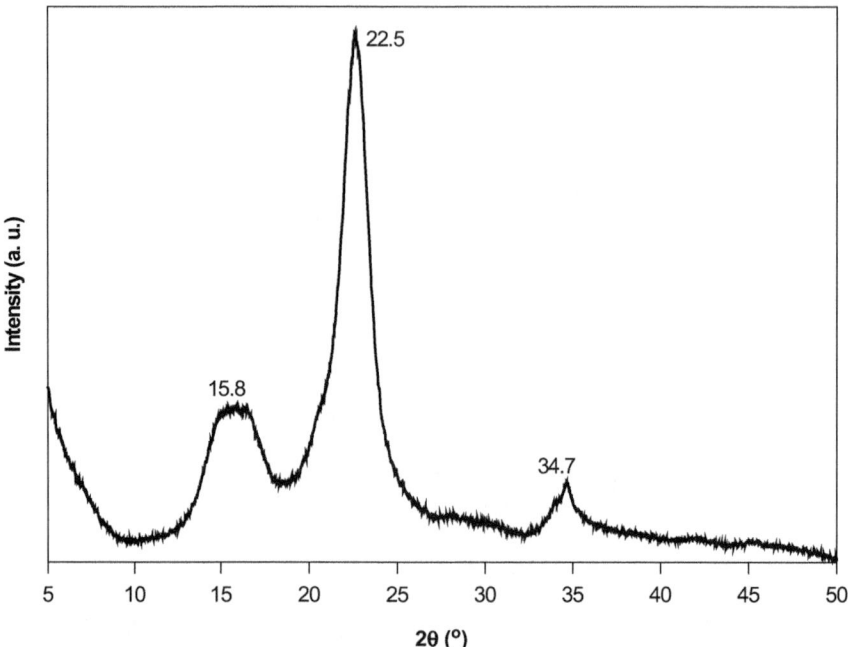

Figure 1.4 Typical XRD pattern of microcrystalline cellulose.

polymerization (DP), for example in wood pulp, it ranges from 300 to 1700 and in cotton and plant fibres, it ranges from 800 to 10 000.[20] Because of these properties of cellulose, its hydrolysis becomes difficult. Furthermore, the presence of lignin (covering polysaccharides), which has the role of protecting the polysaccharides from any chemical or biological attacks, hinders their catalytic conversions. Hence, it becomes indispensable to pre-treat the lignocelluloses before hydrolysis for the removal of lignin and to decrease the crystallinity of cellulose. Other significant aspects in pre-treatment are to avoid the degradation or loss of saccharides and make the overall process economic and environmental friendly. During some of the pre-treatments, the DP of cellulose decreases, which may increase the solubility of cellulose (fractions) in water for efficient hydrolysis.

The available biomass pre-treatment methods are classified roughly into three categories namely: (1) physical (milling, grinding, radiation, ultra-sound), (2) physico–chemical (steam explosion, ammonia fiber explosion, carbon dioxide explosion) and (3) chemical (ozonolysis, alkaline hydrolysis, oxidative delignification, organic solvent extraction, acid hydrolysis, enzyme treatment, and ionic liquid treatment).[21,22]

1.2.1 Physical Treatment

The main purpose of the physical pre-treatment of lignocellulose is to de-crease its particle size and cellulose crystallinity. Use of milling and grinding methods can decrease the size of various biomass to 0.2–2 mm from 10–30 mm.[23] The size reduction of lignocellulose is directly related to the energy consumption and time required for pre-treatment processing. Several reports on the ball milling method show a reduction in size (determined by particle size analyser), crystallinity (determined by XRD, NMR studies) and degree of polymerization (DP; determined by anion-exchanged chromatography) in cellulose and hence, increases in its hydrolysis rate.[24–26] During ball milling, an increase in temperature is seen, which has an effect on decrystallization and for this reason, it becomes important to control the temperatures for reproducible results. Recently, an ultrasound technique has been used to decrease cellulose crystallinity within a short time in the presence of water.[27] It is shown that Avicel cellulose (particle size = 38 μm) can be transformed into 0.1–0.6 μm cellulose with a 12.1% decrease in crystallinity index (without changes in its structure) at 80 °C after subjecting it to ultrasound treatment (optimum amplitude = 40%) for 3 h. It is assumed that when cellulose is exposed to ultrasound, which has a higher energy than the H-bonding energy of cellulose (21 kJ mol^{-1}), it breaks the H-bonding in cellulose to form lower crystalline cellulose.[28] Microwave irradiation of lignocellulosic biomass causes localized heating of lignocellulose leading to disruption of the lignocellulose structure[29] and since it is a harsh process, it leads to very high lignin removal from biomass.[30,31] However, in terms of cost and generation of high temperature during treatment, the process is not efficient on a large scale.[32] Additionally, treatment of cellulose with γ-rays leads to the reduction of DP and crystallinity in cellulose but this process also faces the drawback of high costs.[33]

1.2.2 Physico–Chemical Treatment

In the widely used steam explosion pre-treatment, biomass is treated with saturated steam (6.9–48.3 bar) at high temperatures (160–260 °C) and after a certain time (few seconds to a few minutes), the pressure is suddenly reduced to atmospheric pressure allowing the biomass to undergo explosive decompression.[23,34] This process helps in the removal of hemicellulose and redistribution of lignin. Various factors such as biomass size, moisture content, temperature and time are decisive in designing an effective steam explosion biomass pre-treatment method.[35] Moreover, the addition of mineral acid (H_2SO_4, CO_2; 0.3–3 wt/wt) during steam explosion, improves the efficiency of the pre-treatment process by decreasing the temperature and time required.[36] The steam explosion treatment has advantages including a lower energy requirement than mechanical treatment (70% less energy requirement)[37] and less environmental impact, which allows the process to operate on an industrial scale successfully (Iogen Corporation, Canada; steam explosion using dilute acid pre-hydrolysis of corn stover, barley straw and bagasse[38]). Biomass pre-treatment is also carried out in liquid hot water instead of steam.[39] This process can be done in three types of reactors namely co-current, counter-current and flow-through. In a typical procedure, all the reactors are maintained at desired temperatures (200–230 °C) for a specific time (*ca.* 0.25 h) to achieve efficient pre-treatment.

Ammonia fibre/freeze explosion (AFEX) is another process for lignocellulose pre-treatment where liquid NH_3 is in contact with biomass (typically in a biomass–NH_3 ratio of 1 : 1 to 1 : 2) at moderate temperatures (60–100 °C) and high pressures (17–21 bar) for a specific time and subsequently pressure is reduced to make an explosion.[21,40] This process is reported to be effective for lignocelluloses with a low lignin content (bagasse: 15%, bermudagrass: 5% *etc.*) rather than a high lignin content (nutshell: 30–40%, wood: 18–35%, aspen chips: 25% *etc.*).[41]

To surmount the problems with high temperature (steam explosion) and high cost (AFEX) associated with these methods, a carbon dioxide explosion pre-treatment, which uses carbon dioxide in a supercritical state ($Pc = 73.8$ bar, $Tc = 31.1$ °C), has been developed.[21,42] Moreover, due to the liquid–gaseous state of carbon dioxide under supercritical conditions, it has a lower viscosity and higher diffusivity (than water) and hence, it can easily penetrate inside the architecture of biomass and separate cellulose and hemicelluloses from each other and lignin. This process reduces the prospects of degradation product formation from saccharides because of lower operating temperatures.

1.2.3   Chemical Treatment

Effective lignin removal from lignocellulose can be carried out using ozone treatment to biomass at room temperature and atmospheric pressure (ozonolysis).[43,44] This treatment has a minor effect on the hemicellulose part and no effect on the cellulose part of lignocelluloses. Moreover, during this processing, no toxic chemicals are generated. From an environmental point of view, this process is better since the used ozone can easily be catalytically decomposed at increased temperatures.[45] However, the requirement of a large amount of costly ozone makes the process expensive.

Another way to remove the lignin part is to use alkali at low temperature and pressure (alkaline hydrolysis).[34,46] During this process, in the presence of alkaline reagents such as NaOH, KOH, $Ca(OH)_2$, NH_4OH, removal of acetyl and uronic acid substitution in the hemicellulose part is also possible along with removal of lignin.[47–49] The treatment of lignocelluloses with alkali mainly hydrolyses the intermolecular ester bonds between lignin and polysaccharides to remove lignin leaving behind the free polysaccharides (cellulose, hemicellulose).[23] Use of air/oxygen during alkaline hydrolysis significantly improves the delignification process of biomass with high lignin contents.[47] Nonetheless, the chances of polysaccharides undergoing hydrolysis and oxidation reactions under alkaline conditions make this process vulnerable.

Delignification of biomass can as well be carried out using H_2O_2 as an oxidising agent.[46] This process involves removal of both lignin and hemicellulose at a higher extent from lignocellulose. In this method, oxidising chemicals react strongly with the aromatic rings of lignin and generate aromatic carboxylic acids that can act as inhibitors in the later stages of transformations, hence they need to be removed or neutralized before processing the celluloses.[50]

The organosolv process can also remove lignin from biomass. In this method, several organic solvents (methanol, ethanol, acetone, ethylene glycol, triethylene glycol, tetrahydrofurfuryl alcohol *etc.*) or organic solvents with water in the presence of mineral and organic acids (H_2SO_4, HCl, oxalic acid, salicylic acid, acetylsalicylic acid *etc.*) are used to remove lignin.[51–53] The organic solvents used in this method are required to be recoverable to make the process cost efficient. Besides this, the production of very high quality lignin makes this method competent since this lignin can be consumed further for the synthesis of various valuable chemicals.

Dilute acid (<4 wt%) pre-treatment to lignocellulose is another widely used technique. In general, two pathways are undertaken for this purpose: (1) use of a high temperature (>160 °C) continuous flow process for low biomass loading (5–10 wt%) and (2) use of a low temperature (≤160 °C) batch process for high biomass loading (10–40 wt%).[54] Typically, dilute acids (H_2SO_4, HCl, HNO_3, H_3PO_4, maleic acid, fumaric acid *etc.*) and concentrated acids (H_2SO_4, HCl) are sprayed over the lignocellulose to remove (selectively) hemicelluloses in the form of sugars and soluble oligomers ($2 < DP < 10$).[21,23,46,54,55] This method is used on a commercial scale by BlueFire Renewables and Biosulfurol Energy Limited.[56,57] However, due to acid treatment, some portions of sugars are further transformed into furans (furfural and HMF) and other by-products, which leads to the loss of sugars. Although this pre-treatment process is quite impressive, due to the use of acids, it becomes hazardous and corrosive and thus requires a high capital cost.[23,58]

Biological pre-treatment of biomass is a safer and environmentally friendly method which uses various micro-organisms (white-, brown- and soft-rot fungi) to attack the cellulose part selectively.[18,21] Although milder conditions are applied for biological pre-treatment, nevertheless the rate of biological hydrolysis is very low and requires a long processing time, which is a great disadvantage of this method.[23,59,60]

Recently, the use of ionic liquids (ILs) in the pre-treatment of lignocelluloses has also been studied because they have interesting properties such as high thermal stability, low volatility, high activity at low temperatures *etc.*[61,62] ILs are very effective in the de-crystallization of cellulose and in the cleavage of lignin–hemicellulose linkages. As, for example, the anionic part (Cl^-) of 1-butyl-3-methylimmidazolium chloride, [BMIM]Cl ionic liquid forms new H-bonds with the sugar (glucose) part of cellulose in a 1:1 stoichiometric ratio and thus breaks the earlier H-bonding present in cellulose (Figure 1.5).[19,63] This helps in the dissolution of cellulose in ILs, for example [BMIM]Cl, [BMIM]Fm, [BMMIM]Cl, [AMIM]Cl, [AMIM]Br, [AMIM]Fm, [EMIM]Cl, [EMIM]Ac, [BMPyM]Cl *etc.* have the ability to dissolve 3–39% cellulose with different DP (200–6500) at varying temperatures (45–110 °C).[64,65] It is predicted that small cations are helpful in achieving

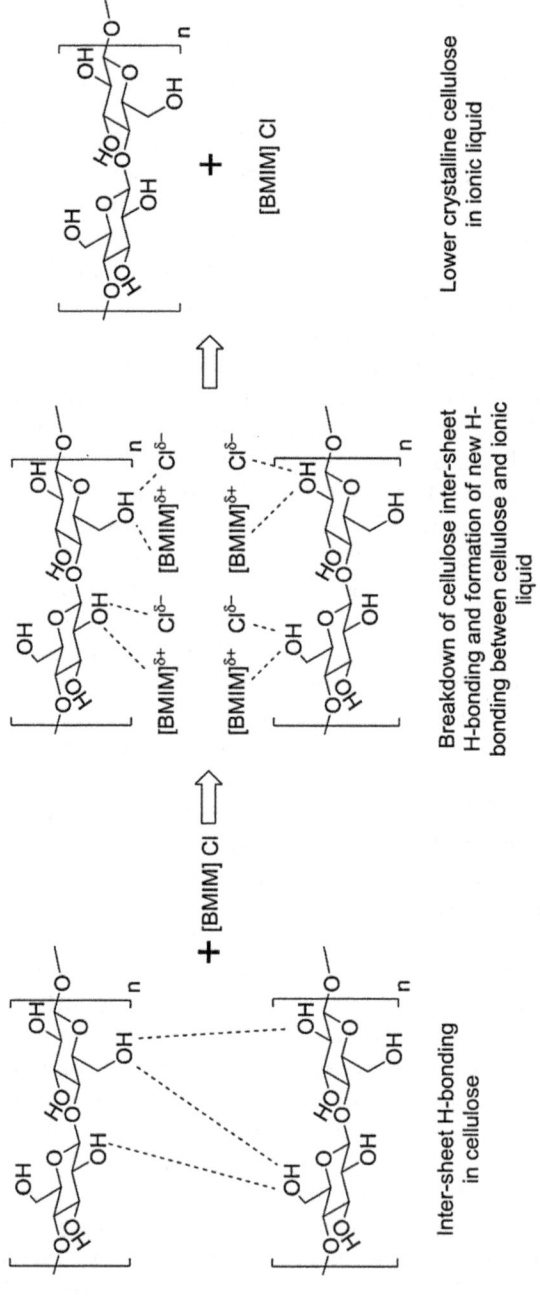

Figure 1.5   Illustration of cellulose H-bond breaking by action of ILs.

efficient dissolution of cellulose since [AMIM] is proven to be better than [BMIM].[66] Although the process may look very attractive for lignocellulose pre-treatment, it can only be possible if the drawbacks (high cost, recycling problems due to homogeneous nature, washing of lignocelluloses *etc.*) of ILs are minimized.[61]

In summary, the choice of pre-treatment method is dependent on whether to remove lignin or hemicelluloses or both. However, the foremost aims of these methods are to decrease the size of particles, reduce the crystallinity of cellulose and separate the cellulose, hemicelluloses and lignin parts from each other for their efficient conversions into value-added chemicals. Although these different methods show different effects on the structure of cellulose, these are still not well studied in hydrolysis reactions.

1.3 Synthesis of C_6 Sugars

As shown in Figure 1.2, glucose (C_6 sugar) can be obtained from various disaccharides (maltose, cellobiose and sucrose (table sugar)) and polysaccharides (starch, cellulose and inulin) by undergoing hydrolysis reactions (addition of one mole of water/two moles of sugars). The presence of acid or base catalysts assists in improving the hydrolysis rates by efficiently cleaving the glycosidic linkages in disaccharides and polysaccharides to form monosaccharides (sugars: glucose, fructose).[67]

Before discussing the catalytic processes for the cellulose hydrolysis into sugars, it is crucial to know the properties of cellulose and the variety of crystal structures (polymorphs) of cellulose, which are identified as, I_α, I_β, II, III and IV based on unit cell parameters.[68] Undoubtedly, these structures play an imperative role in catalysis since these can restrict or enhance the possibility of interactions between cellulose and catalytically active sites. Nature produces most of the cellulose in the form of cellulose I, which is a composite of two crystalline allomorphs *viz.* I_α and I_β in varying ratios depending on the source of cellulose.[16,69] In cellulose I crystalline form, two glucosyl units are arranged in a parallel manner and form one-chain triclinic (I_α) and two-chain monoclinic (I_β) unit cells.[70] It is known that the cellulose I_α form is metastable and upon annealing, it transforms to the more stable cellulose I_β form. Typically, the cell wall of algae and bacterial cellulose contains the I_α crystalline form of cellulose but cotton and wood contain mostly the I_β type of the cellulose crystalline form.[71] Later, treatment to the cellulose I_α and I_β crystalline forms with concentrated NaOH ($\sim$20%) solution (regeneration) and a further water washing (mercerization) transforms cellulose I into the thermodynamically more stable cellulose II in an irreversible way.[69,72] The cellulose II crystalline form (two-chain monoclinic unit cell) consists of two glucosyl units in an antiparallel arrangement. Treatment of both cellulose I and II with amines or liquid ammonia (swelling) and further removal of the swelling agent anhydrously produces the cellulose III crystalline form. Cellulose III is a highly unstable crystalline form and can be sub-classified into cellulose III_I and cellulose III_{II}

depending on its source (cellulose I_α or I_β). In cellulose III, two glycosyl units are present in parallel forming a one-chain monoclinic unit cell. Further, when cellulose III is treated with glycerol at 206 °C, it forms the cellulose IV crystalline form. In general, cellulose IV is the disordered form of cellulose I.

Several reagents such as water, organic solvent, alkali and acids have the ability to swell cellulose and, based on the penetration ability of reagents in the structure of cellulose, swelling can be classified as inter-crystalline or intra-crystalline.[70] If the reagent only affects the amorphous part of cellulose leaving the crystalline part intact then it is called inter-crystalline swelling and if the reagent affects cellulose completely (amorphous and crystalline part) then it is called intra-crystalline swelling. Due to a hydrophilic cellulose surface (the presence of –OH groups), water molecules interact with cellulose *via* H-bond formation.[73] This leads to the inter-crystalline swelling of cellulose. Because of restrictive swelling due to water, and since water is used as a solvent in hydrolysis reactions, it can be expected that it would be difficult to hydrolyse cellulose. Generally, organic solvents have less ability to swell cellulose (inter-crystalline) than water and the effectiveness of organic solvent swelling depends on the higher H-bonding ability and higher polarity of the solvent.[70] In contrast, alkali and mineral acids can strongly affect amorphous and crystalline structures leading to intra-crystalline swelling. This gives an idea that the combined effect of water (as hydrolysis reactant and solvent) and mineral acids or alkalis can pronounce the hydrolysis rates. However, in most of the hydrolysis works, researchers have used commercially (Aldrich, Merck *etc.*) available microcrystalline (Avicel) cellulose (partial depolymerized form of I_α cellulose by mineral acids[74]). Besides this, a few researchers have also used several types of isolated celluloses (pre-treated using acid, alkali, supercritical or any other method as discussed in Section 1.2) for the synthesis of sugars (glucose) for which the type of cellulose is not known. All this makes it difficult to really compare the results of all the reactions with each other since, depending on the type of polymorph of cellulose used in the work, activities of the catalysts will vary. Moreover, results obtained with a particular catalyst in the hydrolysis of a particular cellulose are mostly not reproducible with a change of the source of cellulose as the structures of these celluloses are different. In some reports, it has been shown that supercritical ammonia treatment to cellulose I produces cellulose III, which has more enzymatic accessible sites than cellulose I and thus enhances the hydrolysis activity.[75,76] However, more detailed investigations are necessary to correlate the catalytic activity with cellulose structures.

Considering these discussions on the changes in cellulose structures after isolation (pre-treatment), it would be beneficial to use lignocelluloses without any pre-treatment for sugar synthesis as cellulose will be present in mostly the I_α form, which is easier to hydrolyse than a few other cellulose structures. Moreover, there is a possibility that results obtained with one lignocellulose cannot be reproduced with other lignocelluloses (at least derived from the same plant species). Nonetheless, very few catalytic reports are known to use raw lignocelluloses for the conversion into chemicals. However, unlike academic research, industries prefer to use lignocelluloses instead of isolated polysaccharides to study the effects of catalysts on their hydrolysis. The academic research is mainly done with isolated polysaccharides because as mentioned earlier (Section 1.2), in the native form, cellulose is present as a very rigid, crystalline material and hence it would be challenging to hydrolyse it. Ideally, when lignocelluloses or cellulose are used as substrates that are not soluble in most of the reaction media (mainly water), all the catalytic systems become heterogeneous. But to simplify the discussions, depending on the solubility of the catalyst in the reaction medium, discussions are made for homogeneous as well as heterogeneous catalytic methods.

1.3.1 Use of a Homogeneous Catalytic System

1.3.1.1 Homogeneous Acid Catalysed

Chemical transformations of substrates using homogeneous catalysts are always attractive since these methods prevail over the mass transfer limitations to allow efficient interactions between substrates and catalysts (active sites). An early process (before 1960) depicts that the synthesis of glucose can be achieved from starch in the presence of H_2SO_4 and HCl.[77] Later, simple substrates such as maltose and cellobiose were used to study the reaction mechanism and kinetics in the presence of dilute H_2SO_4 to yield glucose.[78] The activation energy calculated for these substrates hydrolysis was found to be 132–137 kJ mol^{-1} and it was also observed that the rate of hydrolysis increased with an increase in temperature. The results were confirmed by another research group by observing a similar activation energy of 133 kJ mol^{-1} in cellobiose hydrolysis.[79] However, it was seen that with the use of mineral acids such as HCl and H_2SO_4, large amounts of degradation products can be formed. To reduce this possibility, dilute H_3PO_4 was used for starch hydrolysis into glucose although H_3PO_4 shows a lower activity than HCl.[80]

The studies were also done with non-edible substrates instead of the edible substrate starch. The first process for the synthesis of glucose from linen (textile) using concentrated H_2SO_4 was reported as early as 1819.[81] In 1937, hydrolysis of cellulose was reported using 40 wt% HCl at room temperature, however, lower yields were observed and also the reaction took a longer time to complete.[82] In 1931, the process of cellulose hydrolysis into sugars was commercialized (Scholler process) based on the two step method wherein in the first step, in the presence of a mineral acid (0.5 wt% H_2SO_4), hydrolysis of wood wastes at 170 °C was carried out to yield sugars and oligomers. In the second step, fermentation was carried out to achieve 50% yield for sugars.[2] Around the same time (1935–45) in Russia, using dilute mineral acids and fermentation methods, many industrial processes were developed for the hydrolysis of softwood (corn, grain, molasses) and hardwood into sugars, ethanol and furfural.[83] As mentioned earlier, the efficiency of releasing glucose in the solution depends on the structure of cellulose, temperature, time and concentration of acids. With an optimization of reaction conditions, a maximum of *ca.* 70% glucose formation was reported using dilute H_2SO_4 from corn stover.[84] However, at the same time, if a reaction is not curtailed at an appropriate time, then further conversions of glucose into by-products is feasible due to condensation and degradation reactions.[85,86] To avoid the use of higher temperatures (for curbing the degradation reactions of glucose), concentrated acid catalysed low temperature (room temperature) methods were developed at ambient pressures. But the use of concentrated acids again promotes sugar degradation reactions. Nevertheless, with the development of new reactors (continuous, concurrent batch *etc.*), it is possible to achieve better yields of glucose. But, due to the higher costs of these reactors and the severe problem of corrosion associated with the use of concentrated acids, the commercial viability of this method is in jeopardy. Based on recent developments in this research area, BC International Corporation (BCI), USA, uses agricultural wastes such as rice husk, corn stover, bagasse *etc.* to yield sugars and subsequently ethanol (by use of microorganisms).[70] The concentrated acid hydrolysis method is also very well known for obtaining sugars from raw biomass. In 1948, the Japanese commercialized the method using concentrated H_2SO_4 and membranes to separate the acid from sugars. A few attempts were also made to carry out hydrolysis in a stepwise manner. For instance, first, a dilute acid treatment is given to biomass to remove the hemicellulose part and then the cellulose and lignin parts are subjected to concentrated acid treatments. This ensures that the degradation products obtained from hemicelluloses under concentrated acid conditions can be avoided. Moreover, during dilute acid hydrolysis of hemicelluloses, loosening of the cellulose structure occurs, which in turn makes it easy to undergo hydrolysis. Arkenol's process is based on this technology wherein 20–30% and *ca.* 70% sulfuric acid concentrations are used to obtain high yields of C_5 and C_6 sugars. In the last few years (since 2012), along with US DOE, Masada Resource Group and Arkenol have been working to commercialize dilute–concentrated acid technology.[70] It is also mentioned in the literature that the presence of metal salts (LiCl, $CaCl_2$ *etc.*) along with acids helps in achieving a higher hydrolysis rate by the swelling effect as discussed earlier. The same effect is also observed with the use of concentrated acids in higher concentrations (>50 wt%).[87,88] The swelling effect is thought to break the hydrogen bonding and gives the catalyst access to internal glycosidic bonds, which ensures an improvement in the rate of hydrolysis. Additionally, numerous studies on the use of liquid mineral acids (H_2SO_4, HCl, H_3PO_4, H_2CO_3 *etc.*) and organic acids (oxalic acid, maleic acid, fumeric acid, *p*-toluenesulfonic acid *etc.*) have been done with varying concentrations (0.05–40 wt%) to achieve higher yields of glucose in the hydrolysis of isolated cellulose carried out at 100–260 °C for a few minutes to several hours.[14,20,67] Use of gaseous HCl for pre-treatment of wood chips and a further hydrolysis in the presence of dilute HCl yields high amounts of sugars (glucose = 80%, xylose = 95%).[89]

It is reported that the acid hydrolysis of cellulose follows first order kinetics and the reaction rate is dependent on several factors such as temperature, acid concentration, physical state of cellulose (polymorph) *etc.*[68,73] In the acid catalysed cellulose hydrolysis reactions, initially, a rapid decrease in the degree of polymerization (DP) is observed. However, after reaching a certain value (depending on cellulose's physical state), the DP value remains almost constant and this phenomenon is termed as 'levelling-off degree of polymerization (LODP)'.[90] This phenomenon can be explained based on the fact that first, the easy to hydrolyse amorphous part of cellulose undergoes rapid conversion into oligomers and sugars and then the difficult to hydrolyse crystalline part starts converting at a much slower rate. When cellulose is heated in the presence of dilute or concentrated mineral acids (HCl, H_2SO_4), it yields glucose (hydrolysis product) as a primary product along with secondary (degradation) products: 5-hydroxymethylfurfural (HMF), formic and levulinic acids, humins (re-polymerization or condensation products) *etc.* The kinetic studies have revealed that in the presence of H_2SO_4, the rate of cellulose hydrolysis into glucose is always higher (Douglas Fir: 1.73×10^{19} per min, Solka Floc: 1.22×10^{19} per min, filter paper: 1.22×10^{19} per min and municipal solid waste: 1.16×10^{19} per min) than the rate of glucose degradation reactions (Douglas Fir: 2.38×10^{14} per min, Solka Floc: 3.79×10^{14} per min, filter paper: 3.79×10^{14} per min and municipal solid waste: 4.13×10^{15} per min).[91–93] It is known that the activation energies for cellulose hydrolysis reactions are higher (172–180 kJ mol^{-1}) than the glucose degradation reactions (137–142 kJ mol^{-1}). Considering these somewhat contradictory results, it is expected that in the acid hydrolysis of cellulose, glucose degradation products will always be observed. This might be another reason to observe the LODP effect since glucose may undergo repolymerization (condensation) reactions (to achieve equilibrium between hydrolysis and condensation reactions).

Besides mineral acids, homogeneous heteropoly acids (HPAs) are also presented as effective catalysts in the hydrolysis of cellobiose and cellulose. Silicotungstic acid ($H_4SiW_{12}O_{40}$), phosphomolybdic acid ($H_3PMo_{12}O_{40}$) and silicomolybdic acid ($H_4SiMo_{12}O_{40}$) are used for cellobiose hydrolysis reactions to yield 42–53% glucose at 150 °C.[94] Moreover, the cellulose hydrolysis reaction has also been reported using $H_3PW_{12}O_{40}$ and $H_3BW_{12}O_{40}$ catalysts to produce 18–77% of glucose and other reducing sugars.[94–97]

Based on the kinetic data obtained with various substrates and many studies, the cellulose hydrolysis mechanism is proposed. The hydrolysis of cellulose happens by the action of a proton (H^+)/H_2O or conjugated acid (H_3O^+) at the glycosidic O-linkages *via* formation of a positively charged acyclic or cyclic intermediate to yield oligomers and further glucose (Figure 1.6).[98]

Although several researchers have shown that homogeneous acids can be used as catalysts for the synthesis of C_6 sugars (glucose, fructose) from cellulose (saccharides), due to various practical problems such as acid corrosion of reactors (acid concentration = 0.05–40 wt%), difficulty in catalyst

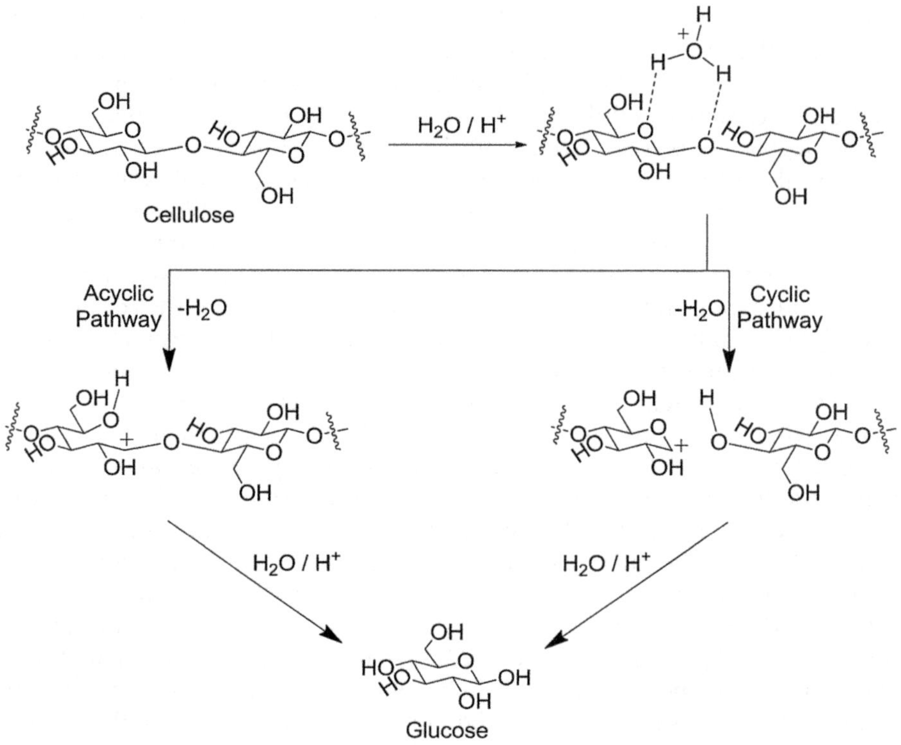

Figure 1.6 Mechanism of acid catalysed cellulose hydrolysis into glucose.

and product recovery, poor catalyst recyclability (solubility of both, product and catalyst in water), formation of degradation products (at high temperatures and using concentrated acids), generation of neutralization wastes (salt formation), toxicity, health hazards *etc.*, these catalysts may not be suitable for hydrolysis. However, due to a lack of any alternate efficient method, they are still used on an industrial scale.

1.3.1.2 Homogeneous Base (Alkali) Catalysed

Cellulose is also hydrolysed using alkalis in two possible pathways: (1) endwise degradation, which is also known as peeling and (2) hydrolysis of glycosidic bonds.[68,70,73] The peeling effect literately means peeling off monomers from reducing ends of cellulose below 100–140 °C. Thus, this process eventually will reduce the chain length and will also hydrolyse cellulose completely into glucose. Nonetheless, this process is too slow and does not continue for long. However, if this process was fast or continued until complete hydrolysis was achieved then it would have been practically impossible to subject lignocelluloses to the alkali pre-treatment/Kraft process (Section 1.2.3). Researchers have, however, devised a method by which the peeling effect can be controlled. This can be done by oxidising the reducing ends of cellulose and also by reducing the hemiacetal group of cellulose. If alkali treatment is given above 140 °C, along with peeling, cleavage of glycosidic bonds will start initiating a quick decline in chain length (DP). Since cleaving glycosidic bonds is an arbitrary process, several new reducing ends will be generated and the peeling process will gain momentum. The mechanism of peeling and alkali hydrolysis is illustrated in Figure 1.7. As observed, the mechanistic pathway studied with the help of model compounds proposes the involvement of S_N1, $S_N$icB (2) (nucleophilic substitution by an internal nucleophile *i.e.* conjugate base of C2 hydroxyl

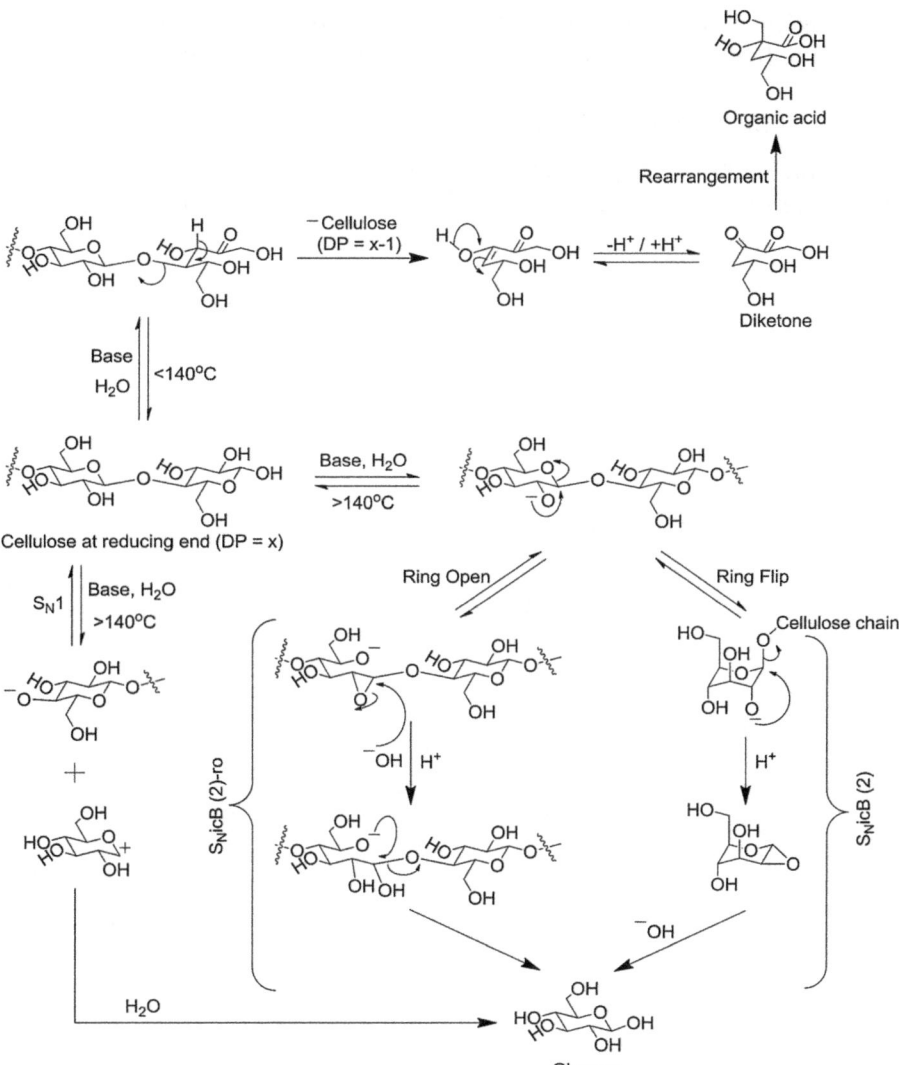

Figure 1.7 Mechanism of base catalysed hydrolysis of cellulose into glucose.

group) and S_NicB (2)-ro (nucleophilic substitution *via* intermediate ring opening) pathways. As shown in Figure 1.7, to carry out the S_NicB (2) mechanism, first the pyranose ring structure needs to be flipped, but this ring flipping is unfeasible in the case of crystalline cellulose because of severe hydrogen bonding, which makes the structure very rigid. Hence, it is probable that the alkali catalysed cellulose hydrolysis may happen *via* a S_N1, S_NicB (2)-ro mechanism. Once the glucose is formed, under alkaline conditions it can undergo an isomerization reaction to yield fructose and mannose (Section 1.5) and thus can hamper the yields of glucose. Although cellulose hydrolysis can be achieved using alkalis, it is also possible for cellulose to undergo oxidation reactions (done purposefully to reduce the peeling off effect) in the presence of alkalis to yield oxycelluloses or carbon dioxide and water (complete oxidation) depending on the severity of the process.

1.3.1.3   Enzyme Catalysed

The actions of enzymes on cellulose hydrolysis are explained by 'inversion' and 'retention' mechanisms.[70,99] As illustrated in Figure 1.8, in both

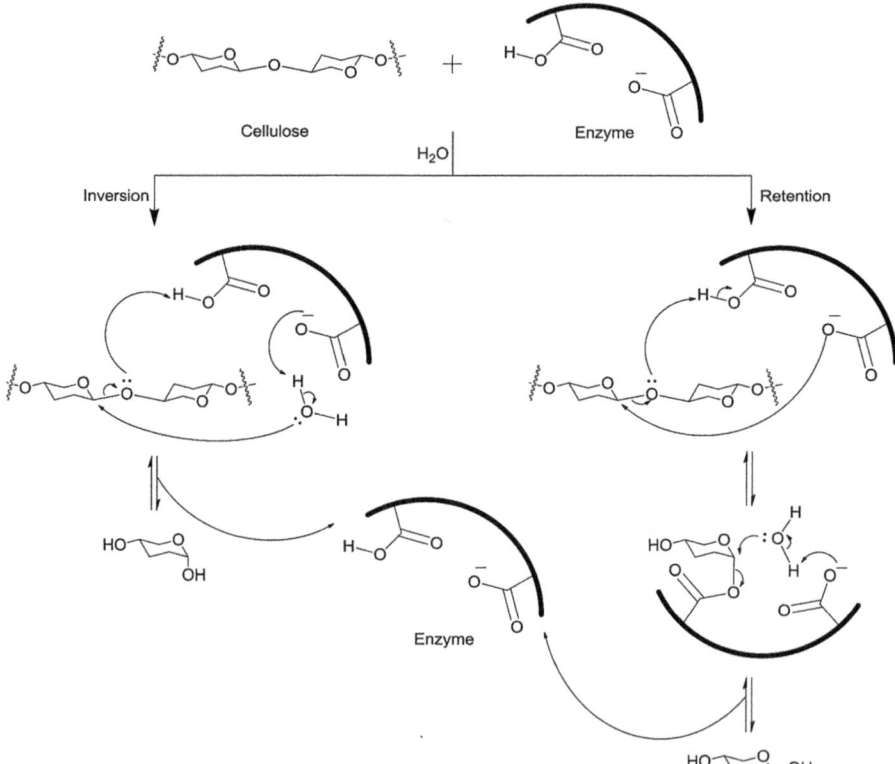

Figure 1.8   Mechanism of the enzyme catalysed hydrolysis of cellulose into glucose.

mechanisms, cellulose hydrolysis can be carried out with the help of two carboxylic acid groups, wherein one acts as a proton donor (acid) and the other acts as a nucleophile (base).

It is observed that cellulose hydrolysis in the presence of enzymes occurs *via* two subsequent pathways: (1) dissolution of crystalline cellulose (physical disruption) and (2) hydrolysis of disrupted cellulose glycosidic linkages to form sugar.[100] In April 2004, Iogen Corporation of Canada started production of bio-ethanol from wheat straw *via* the enzymatic hydrolysis of cellulose into sugars and their further fermentation into ethanol. An ethanol producer company from USA, POET, also announced in 2009 the production of bio-ethanol from cellulose (corn cob) in their South Dakota facility under Project Liberty. Abengoa has also shown a possibility for the production of ethanol from corn Stover *via* sugar formation. In 2013, Raízen Energia SA of Brazil (joint venture of Shell and Cosan) announced the production of bio-ethanol *via* sugar formation using an enzymatic process. In October 2004, taking into account the high cost of enzymes, Novozymes (Denmark) and Genencor (USA) industries have announced a modified cellulase production method using genetically modified organisms and they showed that the cost of the cellulase enzyme can be reduced to $0.1–0.2 from $5 (cost in 2001).[70] This has definitely improved the economics of the enzymatic hydrolysis process and hence, after 2004, several facilities started using enzymatic routes for the conversion of lignocelluloses into sugars and further into ethanol. In these processes, cellulase enzyme (endoglucanase, exoglucanase, β-D-glucosidase *etc.*) are used to convert cellulose into sugars.[101] Due to the specific role of each enzyme, for example while endoglucanase cleaves the internal bonds of cellulose, the exoglucanase and β-D-glucosidase give complete conversion of cellulose into sugars, a mixture of these enzymes is required (endoglucanases and exoglucanases for the hydrolysis of cellulose into cellobiose and β-1,4-glucosidases for the conversion of cellobiose to glucose). One major problem that affects the hydrolysis of cellulose is the inhibitory effect that cellobiose causes on β-1,4-glucosidases. This problem is pertinent with the cellulase derived from *Trichoderma viride* which has a high efficiency for cellulose hydrolysis to cellobiose. However, if the β-1,4-glucosidases are obtained from any other source then it can partially nullify this inhibitory effect. A few other studies have also shown that compared to ethanol, cellobiose is a potent inhibitor. Although, as discussed, several industrial processes have been developed based on enzymes, other drawbacks due to their homogeneous nature persist.[102] To overcome these problems, researchers have worked on the immobilization of enzymes. In this method, several active enzymes are entrapped on an inert and solid material to improve their stability as well as separation ability. It is reported that the cellulase enzymes (celB and β-glucosidase) covalently immobilized on polystyrene or calcium alginate gel particles are reused several times in glucose formation from cellulose.[103,104]

1.3.1.4   Ionic Liquid (IL) Catalysed

Cellulose hydrolysis reactions are also catalysed by ionic liquids (ILs) with acidity. Since the role of ILs has been recognized in the pre-treatment of lignocelluloses (as discussed in Section 1.2.3) for quite a few years and their effect on breaking hydrogen bonds is well studied, it was thought that these ILs, if functionalized with acidic groups, may facilitate the hydrolysis of cellulose. Moreover, the ability of ILs to dissolve cellulose (even crystalline), which is a major problem in achieving higher yields of glucose with other catalytic systems, is also an interesting phenomenon that can help in observing higher glucose yields with ILs. Additionally, the advantage of ILs with distinctive properties such as high thermal stability, low vapour pressure, tuneable polarity *etc.* compared to molecular solvents can be tuned according to the requirements of the reaction. Considering this, first, ILs were used along with mineral acid (H_2SO_4) to achieve higher yields of glucose. In this way, ILs were responsible for dissolving crystalline cellulose and further hydrolysis was catalysed by H^+ from mineral acid. Recently, studies were performed using cellulose + [C_4mim]Cl solution along with H_2SO_4/cellulose with varied mass ratios for several minutes to hours to achieve *ca.* 50% glucose yield.[105] It is also claimed that in the presence of cellulose and [BMIM]Cl, the rate of hydrolysis catalysed by maleic acid was enhanced. The catalytic activity of various acids in the presence of [BMIM]Cl was observed as follows:[106] H_3PO_4 < maleic acid < H_2SO_4 < HNO_3 < HCl.

To prevail over the use of mineral acids, later acidic ILs [1-(1-propylsulfonic)-3-methylimidazolium chloride and 1-(1-butylsulfonic)-3-methylimidazolium chloride] with a –SO_3H group attached *via* the propyl chain to the cation were shown to hydrolyse cellulose into sugars at 70 °C in water.[107] Soon after, several research groups studied the effects of various types of acidic ILs on the hydrolysis of cellulose.[108–110] However, like mineral acids, ILs also face drawbacks such as a homogeneous nature, recyclability and high costs, which make the overall process commercially unfavorable.[111]

Overall, the use of homogeneous catalysts in cellulose hydrolysis to yield glucose has long been known and has been practiced on an industrial scale with renewed interest. However, these methods face quite a few drawbacks as discussed. Realising this, researchers have recently started using heterogeneous catalysts which may surmount the problems (non-corrosive, easy recovery and reusability of the catalyst, elimination of neutralization process *etc.*) faced while using homogeneous catalysts. Moreover, knowing that nearly 80% of the current industrial processes use heterogeneous catalysts, it is interesting for researchers to generate knowledge on the conversion of saccharides into sugars.

1.3.2   Use of a Heterogeneous Catalytic System

As discussed, it will be beneficial to use heterogeneous (solid) catalysts in the hydrolysis of cellulose, because they can be easily separated from the reaction mixture *via* filtration and their acidity (strength) can be tuned as per the requisite of the reaction.[8,112] Synthesis of glucose (C_6 sugar) from various feedstocks (mostly from cellulose) using various heterogeneous catalysts can be subcategorised as (1) solid acid catalysed and (2) supported metal catalysed. The details on these processes are discussed next.

1.3.2.1 Solid Acid Catalysed

Solid acid catalysts defined as 'solid materials having acidic properties because of the presence of either Brønsted acid sites or Lewis acid sites or both' are useful in many reactions. Solid acid catalysts such as zeolites, ion-exchanged resins, functionalized mesoporous silicas, functionalized carbons, functionalized and supported metal oxides, heteropoly acids *etc.* (Figure 1.9) can influence the hydrolysis reaction of disaccharides (cellobiose, maltose and

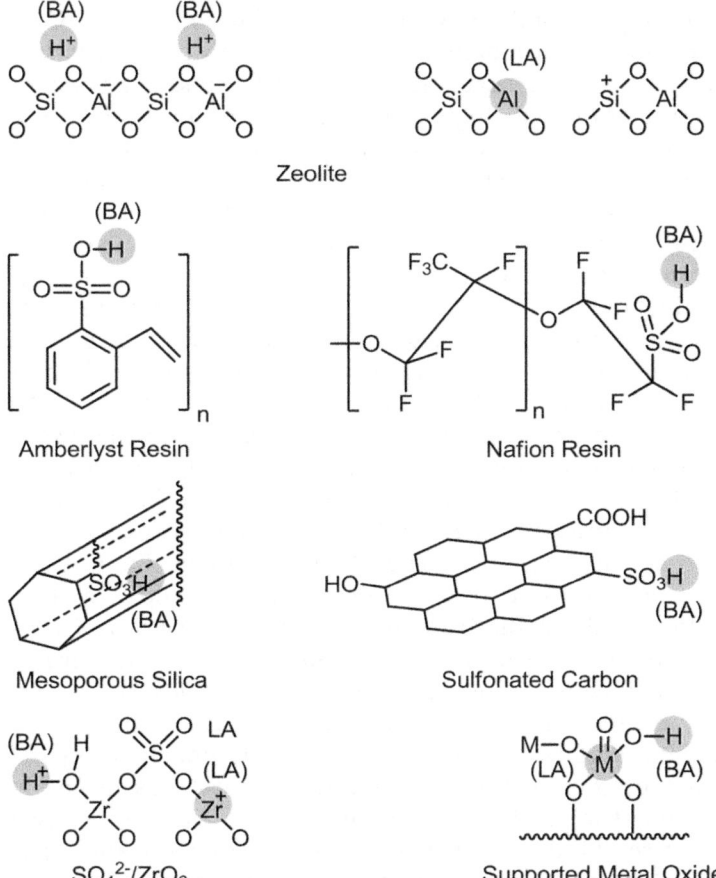

Figure 1.9 Solid acid catalysts with their typical acid sites. (BA) indicates Brønsted acid sites and (LA) indicates Lewis acid sites.

sucrose) and polysaccharides (cellulose, starch, and inulin) into sugar monomers (glucose and fructose). Solid acids have advantages over homogeneous acids (typically mineral acids) in terms of selectivity, stability, recovery and reusability. They can be used as a direct replacement to the homogeneous acids as their acid strengths can be manipulated for achieving optimum concentration of protons in the solution. They also have lower operating costs and avoid any corrosion of reactors as after removal of the catalyst from the solution by simple filtration, the solution becomes neutral. However, they have the drawback of diffusion limitations and thus can slow down the reaction compared to their homogeneous counterparts. Moreover, solid acids can easily be poisoned by water (by forming a solvation layer around acid sites by hydrogen bonding) or they degrade during the hydrothermal treatment involved under biomass conversion conditions and consequently, it becomes absolutely critical to design and develop solid acids that can tolerate water poisoning and are hydrothermally stable.

In the early studies, as an alternative to homogeneous acids, zeolites, a class of solid acid catalysts, were used in the conversion of various disaccharides (maltose, cellobiose and sucrose) and polysaccharides (inulin, starch and cellulose) to sugars (glucose and fructose) in a water medium. Zeolites are microporous, high surface area aluminosilicates with a framework consisting of AlO_4 and SiO_4 tetrahedra linked with each other *via* oxygen, are naturally available and have been known for over 200 years following the discovery of stilbite.[113] Commercially synthesized zeolites with specific acidity, tuneable hydrophilicity, channel structure, textural property, varying porosity, high thermal stability and Si/Al ratio find much application in oil refining and petrochemistry. Their porous structure developed by 1D, 2D or 3D channel formation can accommodate various substrates depending on the type of zeolites and thus provide the sites for obtaining shape selective catalysis. In the early works, faujasite (HY, Si/Al = 2.4, 27, 55, 110) zeolite was employed in the hydrolysis of sucrose and the fact was established that after dealumination (higher Si/Al ratio), the reaction rate enhances.[114] Later, in another report, it was shown that the use of sucrose as a substrate in the presence of HY (Si/Al = 15) zeolite yields glucose and fructose (invert sugars) at 85 °C.[115] The HY (Si/Al = 15) zeolite is also used as a catalyst for the hydrolysis of other disaccharides (maltose and cellobiose). When maltose and cellobiose are treated at 150 °C, nearly 90% conversion is achieved to yield ∼83% and ∼88% of glucose, respectively. It is suggested that the catalytic activity is dependent on the ease of adsorption of substrates *via* two oxygen atoms (ring oxygen and oxygen attached to anomeric carbon) on the catalyst surface. Thereafter, the substrate undergoes hydrolysis over acids sites. Moreover, hydrolysis of inulin and starch has also been reported with HY catalyst to yield 92% fructose and 95% glucose, respectively. The influence of the Si/Al ratio in Hβ is proved with the results that with an increase in Si/Al ratio, the Hβ zeolite becomes more hydrophobic, which promotes an easy maltose adsorption on the catalyst surface and thereby improves the catalytic activity.[116] This phenomenon explains the observation of higher amounts (80%) of glucose with Hβ (Si/Al = 50) compared to Hβ (Si/Al = 12.5) (43%) under similar reaction conditions. The use of dealuminated H-form of mordenite (HMOR) (Si/Al = 12) catalyst to produce 66% glucose from 70% maltose conversion at 130 °C is also known. Later, the use of zeolites (HMOR and Hβ) in the hydrolysis of starch was shown to yield 18% glucose.[116] However, due to a lower hydrothermal stability, catalysts undergo deactivation *via* leaching of their active component in water. When cellulose is used as a substrate, due to its insolubility in most of the solvents, the reaction system becomes heterogeneous with respect to the substrate itself. This eventually affects the reaction rate of cellulose hydrolysis in the presence of solid acid catalysts. However, as discussed earlier in this chapter (Section 1.2), pre-treatment of cellulose reduces its crystallinity and helps in solubilizing part of it in hot water. It is shown that the ball milling of microcrystalline cellulose (Avicel) (crystallinity = 75%) using ZrO_2 balls (mass = 1.8 kg, diameter = 2 cm) for 48 h (spinning speed = 60 rpm) completely diminishes the cellulose crystallinity leading to formation of amorphous cellulose as proven by XRD technique.[117] Subsequently, this treated cellulose is hydrolysed over several zeolites such as HMOR (Si/Al = 10), Hβ (Si/Al = 12, 75) and HZSM-5 (Si/Al = 45) at 150 °C. Among all zeolites, Hβ (Si/Al = 75) showed the best catalytic activity (glucose yield = 13%) due to its higher hydrophobicity compared to other zeolites. Next, to improve the yields with zeolites, reactions were conducted in IL medium as ILs can solubilize cellulose (Figure 1.5). It is seen that with the use of a [BMIM]Cl/H_2O system along with HY catalyst, a *ca.* 50% glucose yield can be obtained at 130 °C within 2 h.[118] However, when conventional heating is replaced with microwave heating, a *ca.* 37% glucose yield at 100 °C within just 0.13 h was observed.[119] Although zeolites are active in these reactions, lower hydrothermal stability and leaching of Si and Al in the solution hamper their use. Thus, it is important to synthesize zeolites with higher thermal stability to help achieve recyclable activity in cellulose hydrolysis reactions. Recently, it has been proposed that to increase the stability of zeolites (by avoiding desilication) in hot water, the presence of Al in both forms, lattice and extra framework is helpful.[120] It is suggested that lattice Al counteracts the hydrolysis of framework bonds and extra framework Al prevents the solubilization of the framework. Moreover, the authors claim that with a decrease in the Si/Al ratio, an increase in structure stability is possible because of the higher Al content. When H-form of ultra stable Y (HUSY) is subjected to mild steam treatment, the concentration of extra framework Al increases on the external surface of the zeolite, thereby giving it a protective cover, which as mentioned earlier, helps in reducing the solubility of the zeolite.

Typically, (cation) ion-exchanged resins with acidity due to the presence of –SO_3H groups are used in many acid catalysed organic transformations. Though they are stable in various organic solvents, their hydrothermal stability is not very high (<150 °C). It is shown that with an increase in the degree of resin cross-linkages, the access of substrate molecules towards the catalyst active centre is more restricted, decreasing the catalytic activity.[116]

Use of 2% cross-linked resin (Dowex 50x2-100) can afford complete hydrolysis of maltose to yield *ca.* 95% glucose within 26 h at 120 °C while 8% cross-linked resin (Dowex 50x8-100) catalyst requires 130 °C for complete maltose hydrolysis.[116] Amberlyst-15 resin has a very high acid amount (4.65 mmol g^{-1}) compared to Nafion-silica resin (0.17 mmol g^{-1}) and hence, Amberlyst-15 showed higher sucrose hydrolysis activity (conversion = 88%, glucose yield = 88%, fructose yield = 87%) than Nafion-silica (conversion = 28%, glucose yield = 28%, fructose yield = 26%) under identical reaction conditions (80 °C, 4 h).[121] These findings are confirmed by another study carried out with Amberlyst-15 (78%), Nafion NR-50 (42%) and Nafion SAC-13 (29%) catalysts for the sucrose hydrolysis reaction at 80 °C.[122] However, it should be considered that this activity can be manipulated by charging varying quantities of catalysts in the reaction. Some other reports also discuss the use of ion exchange resins such as Amberlite A120 and Amberlite 200 for the hydrolysis of sucrose to yield 82–98% sugars at 80 °C.[123,124] When maltose is used as a substrate, Amberlite 200 resin gives 95% sugar formation at 80 °C. The same catalyst, however, shows lower activity in cellobiose conversion (19%) into sugars (15%).[124] At the same time, with a slight increase in temperature to 90 °C and use of Amberlyst-15 catalyst, a maximum of 61% glucose formation with 62% cellobiose conversion was reported.[125] Amberlyst-15 catalyst was also evaluated in starch hydrolysis at 130 °C giving 25% glucose along with 12% maltose.[121] Although strongly acidic resin catalysts (Amberlyst-15, Nafion NR-50) showed high activity in the hydrolysis reactions of sucrose, maltose and starch, they showed low activity when used in the hydrolysis of microcrystalline cellulose.[126] It is suggested that obtaining reproducible activity with resin catalysts in the cellulose hydrolysis reaction is difficult since this process requires temperatures higher than the degradation temperatures of these materials. To overcome these problems, use of ILs along with resins has been suggested by a few research groups. In one report, use of an acidic resin, NKC-9 along with [C$_4$min]Cl ionic liquid, was shown to yield 39% glucose.[119] A few other reports also claimed the formation of the hydrolysis product, glucose (35 and 83%), from cellulose using a combination of resin catalysts (Nafion-50 and Dowex 50wx8-100, respectively) and ILs ([BMIM]Cl and [EMIM]Cl, respectively).[127,128]

In the early 1990s, siliceous mesoporous silicas such as FSM-16 (Folded Sheet Mesoporous/Material), SBA-15 (Santa Barbara Amorphous) and MCM-41 (Mobil Corporation/Composite Mesoporous/Material) with large pore openings (>2 nm) were invented. The discovery of these materials paved the way to overcoming the diffusion limitations faced by bigger molecules to enter the smaller (micro, <2 nm) pores of zeolites.[129–131] Though these materials had a 1D channel structure, subsequent developments in the synthesis of many other mesoporous materials with 2D and 3D channel structures have been successful. Characteristically siliceous materials do not have any acidity (or very weak acidity due to surface –OH groups), though by incorporation of heteroatoms such as Al, Ga *etc.*, acid sites can be generated on these materials. Moreover, anchoring sulfonyl groups ($-SO_3H$) on the surface *via* one-pot or grafting methods also gives rise to acidity in these materials. Because of a large pore diameter and acidity, these materials (MCM-41, SBA-15, FSM-16 and HMM (Hybrid Mesoporous Materials))[132] are largely used in several acid catalysed reactions such as etherification, esterification *etc.* carried out in organic media.[133,134] Nevertheless, when these catalysts are used in the presence of water, leaching of $-SO_3H$ groups is observed, which makes these catalysts unsuitable for hydrolysis or dehydration reactions. Considering these facts, researchers have used hot water treated (to remove loosely bound $-SO_3H$ groups) sulfonic acid functionalized phenylene-bridged mesoporous silica (Ph-HMM) and ethylene-bridged mesoporous silica material (Et-HMM) catalysts in the sucrose reaction.[121] These materials with an acid amount of 0.31–0.90 mmol g^{-1} showed very high catalytic activity for sucrose conversion (81–90%) into glucose and fructose (81–90%) at 80 °C. Moreover, sulfonic acid functionalized FSM-16 (acid amount = 1.11 mmol g^{-1}) synthesized by a post-synthesis grafting method was also used to produce glucose and fructose from sucrose.[121] Later, these catalysts were employed for the hydrolysis of starch at 130 °C to obtain ~67% glucose. Another report compared the activities of silica materials after surface functionalizations with various acid groups such as butylcarboxylic, propylsulfonic and arenesulfonic acids in cellobiose hydrolysis.[135] It was shown that both propylsulfonic and arenesulfonic acid functionalized silica materials are able to convert >90% of cellobiose at 175 °C, while butylcarboxylic acid functionalized silica exhibited only 15% cellobiose conversion. The ambiguity in results is because of fewer acid sites generated in water for butylcarboxylic acid functionalized silica (pH = 4.90) compared to the other two silica materials (pH = 2.67–2.89).

This means that pH below 4.0 is mandatory for hydrolysis of cellobiose. Moreover, an almost similar activation energy (110–138 kJ mol^{-1}) for the cellobiose hydrolysis reaction was reported in the presence of all silica catalysts[135] and eventually, the values matched well with those reported earlier for H_2SO_4 (110 ± 29.6 kJ mol^{-1}) and maleic acid (114 ± 9.3 kJ mol^{-1}).[136] The good correlation of activation energy values between silica catalysts and homogeneous catalysts revealed that mass transfer limitation is not playing any significant role and the reactions are guided only by hydrated protons. These results are interesting and subsequent studies showed that dispersion of solid acids in water eventually forms H_3O^+ species in bulk water, which are actually active in the hydrolysis reaction as explained in Figure 1.10.[137] These facts elucidate the capability of many solid acid catalysts with a pore diameter <2 nm to hydrolyse bigger molecules such as cellulose or hemicelluloses. Studies have also confirmed that at 175 °C, glucose undergoes only 15% conversion, implying that at lower temperatures, glucose is stable.[135]

For numerous years, inexpensively available carbon materials (activated carbon, char, graphite, nanotubes *etc.*) have attracted a lot of attention as adsorbents, catalysts or supports due to their large specific surface area,

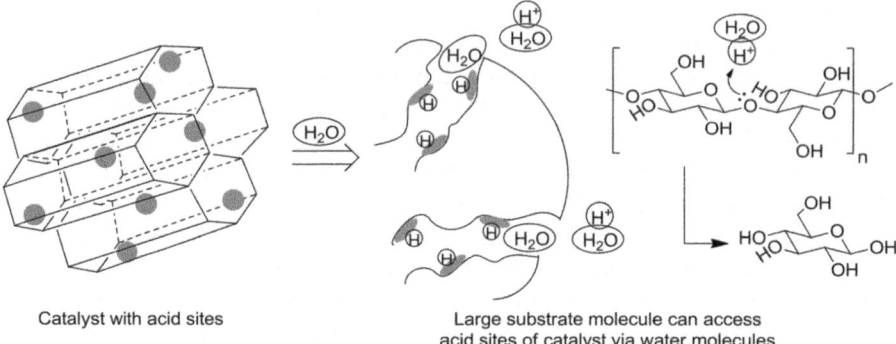

Catalyst with acid sites

Large substrate molecule can access
acid sites of catalyst via water molecules

Figure 1.10 Possible pathway for the formation of H_3O^+ in bulk water to catalyse the hydrolysis of saccharides.

high porosity, excellent electron conductivity, relative chemical inertness *etc.*[138] These excellent materials with the incorporation of acidic functionality can enhance their performance in several acid catalysed reactions.[139,140] The acidity can be incorporated in these materials by washing the carbon with H_2SO_4 or grafting sulfonic acid groups on its surface. Additionally, activated carbons may have –OH and $-CO_2H$ groups on their surface, which can give rise to weak acidity. Sulfonated activated carbon ($AC-SO_3H$) prepared by treatment of AC with concentrated H_2SO_4 at 150 °C under an Ar gas flow generates high acidity (1.63 mmol g^{-1}), which is responsible for achieving a $\sim 90\%$ glucose yield in a starch hydrolysis reaction carried out at 120 °C.[117] Soon after, various functional groups such as $-SO_3H$, $-CO_2H$ and –OH bearing amorphous carbon materials ($CH_{0.64}O_{0.49}S_{0.032}$) were synthesized by sulfonation (fuming H_2SO_4, 120 °C, 10 h, N_2 gas flow) of carbon obtained from partial carbonization of cellulose (400 °C, 1 h, N_2 gas flow).[125] During cellobiose hydrolysis, it was suggested that the –OH groups present on the carbon surface form strong H-bonds with cellobiose glycosidic oxygen, allowing it to adsorb on the carbon surface. Subsequently, the $-SO_3H$ (acid density = 1.5 mmol g^{-1}) group on the carbon surface catalyses the hydrolysis of this anchored cellobiose easily to yield a high amount of glucose (81–83%) at 90 °C. A similar influence of the catalyst surface property is also valid in the hydrolysis of cellohexaose to yield glucose.[125] Due to very minor leaching of $-SO_3H$ groups from this catalyst, it showed recyclable activity in five runs. Another report showed that partial carbonization of polyvinyl chloride (PVC) at 400 °C in a N_2 atmosphere and its subsequent sulfonation using fuming H_2SO_4 produced a carbon material (PVC-AC-673) with $-SO_3H$ group bonded carbon sheets linked *via* rigid sp^2 bonds and flexible aliphatic hydrocarbons linked *via* sp^3 bonds.[141] It was proposed that the additional flexible aliphatic hydrocarbons along with planar carbon sheets overcome the diffusion limitations of reactant molecules and hence enhance its catalytic activity to yield 30% glucose from cellobiose. Later, $AC-SO_3H$ catalyst was checked in a ball-milled cellulose hydrolysis reaction to yield superior amounts of glucose (41%).[117] The better catalytic activity of AC-SO$_3$H is ascribed to the fact that this catalyst has higher hydrophobic graphene planes and strong acidic −SO$_3$H surface functionalization. Furthermore, AC-SO$_3$H catalyst is also shown to be recyclable in three runs without losing its activity. Amorphous carbon bearing −SO$_3$H, −CO$_2$H and −OH groups (CH$_{0.62}$O$_{0.54}$S$_{0.05}$) is known for the successful hydrolysis of microcrystalline cellulose into 64% oligomers (β-1,4-glucan; DP = 2–4) and 4% glucose at 100 °C.[126] Moreover, the carbon catalyst is shown to be recyclable 25 times with similar activity despite leaching of −SO$_3$H groups after the first run. It is suggested that the better activity of this carbon catalyst is due to the strong adsorption of the substrate on the catalyst surface, which also reduces the activation energy barrier (110 kJ mol^{-1}) for the cellulose hydrolysis reaction compared to H$_2$SO$_4$ (170 kJ mol^{-1}).[126,142] Later, from various mathematical and statistical calculations (artificial neural network (ANN) model and response surface methodology (RSM)), authors tried to understand the correlation between reaction physical phenomenons.[143] It is concluded that in the hydrolysis of cellulose, the amorphous carbon catalyst follows the same pathway that concentrated H$_2$SO$_4$ follows. One report suggests that due to the presence of aromatic rings in mesoporous carbon nanoparticles (MCN), it forms strong CH−π interactions with hydrogens of glucans (glucose polysaccharide), because of which higher adsorption of glucans on these materials is possible.[144] This observation was confirmed by GPC, ^{13}C Bloch Decay NMR and MALDI-TOF-MS analysis. Moreover, it was found that with an increase in glucan chain length, the adsorption free energy decreases (*ca.* 1.67 kJ mol^{-1} with each glucose unit in glucan series) due to a higher degree of CH−π interactions. Three sulfonated carbon catalysts with −SO$_3$H, −CO$_2$H and −OH groups were synthesized from bamboo (BC-SO$_3$H-1), cotton (BC-SO$_3$H-2) and starch (BC-SO$_3$H-3) and were used as catalysts in a microcrystalline cellulose (DP = 200–1000) hydrolysis reaction under microwave heating.[145] Because of the use of a microwave power of 350 W (cellulose crystallinity reduction from 75% to 58%) in the presence of BC-SO$_3$H-1 catalyst, the highest amount (24%) of reducing sugar formation (glucose = 17%, cellooligomers = 7%) was observed. In contrast to this, only 5% reducing sugar (glucose = 3%, cellooligomers = 2%) formation was observed under conventional heating (oil bath) at 90 °C. Other carbon catalysts (BC-SO$_3$H-2 and BC-SO$_3$H-3) afforded 28% (glucose = 20%, cellooligomers = 8%) and 13% (glucose = 5%, cellooligomers = 8%) of reducing sugars, respectively. Leaching of −SO$_3$H groups in the solution (fresh = 1.87 mmol g^{-1}, spent = 1.12 mmol g^{-1}) was, however, observed for the BC-SO$_3$H-1 catalyst, which led to a decrease in the catalytic activity in recycle runs (reducing sugar yields; 1st run: 24%, 2nd run: 18%, 3rd run: 16%, 4th run: 15%). Leaching is a major problem with any −SO$_3$H functionalized catalyst and care should be taken to confirm that the activity observed is due to anchored −SO$_3$H groups and not because of leached out −SO$_3$H groups. The influence of sulfonation temperature during the synthesis of sulfonated activated carbon was also studied and it was shown that compared to 150, 200, 280 and 300 °C, 250 °C treatment gave the best yields of glucose (63%) from ball milled cellulose.[146] Moreover, the influence of various sources of sulfonated carbons (sulfonation temperature = 250 °C), such as acetylene carbon black (ACB), multi-wall carbon nanotube (MWCNT), cell-carbon (carbonization of microcrystalline cellulose), coconut shell active carbon (CSAC), resin carbon (carbonization of mesoporous resin) and mesoporous carbon (CMK-3), was also investigated.[146] It was found that the sulfonated CMK-3 catalyst was highly active due to its mesoporous nature and yielded the highest amount of glucose (75%). However, all other sulfonated carbons showed lower activity (glucose yield = 15–55%) under similar conditions. Execution of both a strong Brønsted acid site presence and favourable glucan adsorption on the catalyst was explored by using sulfonated silica/carbon nanocomposite catalyst.[147] The composite catalyst with a particular composition for Si/C (33 wt%/66 wt%) and particular carbonization temperature (600 °C) showed the best catalytic activity for ball milled cellulose hydrolysis (conversion = 61%, glucose yield = 50%) compared to other catalysts. This higher catalytic activity is explained by the presence of strong Brønsted acidic sites in the composite catalyst and its ability to adsorb glucan strongly. It is suggested that the silica may provide better mechanical and thermal stability to the catalyst. Nevertheless, when the composite catalyst was reused, it showed a slow but continuous decrease in activity due to leaching of $-SO_3H$ groups. From the discussions on carbon catalysts, it is evident that, to achieve higher activity, few hydrophilic groups on the surface are necessary to attract saccharides otherwise they will be repelled by the hydrophobic surface of carbon.

Metal oxides characteristically give Lewis acid sites and due to their higher thermal stability, they are widely used in several organic transformations and in sugar synthesis. With a layered $HNbMoO_6$ catalyst (1.9 mmol g^{-1}) at 80 °C, almost complete conversion of sucrose is observed to achieve selectively glucose and fructose as products. With a $HTiNbO_5$ nanosheet (0.4 mmol g^{-1}) catalyst, only 42% yield was seen, however.[122] The very high activity observed with the layered $HNbMoO_6$ catalyst is explained by the strong interactions between sucrose and the interlayer of $HNbMoO_6$ (intercalation; proven by XRD and elemental analysis). However, with the same catalyst, the rate of glucose formation decreased when cellobiose was used as a substrate (1.18 mmol g h^{-1}) instead of sucrose (24.1 mmol g h^{-1}). To understand this difference, the authors studied the intercalation ability of both substrates in the presence of $HNbMoO_6$ catalyst. The results displayed higher substrate adsorption in the case of sucrose (1.12 mol%) compared to cellobiose (0.21 mol%) per mole of catalyst. This difference may arise from the orientation of sugars in these disaccharides. In starch hydrolysis, $HNbMoO_6$ catalyst yielded a maximum of 45% glucose at 100 °C. Cellulose hydrolysis was also performed using layered $HNbMoO_6$ catalyst, however very low yields of glucose were observed.[122] Catalytic activity with Nb-W oxide in the hydrolysis of sucrose and cellobiose was also reported.[148] With an increase in W content in the catalyst, the formation of strongly acidic

Nb_3W_7 oxide species was observed. This in turn was responsible for improvements in the glucose yields (65%) from sucrose. However, due to the presence of low acid density in the Nb-W catalyst, it showed lower cellobiose hydrolysis activity. Several other metal oxides such as $SiO_2-Al_2O_3$, SiO_2-ZrO_2, Nb_2O_5 and $Nb_2O_5-PO_4$ have also been evaluated for their activity in sucrose hydrolysis and among them, $Nb_2O_5-PO_4$ yielded the highest amount (62%) of glucose.[124] A bare nano Zn-Ca-Fe oxide catalyst was also shown to be efficient in the hydrolysis of crystalline cellulose to yield 29% glucose.[149] Moreover, due to the paramagnetic nature of the Fe oxide present in the catalyst, it can be easily separated from a reaction mixture. Use of $-SO_3H$ functionalized mixed metal oxide ($CoFe_2O_4-SiO_2$) catalyst was demonstrated in the hydrolysis of cellobiose to yield 50% glucose at 175 °C.[150] In ball milled cellulose hydrolysis, a sulfonated zirconia (SO_4^{2-}/ZrO_2) catalyst was applied but with limited activity (14% glucose).[117] However, the leaching of SO_4^{2-} in water is very well known and hence these catalysts are not recyclable and thus were not evaluated further.

Heteropoly acids (HPA) or polyoxometalates are another class of acid catalysts that have the composition of various oxoacids (transition metal–oxygen anion clusters).[151–153] Structurally, HPAs are of two types, namely Keggin and Dawson with molecular formulae of $XY_xM_{(12-x)}O_{40}$ and $X_2M_{18}O_{62}$, respectively (where X is a heteroatom from the p-block and Y and M are addendum atoms). For example, $H_3PW_{12}O_{40}$ and $H_5PMo_{10}V_2O_{40}$ have Keggin structures and $H_6P_2Mo_{18}O_{62}$ has a Dawson structure. The protonated forms of HPAs are homogeneous in nature, however, partial replacement of protons by larger ions such as Cs^+, Ag^+ *etc.* makes HPAs heterogeneous. These catalysts are used in the cellobiose hydrolysis reaction and their activity is shown to be dependent on their Brønsted acid strength.[94] Homogeneous HPA and metal ion incorporated phosphotungstic acids $(PW_{12}O_{40}^{3-})$ are also used as catalysts in cellulose (ball milled) hydrolysis reactions to understand the influence of Brønsted and Lewis acid sites.[94] From the results obtained with all catalysts, the authors have concluded that stronger Brønsted acid catalysts are more favourable for hydrolysis of β-1,4-glycosidic bonds compared to Lewis acid catalysts. The difference between the activity of Brønsted and Lewis acid sites is because, as mentioned in Figure 1.10, Brønsted acid sites can liberate H^+ to give rise to H_3O^+ which then is available in bulk water. In this way, it can easily interact with the substrate molecule to give rise to protonated oxygen. In the case of Lewis acid sites, the substrate has to interact with them through a lone pair on oxygen (bridged) and only then can the reaction occur. Use of HPA, $H_3PW_{12}O_{40}$ and $Sn_{0.75}PW_{12}O_{40}$ showed ~40% formation of total reducing sugars from ball milled cellulose.[94] In another report, use of $H_3PW_{12}O_{40}$ as a catalyst for the conversion of microcrystalline cellulose was also shown to produce a maximum of 52% glucose at 180 °C.[95] It was discussed that although the $H_3PW_{12}O_{40}$ catalyst is homogeneous in nature, it could be recycled six times after the catalyst was extracted with diethyl ether. However, leaching of Keggin-type $PW_{12}O_{40}^{3-}$ ions was seen in the reaction solution as proven by UV-Vis analysis (absorption bands at 200 nm and 265 nm). In another report, micellar HPA, $[C_{16}H_{33}N(CH_3)_3]H_2PW_{12}O_{40}$ and $Cs_{2.5}H_{0.5}PW_{12}O_{40}$ catalysts were studied for the hydrolysis of starch and cellulose.[154] Higher amounts of glucose formation were demonstrated using micellar HPA catalyst (starch: 82%, cellulose: 39%) compared to $Cs_{2.5}H_{0.5}PW_{12}O_{40}$ (starch: 43%, cellulose: 21%) since the micellar HPA catalyst attracts cellulose to accumulate around the micellar core, which gives cellulose easy access to catalytically active sites.

Although various methodologies have been developed for the hydrolysis of isolated cellulose into glucose in the presence of several solid acid catalysts, it would obviously be beneficial if a methodology was developed wherein sugars are obtained directly from raw biomass without any pre-treatment (without isolating substrates). This will help to reduce the cost of sugar synthesis (by virtue of reducing capital costs for the isolation process) and will also avoid generation of wastes (generated during isolation). Very few processes for the direct hydrolysis of polysaccharides from lignocelluloses into sugars using heterogeneous catalysts have been reported in the literature. It was demonstrated earlier that carbon catalysts bearing $-SO_3H$, $-CO_2H$ and $-OH$ groups have better catalytic influence on the hydrolysis of isolated polysaccharides *via* strong adsorption capability.[126] As an extension to the work, the authors used dried *Eucalyptus* flakes as a substrate for hydrolysis of its cellulose part along with its hemicellulose part into water soluble saccharides.[126] The authors claimed that all the cellulosic material was hydrolysed into water soluble saccharides at 100 °C but a lack of detailed quantitative information on the products' formation is detrimental to studying this catalyst further. However, from earlier results, it can be assumed that most of the products are water soluble oligomers or polymers. Another research group showed the use of recyclable carbonaceous solid acid (CSA) with $-SO_3H$, $-CO_2H$ and $-OH$ groups in the hydrolysis of cellulose and hemicellulose from corn cob to yield C_6 and C_5 sugars.[155]

The discussion on solid acid catalysts reveals that it is possible to hydrolyse several di- and polysaccharides into sugars, although it is essential to develop a catalyst that is stable and recyclable even if it gives lower activity. Moreover, poisoning of solid acid catalysts (for example $-SO_3^-H^+ \rightarrow -SO_3^-K^+$; $-O-Si-O(H)-Al-O- \rightarrow -O-Si-O(K)-Al-O-$) with the impurities present in lignocelluloses (nutrients, waxes) may reduce their activity in recycle runs. To use these catalysts again, treatment of these catalysts would be necessary, which may add the cost to the process and moreover, during these treatments, the catalysts may undergo structural changes by which initial activity may not be retained.

1.3.2.2 Supported Metal Catalysed

As early as 1957, pioneering work was done by Balandin and co-workers for the one-pot synthesis of sugar alcohols (sorbitol) from cellulosic biomass with the help of dilute acid and Ru/C catalyst.[156] The studies were carried out with 1–2 wt% H_2SO_4 and Ru/C at 70 bar hydrogen pressure to yield *ca.* 80% of sorbitol. Later, the same group replaced H_2SO_4 with 0.7 molar phosphoric acid to increase the yield to *ca.* 90%.[157] The increase in yield with phosphoric acid is due to suppression of strong acid (H_2SO_4) catalysed cyclodehydration reactions of sorbitol to give sorbitan.[158] In these studies, homogeneous acids were used to hydrolyse cellulose into glucose and further, its hydrogenation was catalysed by Ru/C catalyst. In 2006, a new method was disclosed to convert cellulose into sugar alcohols (*via* glucose formation) by avoiding the use of homogeneous acids.[159] In the report, it is suggested that by using supported metal catalysts (Pt/Al_2O_3) under hydrogen pressure and in the presence of water, hetero cleavage of hydrogen is possible. The H^+ formed during the process acts as a proton source which helps in the hydrolysis of cellulose to yield glucose. However, under the reaction conditions, it was seen that glucose immediately undergoes a hydrogenation reaction to yield sugar alcohols (sorbitol, mannitol). Around the same time, another group also showed the conversion of cellobiose into sugar alcohols over supported metal catalysts and proposed a similar mechanism.[160] These results also point out that when homogeneous acids and Ru/C catalysts were used in early studies, it is possible that Ru/C was also capable of catalysing hydrolysis reactions. Subsequently, several other reports claimed *in situ* acid site generation over various metals (Pt, Ru, Ni) for the conversion of cellulose.[161,162] It is believed that the support material also has an influence on acid site generation. When a carbon nanotube (CNT) was used to support Ru metal, it enhanced the amount of acid sites and thereby catalytic activity since CNT has a higher ability for hydrogen adsorption.[163] Based on the concept of *in situ* acid site generation, recently, it was shown that a Pt/fibrous 3D carbon catalyst can convert lignocelluloses directly into sugar alcohols *via* a hydrolytic hydrogenation reaction.[164] Though supported metal catalysts are probed further in the conversion of cellulose, the main product formed is sugar alcohol. This is because hydrolysis of cellulose is a rate-determining step and requires harsher reaction conditions than the hydrogenation of sugars to yield sugar alcohols. This means that, independent of the type of supported metal catalyst, used along with hydrogen, they will surely yield hydrogenated products of sugars.

1.4 Synthesis of C_5 Sugars

Four C_5 sugars, namely xylose, arabinose, ribose and lyxose, are present in nature. Invented in 1891, D-ribose is a C-2 epimer of D-arabinose and is present in the backbone of RNA. On the other hand, lyxose is a C-2 epimer of xylose and is available as a component of bacterial glycolipids. Since D-ribose and lyxose are very rarely available in nature and also are not present in plant biomass, discussions on these sugars are not made in this chapter. As shown in Figure 1.2, the abundant sugars, xylose and arabinose can be produced from the hemicellulose part of lignocelluloses. Hemicelluloses are polysaccharides with low molecular weights and are covalently or non-covalently bonded with cellulose and lignin in plant cell walls. They are heteroglycans and are mostly made up of D-xylose, D-mannose, D-galactose, L-arabinose, D-glucose, D-galacturonic acid, glucuronic acid and 4-*O*-methylglucuronic acid *etc.*[165,166] At times, these sugars also have substituents such as acetyl and methyl.[167] Depending on the various sugars linked *via* different fashions with each other, hemicelluloses are named. For example, hemicelluloses made up of xylose and arabinose are termed as arabinoxylan. Based on the origin of hemicelluloses, either from hardwood or softwood or grasses, differences in their composition and linkages are observed and those are summarized in Table 1.1.

Typically, hardwood hemicelluloses and grasses contain a high degree of xylans (main backbone is made up of xylose units with branching after 7–10 units). Based on the substituent present on xylose, their names are derived as arabinoxylan and arabinoglucoronoxylan. On the contrary, softwood hemicelluloses are largely galactoglucomannans but along with them, some small portion of xylan is also present. Due to the random and branched structure of hemicelluloses, they do not have any extensive hydrogen bonding (like in cellulose), which makes hemicelluloses amorphous in nature. These amorphous hemicelluloses during isolation from lignocellulosic materials undergo various transformations such as partial hydrolysis (due to high temperatures and pressures) to yield low molecular weight fractions and saponification during alkali treatment. Hence, upon hydrolysis of isolated hemicelluloses, besides the expected products, a few other products might also be detected although on a ppm level. Moreover, due to their varying compositions and linkages, it becomes crucial to develop a universal catalytic method that can hydrolyse various types of hemicelluloses either present in isolated form or present in lignocelluloses (without separation from cellulose and lignin) in an efficient way. It is important to mention that oligosaccharides ($2 < DP < 7$) of hemicelluloses (obtained upon hydrolysis)

Table 1.1 Summary of linkages and residues present in various types of hemicellulose.

Type	Linkages	Components
Softwood hemicellulose		
Arabinogalactan	1,3-, 1,6-, 1,4-, 1,5-	Arabinose, galactose, glucuronic acid
Galactoglucomannan	1,4-, 1,6-	Galactose, mannose, glucose, acetyl substitution
Arabino-4-*O*-methylglucuronoxylan	1,4-, 1,2-, 1,3-	Arabinose, xylose, glucuronic acid, methyl substitution
Hardwood hemicellulose		
Glucomannan	1,4-	Glucose, mannose
Arabinoxylan/ arabinoglucuronoxylan	1,4-, 1,3-, 1,2-, 1,2/3-	Xylose, arabinose, glucuronic acid, other substitution (acetyl, methyl, feruloyl, coumaroyl)
O-Acetyl-4-*O*-methyl-glucuronoxylan	1,4-, 1,2-	Xylose, glucuronic acid, other substitution (acetyl, methyl)

are termed as, "functional food" or "dietary fibres" and added in food to exert health benefits.

Similar to cellulose, for the hydrolysis of hemicellulose, several catalytic systems such as homogeneous and heterogeneous ones were investigated. However, unlike several C_6 sugar based disaccharides, which are naturally and commercially available in abundance, almost no C_5 sugars containing disaccharides are available and hence in the next section, the focus is only on hemicelluloses.

1.4.1   Use of a Homogeneous Catalytic System

For the hydrolysis of hemicelluloses, several mineral acids (diluted and concentrated HCl, H_2SO_4 and H_3PO_4) and organic acids (oxalic acid, acetic acid, maleic acid, trifluoroacetic acid) are known and details on these are extensively described in a few excellent reviews.[12,101,168] Typically, organic acids with lower acid strength (pK_a, 3.0–5.0) compared to mineral acids (pK_a <1) are sufficient to hydrolyse hemicelluloses into their respective sugars since hemicelluloses are amorphous, branched and have lower DP (<250) compared to cellulose. It has been proved that with the higher strength of mineral acids, formation of degraded products (*via* sugars) is possible.[169] This phenomenon is observed because different linkages in hemicelluloses cleave under different conditions and in most cases, sugars obtained under mild conditions are degraded under the harsher conditions used to cleave the remaining hemicelluloses. Use of trifluoroacetic acid, even if it is some-what strong, is useful in these reactions since it has a low boiling point (72.4 °C) and thus can be removed from the reaction mixture easily. This also avoids the neutralization required with other acids, which is a major draw-back for the mineral acid catalysed method.[169] It was also found that, de-pending upon the source of hemicellulose and the linkages present in the hemicellulose, the activation energy for its hydrolysis to form sugar varies (50–199 kJ mol^{-1}) in the presence of mineral acids (HCl and H_2SO_4).[12] Moreover, studies were also done to check the ease with which various substrates undergo hydrolysis using mineral acids and the following trend was ob-served:[170] arabinoside > xyloside > galactoside > mannoside > glucoside.

Studies on the acid hydrolysis of hemicelluloses (arabinogalactan) were also done and it was established that with an increase in acid concentration (pH) and temperature, the hydrolysis rate increased.[171] It was shown that at pH 1 and at 90 °C, complete hydrolysis of arabinogalactan is possible with-out the formation of any degradation products. A first-order kinetic model inclusive of two parallel reactions for obtaining arabinose and galactose was proposed based on modelling and experimental data.

For the complete hydrolysis of hemicelluloses, a variety of enzymes (hemicellulase: xylanases, arabinofuranosidases, ferulic and coumaric acid esterases, acetylxylan esterases, acetylmannan esterases, acetylgalactan esterases, glucuronidases, xylosidases *etc.*) are required and details on the enzymes and their modes of action are summarized in Table 1.2.[101] Basically,

Table 1.2 Summary of the actions of some enzymes on hemicellulose hydrolysis.

Entry no.	Enzyme	Mode of action
1	Xylanase	Break down β-1,4 linkages in hemicellulose from non-reducing end and preferably at a region of no-substitution
2	Arabinanase	Break down α-1,5 linkages in hemicellulose
3	Arabinosidase	Break down α-1,2-, α-1,3- and β-linkages in hemicellulose
4	Ferulic and coumaric acid esterase	Break down arabinose linked ferulic and coumaric acids from cereal arabinoxylan
5	Acetylxylan esterase, acetylmannan esterase, acetylgalactan esterase	Remove acetyl substitution form xylan, galactoglucomannan, arabinogalactan
6	Glucuronidase	Break down α-1,2 linked glucopyranosuric acid or 4-O-methylglucopyranosuric acid side chain substitution from hemicellulose
7	Xylosidase	Break down of β-1,4 linkages in hemicellulose from non-reducing end to release xylose

enzymes are classified into two catergories, endo (intracellular) and exo (extracellular)—and both have different roles in the hydrolysis. Because of the complexity of the structure of hemicelluloses, it is essential to use several isoenzymic forms of xylanase for its extensive hydrolysis. Due to the very selective actions of enzymes, many xylanases are not capable of cleaving glycosidic bonds between xylans with substituent(s). However, to overcome this, it is mandatory to use other enzymes that can cleave the substituents on the xylans first and then xylanase can separate various sugars from these treated xylans. Additionally, as discussed for cellobiose/cellulose hydrolysis, in hemicelluloses also, sometimes cleaving a bond between only two xylose units becomes difficult (due to an inhibitory effect). Particularly, the action of enzymes on hemicelluloses in the paper and pulp industry is very important as it can give a bleaching effect without the use of a harmful chlorine treatment.

The rate of acid hydrolysis of hemicellulose is seen to be dependent on the structure (conformation) of anhydrosugar present in the hemicellulose, *i.e.* pyranose or furanose form and α- anomeric or β-anomeric form.[12,170] Due to the higher ring strain in the furanose form of sugar compared to the pyranose form, the hemicellulose with the furanose form of anhydrosugars undergoes an easy hydrolysis reaction. Moreover, hemicellulose with anhydrosugars in the β-anomeric form undergoes a faster hydrolysis reaction (relative rate for β-anomer of d-xyloside: 9.1) than hemicellulose with the α-anomeric form (relative rate for α-anomer of d-xyloside: 4.5).

1.4.2 Use of a Heterogeneous Catalytic System

The superiority of heterogeneous catalysts over their homogeneous counterparts is discussed earlier (Section 1.3.2) in this chapter, however, the selectivity obtained with enzymes is very difficult to achieve with any other catalytic systems. This is particularly important in the case of hemicelluloses since in this substrate, various linkages are present that require varying conditions to cleave them. Subsequently, there is a possibility that sugars extracted at milder conditions can degrade under harsher conditions. Heterogeneous catalysts with acidic functionality such as zeolites, metal oxides, HPAs, carbons, *etc.*, which are known in cellulose hydrolysis, are also used for the hydrolysis of hemicelluloses.

Recently, a one-pot method for the hydrolysis of isolated hemicellulose derived from softwood (oat spelt, xylan) and hardwood (birchwood, xylan) in the presence of various solid acid catalysts was shown.[172] Typically, oat spelt derived hemicellulose has a composition of xylose $\geq 70\%$, arabinose $= 10\%$ and glucose $= 15\%$. However, birchwood derived hemicellulose with a higher degree of polymerization than oat spelt is made up of $>90\%$ xylose. Hence, it is expected that the concentration of the products formed from these two substrates will differ slightly. The authors have studied the effects of several acid catalysts, such as zeolites (HUSY (Si/Al $= 15$), Hβ (Si/Al $= 19$), HMOR (Si/Al $= 10$)), clay (K10), metal oxides (γ-Al$_2$O$_3$, Nb$_2$O$_5$, SO$_4{}^{2-}$/ZrO$_2$), mesoporous materials (Al-MCM-41 (Si/Al $= 50$), Al-SBA-15 (Si/Al $= 100$)) and HPA (Cs$_{2.5}$H$_{0.5}$PW$_{12}$O$_{40}$), in the reactions.[172] Ultimately, it was found that the HUSY catalyst showed the highest amount of C$_5$ sugar (xylose and arabinose) formation (41% yields) from oat spelt derived hemicellulose in water at 170 °C. In these reactions, the presence of oligomers (disaccharide, trisaccharide, tetrasaccharide and pentasaccharide) was confirmed with the help of a LC-MS technique and, based on these observations, the hydrolysis reaction pathway was suggested with hemicellulose (polymer) converting into soluble and detectable oligomers in a stepwise reaction ($\rightarrow$ penta-saccharide $\rightarrow$ tetrasaccharide $\rightarrow$ trisaccharide $\rightarrow$ disaccharide $\rightarrow$) to yield C$_5$ sugars. Yet, there is always a possibility that tetrasaccharide can cleave in two possible ways: (1) into one mole of trisaccharide and one mole of monosaccharide or (2) two moles of disaccharide. However, due to the very fast reaction rate observed by the authors, it is very difficult to reveal the exact nature of the cleavage. Hence, further studies are required to check the actual pathway of this reaction. Although higher yields are observed with the HUSY catalyst, further characterization of the spent catalyst showed that it undergoes structural changes and hence loses its activity over time. Another research group showed the possibility of the hydrolysis of arabinogalactan with galactose : arabinose : glucoronic acid $= 5 : 1 : 0.08$ and a molar mass of 20 000–100 000 g mol^{-1} derived from larch wood in a one-pot method.[173] Two solid acid catalysts *viz.* Smopex-101 (fibrous, non-porous catalyst; sulfonic acid functionalized polyethene-graft-polystyrene) and Amberlyst-15 (macroporous resin; sulfonic acid functionalized styrene-divinyl benzene) were used by the authors for the hydrolysis of arabinogalactan. It was seen that Smopex-101 (86% yield for arabinose) gave better activity than Amberlyst-15 (50% yield for arabinose) at 90 °C. This result is interesting since Amberlyst-15 has a higher acid amount (4.7 mmol g^{-1}) than Smopex-101

(3.6 mmol g^{-1}) but the activity was reversed. This observation was explained based on the catalyst structure and diffusion limitations. Due to the fibrous nature of Smopex-101, all the sulfonic acid groups distributed on the catalyst surface became easily accessible to the substrate but, due to presence of acid groups in the pores of Amberlyst-15, the same was not possible with the latter catalyst. Subsequently, an enhancement in arabinose formation (95%) was also achieved by increasing the reaction time to 36 h with Smopex-101. Lower yields of galactose compared with arabinose are possibly due to an easy cleavage of branched arabinose linked *via* β-1,3 linkages on the galactose backbone. The ease of hydrolysis is also dependent on the source from which arabinogalactan is isolated since in the case of larchwood (which was used in the study), the percentage of branching is very high compared with woody plants and the coffee bean. In soybean seed derived arabinogalactan, branching is at the β-1,3-, α-1,4-, and α-1,5- linkages instead of the β-1,3- or β-1,6- linkages typically observed in other arabinogalactans.[166] This difference in branching may affect the catalyst activity in hydrolysis reactions as they may pose the problem of steric hindrance. In another study, a sulfonated biochar catalyst, prepared by pyrolysis of wood chips at 400 °C for 1 h in a nitrogen flow, followed by sulfonation using sulfuric acid, was also used for the hydrolysis of softwood (derived from locust bean gum) and hardwood hemicelluloses (derived from birchwood).[174] It was reported that hemicellulose conversion can reach up to 90% in the presence of the sulfonated biochar catalyst at 120 °C while, under similar reaction conditions, sulfonated activated carbon showed lower catalytic activity (hemicellulose conversion = 50%). This is due to the fact that sulfonated activated carbon has fewer strong acid sites (2.59 mmol g^{-1}) and lower xylan adsorption capacity ($\sim$145 mg xylan per g catalyst at 23 °C) than sulfonated biochar (acid density = 5.65 mmol g^{-1}, xylan adsorption capacity $\sim$390 mg xylan per g catalyst at 23 °C). However, the sulfonated biochar catalyst exhibited a drastic reduction in hemicellulose conversion activity (1st run: $\sim$80%, 2nd run: $\sim$40%, 3rd run: $\sim$10%, 4th run: 0%) in recycle runs due to leaching of sulfonic acid groups (acid density reduction = 3.66 to 0.0 mmol g^{-1}) and its structure deformation (surface area reduction = 365 to 5.3 m^2 g^{-1}, pore volume reduction = 0.20 to 0.004 cc g^{-1}, pore radius reduction = 10.5 to 0.0 Å). Various resins (Amberlyst-70, Amberlyst-35, D5081 and D5082), sulfonic acid functionalized silica gel and zeolites (HZSM-5; Si/Al = 50 and 80, H-faujasite; Si/Al = 5.1 and H-ferrierite; Si/Al = 55) were tested in the hydrolysis of beechwood derived hemicellulose (*ca.* 75% xylose, 2–5% glucose) to achieve C_5 sugars (xylose and arabinose) as products.[175] Amongst all these catalysts, the resins showed a better activity (55–80% of C_5 sugars) at 120 °C because of a higher acid amount. The authors also studied the role of acid density on the catalytic activity and found a direct correlation between activity and acid density. Nonetheless, as discussed earlier for cellulose hydrolysis (Section 1.3.2.1), the resins lost their activity after the first use due to leaching of –SO$_3$H groups. Among the various zeolite catalysts, H-ferrierite offered the best activity for the formation of the highest amount of C_5 sugars

(41%) at 140 °C, due to its higher acid strength compared to other zeolites.[175] Moreover, the catalyst showed recyclable activity up to four runs with a marginal decrease in yields. This indicates that to achieve a recyclable higher activity, it is necessary to have strong acid sites on the catalyst. The use of various acid silicoaluminophosphate (SAPO) catalysts was shown in the hydrolysis of hemicellulose.[137,176] Although the final product in the reaction is furfural, which can be obtained directly from hemicellulose *via* a hydrolysis–dehydration reaction, the results suggested that SAPO catalysts can only be used for the hydrolysis of hemicellulose to produce sugars if the reaction conditions are optimized. This catalyst is interesting to study further since it is stable under the reaction conditions and showed reproducible activity for eight runs at least. This high stability may arise due to the presence of phosphorus in the catalyst, which is responsible for making the catalyst more hydrophilic compared to zeolites. This hydrophilicity is also responsible for better catalyst (active site) and substrate interactions. Supported metal oxide catalysts might also be promising for sugar synthesis from hemicelluloses since they were proven as effective catalysts for hydrolysis–dehydration reactions of hemicellulose.[177] The hemicellulose extracted from *Miscanthus* grass was also used as a substrate for the synthesis of sugars.[178] The extracted solution consists of hemicellulose with a molecular weight of 2008 g mol^{-1} (analysed by GPC), DP of 15, polymer chain length of 7 nm and chain radius of 2 nm (calculated from molecular dynamics simulations). It was demonstrated that hydrothermally treated sulfonated mesoporous carbon nanoparticles (HT$_5$-HSO$_3$-MCN) with a large number of weakly acidic phenolic –OH groups helped in the adsorption of hemicellulose on the catalyst surface effectively *via* H-bond formation. This in turn was responsible for achieving better catalytic activity. A very high amount of xylose formation (74%) was reported from extracted hemicellulose solution at a buffered pH of 3.9 (using sodium acetate buffer solution) at 150 °C. Moreover, the HT$_5$-HSO$_3$-MCN catalyst was shown to provide a similar activity in three recycle runs after washing (stepwise washing with 0.1 M NaOH, 1 : 1 (*v/v*) ethanol and water, only ethanol, only water, 4 M HCl, only water) and drying. The complicated washing and drying procedures may hamper their further use in these reactions. It is thus necessary to devise a simple recycling process.

Additionally, some of the developed processes confirm the use of lignocelluloses directly for the synthesis of C$_5$ sugars in the presence of heterogeneous catalysts. In one report, the selective conversion of the hemicellulose part from bagasse into C$_5$ sugars (xylose and arabinose) over HUSY (Si/Al = 15) catalyst was shown.[172] At 170 °C, 45% of C$_5$ sugar formation was reported while from XRD studies it was shown that the cellulose remained unreacted. Sulfonated allophane catalyst (amorphous aluminosilicate) was also screened for its hydrolysis activity with bamboo, Japanese cedar and rice straw.[179] Under the optimized reaction conditions (150 °C), the hemicellulose part was converted selectively to yield 41% xylose from the bamboo powder, 8% xylose and 6% mannose from Japanese cedar powder and 8% arabinose and 2% xylose from rice straw. The product distributions are solely dependent on the type and concentration of hemicellulose present in these lignocelluloses. Very recently, the selective hydrolysis of hemicelluloses from bagasse, rice husks and wheat straw into sugars and furfural was demonstrated over SAPO type catalysts.[180] Additionally, a few reports showed that under the operating conditions used, carbon catalysts with various surface functionality can hydrolyse both the hemicellulose and cellulose part of the raw biomass (*Eucalyptus* flakes and corn cobs) into C_5 and C_6 sugars.[126,155] It was observed that heterogeneous catalysts can catalyse hydrolysis reactions of hemicelluloses. Moreover, depending on the type and source of hemicelluloses, catalytic activity varies. This brings our attention to the fact that it is vital to design and develop a catalyst that can hydrolyse hemicelluloses derived from different sources with good selectivity and recyclability.

1.5   Synthesis of Sugars *via* an Isomerization Reaction

Besides obtaining sugars from disaccharides and polysaccharides, it is also possible to obtain them from other monosaccharides by virtue of isomerization reactions. During hydrolysis reactions of disaccharides and polysaccharides, it is possible for sugars to undergo isomerization reactions. For example, it is possible to obtain fructose from glucose once the latter is formed during hydrolysis of various disaccharides and polysaccharides. Similarly, the formation of xylulose from xylose is observed. Nevertheless, in this section, we will discuss the isomerization reactions of sugars (monosaccharides). The formation of two ketonic sugars *viz.* fructose (C_6) and xylulose (C_5) from their respective aldonic sugars, such as glucose and xylose, is known to occur *via* an isomerization reaction. In the presence of a base, aldose sugars transform into ketose sugars *via* formation of an enediol intermediate and *vice versa* and this transformation was first discovered in 1885 by Cornelis Adriaan Lobry van Troostenburg de Bruyn and Willem Alberda van Ekenstein and hence is also known as the Lobry de Bruyn–van Ekenstein transformation (Figure 1.11).[181] The isomerization of glucose into fructose is a very important industrial reaction since high fructose syrup (HFS) is used as a sweetener. It is known that fructose has a high degree of relative sweetness (110) compared to glucose (74) and sucrose (100). However, since the glucose to fructose isomerization reaction is equilibrium limited, it is necessary to optimize the reaction conditions to achieve higher yields of fructose.

Homogeneous bases, such as NaOH, KOH, sodium aluminate *etc.*, are frequently used for the isomerization of glucose to achieve 30–50% of fructose yields.[182–184] The reaction proceeds *via* formation of a carbanion by extraction of a H from the C2 carbon by the anionic part (OH⁻, aluminate ion *etc.*) of bases to form an enolate ion. Further, this enolate ion in the presence of water forms an enediol or can cause an inversion to yield mannose (Figure 1.11). The restricted yields obtained in homogeneous systems may be attributed to the fact that fructose, once formed, is immediately

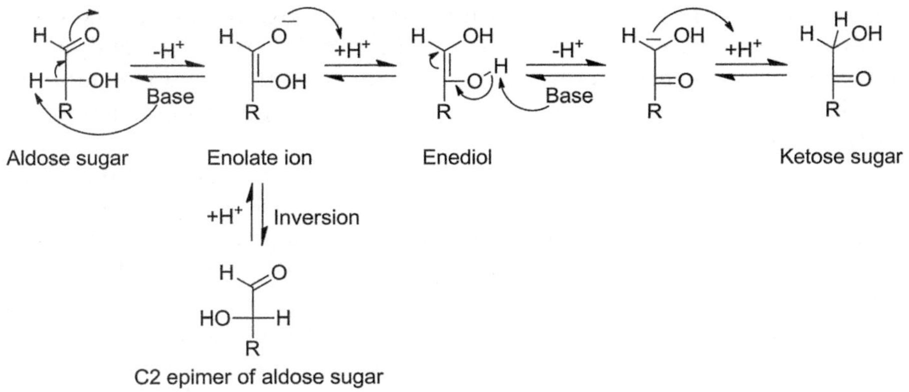

Figure 1.11 Base catalysed isomerization of aldol sugar into ketose sugar.

reverted back into glucose and thus equilibrium is achieved. To improve the yields of fructose, it is necessary to shift the equilibrium towards the right hand side. When a resin is used along with a homogeneous base (NaOH), a slight improvement in yield (57%) is observed.[185] Subsequently, hetero-geneous base catalysts are also known to catalyse this reaction as free hydroxyl anions (OH^-) are liberated in the bulk water from these catalysts. It was reported that aluminate resin with a low hydroxide content can produce very high yields (72%) of fructose from 90% glucose conversion at 2 °C (1007 h).[186] The formation of a stable complex between fructose and aluminate ion (on the catalyst) helps in achieving higher yields as the reverse reaction of fructose to glucose is thus restricted. These results match well with earlier work when aqueous sodium aluminate (68%) was used.[187] It is also possible to obtain mannose as a product when glucose is treated with base, however, the rate of inversion is extremely low compared to enediol formation to yield fructose. Further, in the glucose isomerization reaction, a series of alkali and alkaline earth metals (Li, Na, K, Cs, Ca and Ba) exchanged zeolites (A, X and Y) were screened at 95 °C and the results showed lower conversion (7–34%) with very low fructose formation (4–22%).[188] The same report also presented the use of hydrotalcite (HT) catalyst with Mg/Al = 3 (atomic) for obtaining 42% glucose conversion with 25% fructose formation. Another report also showed that carbonate and hydroxide forms of HT can produce 14% fructose from 15% glucose conversion at 90 °C.[189] When HT was used for the glucose isomerization reaction in *N,N*-dimethylformamide (DMF) solvent at 100 °C, 38% fructose was formed at 62% glucose conversion.[190] Use of HT was also shown by other research groups to yield fructose, however, the main aim was to achieve higher oxidation and hydrogenation activities in the reaction.[191] It can be stated from these reports that when HT is used for isomerization, typically, a high selectivity for fructose formation is achievable, which can be again explained by the possible formation of magnesium aluminate ($MgAl_2O_4$) or ($M^+AlO_2^-$) during the synthesis of HT. Moderate yields of 21–39% for fructose were also reported in the presence of various

metallosilicates, namely titanosilicates, sodium yttrium silicate, alkali calcium silicate and calcium silicate at 100 °C.[192] Although many catalysts are
known for the isomerization reaction, depending on the reaction conditions
(at high temperatures), glucose also may undergo degradation reactions (to
acetic, lactic and propionic acids).

Very recently, a promising pathway for glucose isomerization into fructose
via intramolecular 1,2-hydride shift mechanism was shown using Lewis acid
catalysts (Figure 1.12).[193] Since the process is developed with a Lewis acidic
catalyst, it eventually stops sugar degradation reactions. The Sn-β and Ti-β
catalysts having Lewis acidity can convert glucose (55% and 26%, respectively) into fructose (32% and 14%, respectively) in water at 110 °C.[193] Along
with fructose, formation of mannose is also identified (5–9%) as an epimerization (inversion) product in the reaction mixture. Additionally, Sn-β
catalyst was demonstrated to be stable and recyclable four times without
losing its activity. In an extensive study, with the help of ^{1}H and ^{13}C NMR
analysis of isotope labelled glucose, the authors have proven that in the
presence of Lewis acidic Sn-β catalyst, an intramolecular 1,2-hydride shift
mechanism is operated, which is not possible under alkaline conditions.[194]
Additionally, diffuse reflectance UV-Vis analysis suggested that a partially
hydrolysed Sn framework [(SiO)$_3$Sn(OH)] is the active species in the Sn-β
catalyst. The phenomenon of aldose sugar isomerization in the presence of a
Lewis acid catalyst is also valid in the case of xylose (C$_5$ sugar) for formation
of xylulose. The presence of Sn-β catalyst showed 85% xylose conversion with
18% xylulose formation (isomerization) and 5% lyxose formation

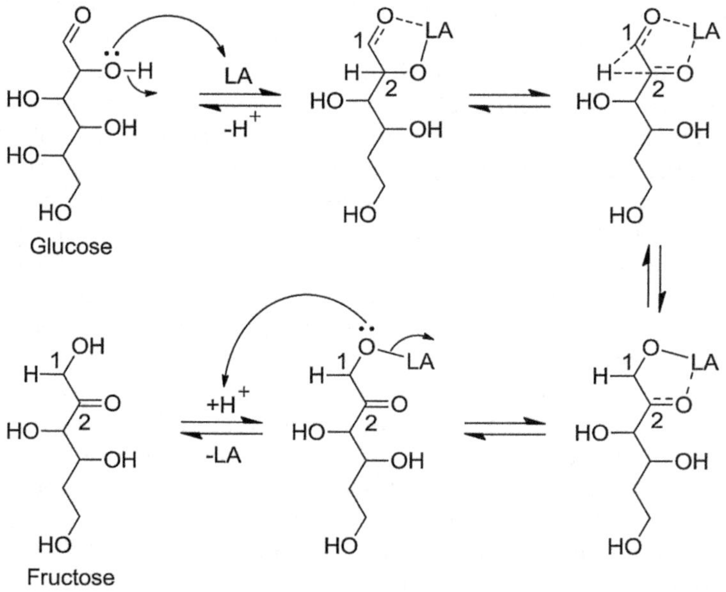

Figure 1.12   Lewis acid catalysed isomerization of glucose into fructose *via* intramolecular 1,2-hydride shifting. 'LA' stands for Lewis acid sites.

(epimerization) at 110 °C in water.[195] But the authors have not pointed out the rest of the products. Considering this concept, another research group has prepared Sn incorporated mordenite framework inverted (MFI) and beta (BEA) zeolites and used them for sugar (glucose and xylose) isomerization reactions at 90 °C.[196] The small pore diameter (0.56 nm) of the MFI zeolite restricted the access of larger sugar molecules (glucose: 0.73 nm, xylose: 0.66 nm) to the metal centre present inside the MFI zeolite, which thus showed lower isomerization activity (glucose conversion = 9%, fructose yield = 4% and xylose conversion = 40%, xylulose yield = 19%). On the contrary, the BEA zeolite with a larger pore diameter (0.77 nm) can easily overcome the diffusion limitations for sugars thereby offering better isomerization activity (glucose conversion = 65%, fructose yield = 34%, mannose yield = 17% and xylose conversion = 81%, xylulose yield = 24%). Due to the presence of Lewis acid sites in SAPO-44, glucose isomerization into fructose is also possible.[86] Since SAPO contains both Brønsted and Lewis acid sites, one-pot synthesis of HMF from glucose *via* fructose formation is possible.[86] In yet another report, use of HT was shown to achieve the acyclic form of glucose to yield higher concentrations of sugar alcohols (sorbitol and mannitol) over supported metal catalysts.[191]

1.6 Conclusions and Outlook

With the onset of renewed interest in lignocellulose utilization for a sustainable future, it is possible to generate fuels and chemicals from this renewable and abundantly available resource. Though several efforts have been devoted since the 1900s and in recent years to the synthesis of fuels (bioethanol, bio-diesel), it is perceived that the synthesis of chemicals from this resource is highly attractive. The estimated worldwide production of bio-fuels (bio-diesel and bio-ethanol) is around 2000 thousand barrels per day, which in comparison to the estimated worldwide production of crude oil is around 2.5% (crude oil production, 79 000 thousand barrels per day).[197] This gives the idea that it is practically impossible to match the requirement of world fuel demand through biomass processing. Looking at this scenario, it would be crucial to develop methods for the synthesis of sugars and other chemicals as they can act as platform chemicals to yield many other industrially important, valuable chemicals such as furans, sugar alcohols, sugar acids, glycols *etc.*

Conversion of plant derived lignocellulosic biomass or lignocelluloses containing cellulose, hemicelluloses and lignin is crucial considering its non-edible property (for humins) compared to food biomass such as starch and various disaccharides. In two possible ways, polysaccharides, cellulose and hemicelluloses, the components of lignocelluloses, can be converted, first, *via* isolated polysaccharides after pre-treatment to extract them from lignocelluloses and second, *via* direct hydrolysis of lignocelluloses. For the first path, pre-treatment has to be done with lignocelluloses to separate cellulose, hemicelluloses and lignin from each other as they are linked together through covalent or non-covalent bonding. However, during isolation procedures typically done by acids, bases, steam *etc.*, changes in the structures of these polysaccharides can be observed. In particular, these polysaccharides may undergo such changes as reduction of the chain length, increase in reducing ends, loosening of structures, loss of crystallinity *etc.*, which are helpful in the hydrolysis of cellulose and also hemicelluloses. The other changes such as oxidations, acylation, incorporation of impurities *etc.*, may, however, hamper the hydrolysis of these polysaccharides. Hence it is essential to choose the correct pre-treatment method for efficient isolation. For the second path, lignocelluloses are used directly without giving any pre-treatment. This pathway is beneficial for avoiding the contamination of substrates with impurities (caused during pre-treatment by use of external reagents) and changes in structures possible during pre-treatment procedures. However, due to the native structure of lignocelluloses, it is difficult to hydrolyse the polysaccharides. Moreover, the other components present in lignocelluloses, such as lignin, waxes and nutrients, may poison the catalysts. It is perceived that lignin can strongly adsorb on the catalyst surface through various functional groups present in lignin (–OH, –CHO, –COOH, =O *etc.*). Another possibility is an exchange of H^+ with M^+ (nutrients such as, K, Mg, Ca *etc.*) in acidic catalysts, hampering their activity.

Although for the conversion of polysaccharides, many pathways are known, such as thermal (pyrolysis, gasification, application of supercritical water), thermo-chemical (acid, base), or biochemical (enzyme) *etc.*, it is obvious that thermo-chemical and biochemical (enzyme) pathways are preferred to yield the maximum amounts of sugars as with thermal treatments, degradation products are predominant. Mineral acid based methods have been known since 1819 and hence much work is done to optimize the reaction conditions for achieving the highest possible sugar yields. As mentioned in Section 1.3.1.1, many industrial processes use mineral acids for yielding sugars that are later converted into fuels. Base catalysed methods are also known to hydrolyse polysaccharides but the possibility of substrate oxidation and isomerization of the sugars formed may restrict the use of bases on an industrial scale. As with mineral acids, enzyme based biochemical methods have also been known for a long time and they have the advantage of achieving very high yields because of the selective action of enzymes. However, in all these methods, it is important to maintain the pH (either acidic or alkaline), which inherently increases the capital cost as the reactor walls are prone to corrosion. Moreover, due to their homogeneous nature, it is difficult to recover the catalysts for reuse. Considering these drawbacks, recently, heterogeneous catalysts such as solid acids and supported metal catalysts have been employed in these reactions. It is observed that solid acid catalysts can, like mineral acids, promote the reaction *via* H^+, which is available on the surface of the catalyst from Brønsted acid sites. Although solid acids are active in these reactions, they are prone to degradation/decomposition during the course of the reaction because of the active sites leaching in the solution or morphological changes (during hydrothermal treatment). It is widely believed that hydrophilic catalysts may help in avoiding these drawbacks and also may overcome diffusion limitations by enhancing the interactions between the catalyst and the substrate. This enhancement may arise due to hydrophilic interactions between hydrophilic surface –OH groups and hydrophilic –OH groups in substrates. The possibility of using supported metal catalysts is also shown by generating *in situ* acid sites by splitting molecular hydrogen on the metal surface. Though heterogeneous catalysts are known for these reactions, it would be a real challenge to develop efficient and reusable catalysts and their application on a larger scale.

Worldwide, several governments, private entities and industries are sponsoring various projects on the development of new environmentally benign methods for the conversion of lignocelluloses into chemicals. In the USA, USDA and DOE support most of the biomass related programs on a large scale with NREL being a prime example. Likewise, the EU funds many projects on biomass valorization. In many other countries, their respective ministries and government organizations fund many projects for developing new methods to valorize biomass. Various multi-national corporations such as DuPont, SABIC, Dow, Shell, BP, Haldor Topsøe, Coca-Cola, Solvay, BASF, Honeywell UOP, Reliance, Tata Chemicals, Mitsubishi, Sumitomo, Toyota, Arkema, DSM, P&G, Unilever, Clarient and Asashi Chemicals, along with a range of other industries, are looking into the use of biomass derived products in their portfolio in the near future.

The use of aquatic biomass is also being looked into by various research groups for the synthesis of fuels (green diesel) and chemicals, however, the work is at a very nascent stage at least in the case of chemical synthesis. However, this feedstock may prove to be very interesting in the near future.

While there are many opportunities to work on the synthesis of chemicals or fuels from biomass (undoubtedly from non-edible resources to avoid the food *vs.* chemical battle), attention to the following few points is required before any methods or technologies are developed:

- Abundance of resources to produce sufficient amount of product, either bulk, fine or niche
- Development of catalytic systems that are efficient, recyclable and easy to handle
- Removal of impurities so that catalyst and product poisoning is avoided
- Utilization of existing facilities (reactors, delivery, man-power)
- Process economy compared to fossil feedstock based methods
- Application of multiple feedstock such as crop wastes (wheat straw, rice husk, corn cob, corn stover, bagasse *etc.*), forest residues, municipal solid waste *etc.*
- Feasibility to scale up the method for commercial purposes

In conclusion, utilization of biomass has thrown up a lot of challenges requiring the development of novel processes to ensure a sustainable future, if not for tomorrow, then for the future.

References

1. P. N. R. Vennestrom, C. M. Osmundsen, C. H. Christensen and E. Taarning, *Angew. Chem., Int. Ed.*, 2011, **50**, 10502–10509.
2. W. L. Faith, *Ind. Eng. Chem.*, 1945, **37**, 9–11.
3. B. Persson, *Sulfitsprit: forhoppningar och besvikelser under 100 år*, DAUS Tryck & Media, Bjasta, 2007.
4. Food and Agriculture Organization of the United Natiions Statistics Division, http://faostat3.fao.org/compare/E.
5. D. Budny, *The Global Dynamics of Biofuels: Potential Supply and Demand for Ethanol and Biodiesel in the Coming Decade* 3, Brazil Institute, Woodrow Wilson International Center for Scholars, Washington, DC, 2007.
6. R. D. Perlack, *Biomass as Feedstock for a Bioenergy and Bioproducts Industry: The Technical Feasability of a Billion-Ton Annual Supply* ORNL/TM-2005/66 TRN: US200617%%291, 2005.
7. S. V. Vassilev, D. Baxter, L. K. Andersen and C. G. Vassileva, *Fuel*, 2010, **89**, 913–933.
8. R. Rinaldi and F. Schuth, *Energy Environ. Sci.*, 2009, **2**, 610–626.
9. Alibaba Group, http://www.alibaba.com/Products?tracelog = beacon_cate_140704.
10. S. Bilgen, K. Kaygusuz and A. Sari, *Energy Sources*, 2004, **26**, 1119–1129.
11. A. J. Ragauskas, C. K. Williams, B. H. Davison, G. Britovsek, J. Cairney, C. A. Eckert, W. J. Frederick, J. P. Hallett, D. J. Leak, C. L. Liotta, J. R. Mielenz, R. Murphy, R. Templer and T. Tschaplinski, *Science*, 2006, **311**, 484–489.
12. P. Maki-Arvela, T. Salmi, B. Holmbom, S. Willfor and D. Y. Murzin, *Chem. Rev.*, 2011, **111**, 5638–5666.
13. E. Papadopoulou, A. Hatjiissaak, B. Estrine and S. Marinkovic, *Eur. J. Wood Wood Prod.*, 2011, **69**, 579–585.
14. H. Kobayashi and A. Fukuoka, *Green Chem.*, 2013, **15**, 1740–1763.
15. P. Zugenmaier, *Prog. Polym. Sci.*, 2001, **26**, 1341–1417.
16. M. Jarvis, *Nature*, 2003, **426**, 611–612.
17. J. Zakzeski, P. C. A. Bruijnincx, A. L. Jongerius and B. M. Weckhuysen, *Chem. Rev.*, 2010, **110**, 3552–3599.
18. L. T. Fan, M. M. Gharpuray and Y. H. Lee, *Cellulose Hydrolysis*, Springer Berlin Heidelberg, Berlin, 1987, pp. 1–198.
19. R. P. Swatloski, S. K. Spear, J. D. Holbrey and R. D. Rogers, *J. Am. Chem. Soc.*, 2002, **124**, 4974–4975.
20. Y.-B. Huang and Y. Fu, *Green Chem.*, 2013, **15**, 1095–1111.
21. P. Kumar, D. M. Barrett, M. J. Delwiche and P. Stroeve, *Ind. Eng. Chem. Res.*, 2009, **48**, 3713–3729.
22. V. B. Agbor, N. Cicek, R. Sparling, A. Berlin and D. B. Levin, *Biotechnol. Adv.*, 2011, **29**, 675–685.
23. Y. Sun and J. Cheng, *Bioresour. Technol.*, 2002, **83**, 1–11.

24. M. Schwanninger, J. C. Rodrigues, H. Pereira and B. Hinterstoisser, *Vib. Spectrosc.*, 2004, **36**, 23–40.
25. Y. Yu and H. Wu, *AIChE J.*, 2010, **57**, 793–800.
26. M. Yabushita, H. Kobayashi, K. Hara and A. Fukuoka, *Catal. Sci. Technol.*, 2014, **4**, 2312–2317.
27. Q. Zhang, M. Benoit, K. De Oliveira Vigier, J. Barrault, G. Jegou, M. Philippe and F. Jerome, *Green Chem.*, 2013, **15**, 963–969.
28. A. M. Bochek, *Russ. J. Appl. Chem.*, 2003, **76**, 1711–1719.
29. H. Ooshima, K. Aso, Y. Harano and T. Yamamoto, *Biotechnol. Lett.*, 1984, **6**, 289–294.
30. H. Ma, W. W. Liu, X. Chen, Y. J. Wu and Z. L. Yu, *Bioresour. Technol.*, 2009, **100**, 1279–1284.
31. O. Lindroos, *Biomass Bioenergy*, 2010, **35**, 385–390.
32. D. Jackowiak, D. Bassard, A. Pauss and T. Ribeiro, *Bioresour. Technol.*, 2011, **102**, 6750–6756.
33. L. Olsson, M. Galbe and G. Zacchi, Pretreatment of Lignocellulosic Materials for Efficient Bioethanol Production, in *Biofuels*, Springer Berlin Heidelberg, 2007, vol. 108, pp. 41–65.
34. D. M. James, Pretreatment of Lignocellulosic Biomass, in *Enzymatic Conversion of Biomass for Fuels Production*, ed. M. E. Himmel, J. O. Baker and R. P. Overend, American Chemical Society, Washington DC, 1994, vol. 566, ch. 15, pp. 292–324.
35. S. J. B. Duff and W. D. Murray, *Bioresour. Technol.*, 1996, **55**, 1–33.
36. I. Ballesteros, M. J. Negro, J. M. Oliva, A. Cabanas, P. Manzanares and M. Ballesteros, *Appl. Biochem. Biotechnol.*, 2006, **130**, 496–508.
37. M. T. Holtzapple, A. E. Humphrey and J. D. Taylor, *Biotechnol. Bioeng.*, 1989, **33**, 207–210.
38. Iogen Corporation, http://www.iogen.ca/technology/platform.html.
39. W. S. L. Mok and M. J. Antal, *Ind. Eng. Chem. Res.*, 1992, **31**, 1157–1161.
40. F. Teymouri, L. Laureano-Perez, H. Alizadeh and B. E. Dale, *Bioresour. Technol.*, 2005, **96**, 2014–2018.
41. M. Holtzapple, J.-H. Jun, G. Ashok, S. Patibandla and B. Dale, *Appl. Biochem. Biotechnol.*, 1991, **28–29**, 59–74.
42. Y. Zheng, H. M. Lin and G. T. Tsao, *Biotechnol. Prog.*, 1998, **14**, 890–896.
43. W. C. Neely, *Biotechnol. Bioeng.*, 1984, **26**, 59–65.
44. R. Travaini, M. D. M. Otero, M. Coca, R. Da-Silva and S. Bolado, *Bioresour. Technol.*, 2013, **133**, 332–339.
45. J. Quesada, M. Rubio and D. Gomez, *J. Wood Chem. Technol.*, 1999, **19**, 115–137.
46. N. Mosier, C. Wyman, B. Dale, R. Elander, Y. Y. Lee, M. Holtzapple and M. Ladisch, *Bioresour. Technol.*, 2005, **96**, 673–686.
47. V. Chang and M. Holtzapple, *Appl. Biochem. Biotechnol.*, 2000, **84–86**, 5–37.
48. F. Kong, C. Engler and E. Soltes, *Appl. Biochem. Biotechnol.*, 1992, **34–35**, 23–35.

49. M. L. Soto, H. Dominguez, M. J. Nunez and J. M. Lema, *Bioresour. Technol.*, 1994, **49**, 53–59.
50. A. M. Azzam, *J. Environ. Sci. Health, Part B*, 1989, **24**, 421–433.
51. X. Zhao, K. Cheng and D. Liu, *Appl. Microbiol. Biotechnol.*, 2009, **82**, 815–827.
52. H. L. Chum, D. K. Johnson, S. Black, J. Baker, K. Grohmann, K. V. Sarkanen, K. Wallace and H. A. Schroeder, *Biotechnol. Bioeng.*, 1988, **31**, 643–649.
53. T. K. Ghose, P. V. Pannir Selvam and P. Ghosh, *Biotechnol. Bioeng.*, 1983, **25**, 2577–2590.
54. A. Esteghlalian, A. G. Hashimoto, J. J. Fenske and M. H. Penner, *Bioresour. Technol.*, 1997, **59**, 129–136.
55. X. B. Lu, Y. M. Zhang, J. Yang and Y. Liang, *Chem. Eng. Technol.*, 2007, **30**, 938–944.
56. BlueFire Renewables, http://bluefireethanol.com.
57. Biosulfurol Energy Limited, http://biosulfurol-energy.com.
58. M. von Sivers and G. Zacchi, *Bioresour. Technol.*, 1995, **51**, 43–52.
59. R. P. Tengerdy and G. Szakacs, *Biochem. Eng. J.*, 2003, **13**, 169–179.
60. C. A. Cardona and O. J. Sanchez, *Bioresour. Technol.*, 2007, **98**, 2415–2457.
61. A. Brandt, J. Grasvik, J. P. Hallett and T. Welton, *Green Chem.*, 2013, **15**, 550–583.
62. A. M. da Costa Lopes, K. G. Joao, A. R. C. Morais, E. Bogel-Lukasik and R. Bogel-Lukasik, *Sustainable Chem. Processes*, 2013, **1**, 1–31.
63. R. C. Remsing, R. P. Swatloski, R. D. Rogers and G. Moyna, *Chem. Commun.*, 2006, 1271–1273.
64. H. Olivier-Bourbigou, L. Magna and D. Morvan, *Appl. Catal., A*, 2010, **373**, 1–56.
65. J. Holm and U. Lassi, Ionic Liquids in the Pretreatment of Lignocellulosic Biomass, in *Ionic Liquids: Applications and Perspectives*, ed. A. Kokorin, InTech, Europe, 2011, ch. 24, pp. 545–560.
66. H. Zhang, J. Wu, J. Zhang and J. He, *Macromolecules*, 2005, **38**, 8272–8277.
67. C.-H. Zhou, X. Xia, C.-X. Lin, D.-S. Tong and J. Beltramini, *Chem. Soc. Rev.*, 2011, **40**, 5588–5617.
68. H. Kraessig, *J. Polym. Sci., Part C: Polym. Lett.*, 1987, **25**, 87–88.
69. A. O'Sullivan, *Cellulose*, 1997, **4**, 173–207.
70. *Cellulose Science and Technology*, ed. J.-L. Wertz, O. Bedue and J. P. Mercier, CRC Press, Taylor & Francis Group, London, 2010.
71. Y. Nishiyama, J. Sugiyama, H. Chanzy and P. Langan, *J. Am. Chem. Soc.*, 2003, **125**, 14300–14306.
72. M. E. Himmel, S.-Y. Ding, D. K. Johnson, W. S. Adney, M. R. Nimlos, J. W. Brady and T. D. Foust, *Science*, 2007, **315**, 804–807.
73. D. Klemm, B. Philipp, T. Heinze, U. Heinze and W. Wagenknecht, General Considerations on Structure and Reactivity of Cellulose, in *Comprehensive Cellulose Chemistry*, Wiley-VCH Verlag GmbH & Co. KGaA, 2004, ch. 2.3–2.3.7, vol. 1, pp. 83–129.

74. M. Sasaki, T. Adschiri and K. Arai, *J. Agric. Food Chem.*, 2003, **51**, 5376–5381.
75. K. Igarashi, M. Wada and M. Samejima, *FEBS J.*, 2007, **274**, 1785–1792.
76. K. Igarashi, T. Uchihashi, A. Koivula, M. Wada, S. Kimura, T. Okamoto, M. Penttilä, T. Ando and M. Samejima, *Science*, 2011, **333**, 1279–1282.
77. J. M. Thomas, *Angew. Chem.*, 1994, **106**, 963–989.
78. T. E. Timell, *Can. J. Chem.*, 1964, **42**, 1456–1472.
79. O. Bobleter, W. Schwald, R. Concin and H. Binder, *J. Carbohydr. Chem.*, 1986, **5**, 387–399.
80. J. D. Fontana, D. A. Mitchell, O. E. Molina, A. Gaitan, T. M. B. Bonfim, J. Adelmann, A. Grzybowski and M. Passos, *Food Technol. Biotechnol.*, 2008, **46**, 305–310.
81. H. Braconnot, *Ann. Chim. Phys.*, 1819, **12**, 172–195.
82. F. Bergius, *Ind. Eng. Chem.*, 1937, **29**, 247–253.
83. M. L. Rabinovich, *Cellul. Chem. Technol.*, 2010, **44**, 173–186.
84. J. J. McParland, H. E. Grethlein and A. O. Converse, *Sol. Energy*, 1982, **28**, 55–63.
85. A. H. Brennan, W. Hoagland, D. J. Schell and C. D. Scott, High temperature acid hydrolysis of biomass using an engineering – scale plug flow reactor. Results of low testing solids, *Biotechnol. Bioeng. Symp.*, 1986, 53–70.
86. P. Bhaumik and P. L. Dhepe, *RSC Adv.*, 2013, **3**, 17156–17165.
87. J. F. Saeman, J. L. Bubl and E. E. Harris, *Ind. Eng. Chem., Anal. Ed.*, 1945, **17**, 35–37.
88. F. Camacho, P. Gonzalez-Tello, E. Jurado and A. Robles, *J. Chem. Technol. Biotechnol.*, 1996, **67**, 350–356.
89. R. A. Antonoplis, H. W. Blanch, R. P. Freitas, A. F. Sciamanna and C. R. Wilke, *Biotechnol. Bioeng.*, 1983, **25**, 2757–2773.
90. O. A. Battista, *Ind. Eng. Chem.*, 1950, **42**, 502–507.
91. H. Arthur E, The Hydrolysis of Cellulosic Materials to Useful Products, in *Hydrolysis of Cellulose: Mechanisms of Enzymatic and Acid Catalysis*, ed. J. Ross, D. Brown and L. Jurasek, ACS, 1979, ch. 2, vol. 181, pp. 25–53.
92. J.-P. Franzidis, A. Porteous and J. Anderson, *Conserv. Recycl.*, 1982, **5**, 215–225.
93. I. A. Malester, M. Green and G. Shelef, *Ind. Eng. Chem. Res.*, 1992, **31**, 1998–2003.
94. K.-i. Shimizu, H. Furukawa, N. Kobayashi, Y. Itaya and A. Satsuma, *Green Chem.*, 2009, **11**, 1627–1632.
95. J. Tian, J. Wang, S. Zhao, C. Jiang, X. Zhang and X. Wang, *Cellulose*, 2010, **17**, 587–594.
96. Y. Ogasawara, S. Itagaki, K. Yamaguchi and N. Mizuno, *ChemSusChem*, 2011, **4**, 519–525.
97. X. Li, Y. Jiang, L. Wang, L. Meng, W. Wang and X. Mu, *RSC Adv.*, 2012, **2**, 6921–6925.
98. B. Philipp, V. Jacopian, F. Loth, W. Hirte and G. Schulz, Influence of Cellulose Physical Structure on Thermohydrolytic, Hydrolytic, and

Enzymatic Degradation of Cellulose, in *Hydrolysis of Cellulose: Mechanisms of Enzymatic and Acid Catalysis*, ed. J. Ross, D. Brown and L. Jurasek, ACS, 1979, ch. 6, vol. 181, pp. 127–143.

99. E. A. Bayer, H. Chanzy, R. Lamed and Y. Shoham, *Curr. Opin. Struct. Biol.*, 1998, **8**, 548–557.

100. E. T. Reese, R. G. H. Siu and H. S. Levinson, *J. Bacteriol.*, 1950, **59**, 485–497.

101. E. W. Charles, R. D. Stephen, E. H. Michael, W. B. John, E. S. Catherine and V. Liisa, Hydrolysis of Cellulose and Hemicellulose, in *Polysaccharides: Structural Diversity and Functional Versatility*, ed. S. Dumitriu, CRC Press, 2nd edn, 2004, ch. 43, pp. 995–1034.

102. R. M. Wahlstrom and A. Suurnakki, *Green Chem.*, 2015, **17**, 694–714.

103. C. T. H. Tran, N. J. Nosworthy, A. Kondyurin, D. R. McKenzie and M. M. M. Bilek, *RSC Adv.*, 2013, **3**, 23604–23611.

104. C. T. Tsai and A. S. Meyer, *Molecules*, 2014, **19**, 19390–19406.

105. C. Li and Z. K. Zhao, *Adv. Synth. Catal.*, 2007, **349**, 1847–1850.

106. C. Li, Q. Wang and Z. K. Zhao, *Green Chem.*, 2008, **10**, 177–182.

107. A. S. Amarasekara and O. S. Owereh, *Ind. Eng. Chem. Res.*, 2009, **48**, 10152–10155.

108. A. S. Amarasekara and B. Wiredu, *Ind. Eng. Chem. Res.*, 2011, **50**, 12276–12280.

109. F. Jiang, Q. Zhu, D. Ma, X. Liu and X. Han, *J. Mol. Catal. A: Chem.*, 2011, **334**, 8–12.

110. Y. Liu, W. Xiao, S. Xia and P. Ma, *Carbohydr. Polym.*, 2013, **92**, 218–222.

111. S. Zhu, *J. Chem. Technol. Biotechnol.*, 2008, **83**, 777–779.

112. C. J. King, Separation Processes, Introduction, in *Ullmann's Encyclopedia of Industrial Chemistry*, Wiley-VCH Verlag GmbH & Co. KGaA, 2000.

113. A. F. Cronstedt, *Akad. Handl., Stockholm*, 1756, **18**, 120–130.

114. C. Buttersack and D. Laketic, *J. Mol. Catal.*, 1994, **94**, L283–L290.

115. C. Moreau, R. Durand, J. Duhamet and P. Rivalier, *J. Carbohydr. Chem.*, 1997, **16**, 709–714.

116. A. Abbadi, K. F. Gotlieb and H. van Bekkum, *Starch – Stärke*, 1998, **50**, 23–28.

117. A. Onda, T. Ochi and K. Yanagisawa, *Green Chem.*, 2008, **10**, 1033–1037.

118. H. Cai, C. Li, A. Wang, G. Xu and T. Zhang, *Appl. Catal., B*, 2012, **123–124**, 333–338.

119. Z. Zhang and Z. K. Zhao, *Carbohydr. Res.*, 2009, **344**, 2069–2072.

120. T. Ennaert, J. Geboers, E. Gobechiya, C. M. Courtin, M. Kurttepeli, K. Houthoofd, C. E. A. Kirschhock, P. C. M. M. Magusin, S. Bals, P. A. Jacobs and B. F. Sels, *ACS Catal.*, 2015, **5**, 754–768.

121. P. Dhepe, M. Ohashi, S. Inagaki, M. Ichikawa and A. Fukuoka, *Catal. Lett.*, 2005, **102**, 163–169.

122. A. Takagaki, C. Tagusagawa and K. Domen, *Chem. Commun.*, 2008, 5363–5365.

123. I. Plazl, S. Leskovsek and T. Koloini, *Chem. Eng. J. Biochem. Eng. J.*, 1995, **59**, 253–257.

124. M. Marzo, A. Gervasini and P. Carniti, *Carbohydr. Res.*, 2012, **347**, 23–31.

125. M. Kitano, D. Yamaguchi, S. Suganuma, K. Nakajima, H. Kato, S. Hayashi and M. Hara, *Langmuir*, 2009, **25**, 5068–5075.

126. S. Suganuma, K. Nakajima, M. Kitano, D. Yamaguchi, H. Kato, S. Hayashi and M. Hara, *J. Am. Chem. Soc.*, 2008, **130**, 12787–12793.

127. S.-J. Kim, A. A. Dwiatmoko, J. W. Choi, Y.-W. Suh, D. J. Suh and M. Oh, *Bioresour. Technol.*, 2010, **101**, 8273–8279.

128. X. Qi, M. Watanabe, T. Aida and R. Smith, Jr., *Cellulose*, 2011, **18**, 1327–1333.

129. S. Inagaki, Y. Fukushima and K. Kuroda, *J. Chem. Soc., Chem. Commun.*, 1993, 680–682.

130. C. T. Kresge, M. E. Leonowicz, W. J. Roth, J. C. Vartuli and J. S. Beck, *Nature*, 1992, **359**, 710–712.

131. D. Zhao, J. Feng, Q. Huo, N. Melosh, G. H. Fredrickson, B. F. Chmelka and G. D. Stucky, *Science*, 1998, **279**, 548–552.

132. S. Inagaki, S. Guan, T. Ohsuna and O. Terasaki, *Nature*, 2002, **416**, 304–307.

133. J. G. C. Shen, R. G. Herman and K. Klier, *J. Phys. Chem. B*, 2002, **106**, 9975–9978.

134. W. M. Van Rhijn, D. E. De Vos, B. F. Sels and W. D. Bossaert, *Chem. Commun.*, 1998, 317–318.

135. J. A. Bootsma and B. H. Shanks, *Appl. Catal., A*, 2007, **327**, 44–51.

136. N. S. Mosier, C. M. Ladisch and M. R. Ladisch, *Biotechnol. Bioeng.*, 2002, **79**, 610–618.

137. P. Bhaumik and P. L. Dhepe, *Catal. Today*, 2015, **251**, 66–72.

138. E. Lam and J. H. T. Luong, *ACS Catal.*, 2014, **4**, 3393–3410.

139. M. Toda, A. Takagaki, M. Okamura, J. N. Kondo, S. Hayashi, K. Domen and M. Hara, *Nature*, 2005, **438**, 178.

140. M. Okamura, A. Takagaki, M. Toda, J. N. Kondo, K. Domen, T. Tatsumi, M. Hara and S. Hayashi, *Chem. Mater.*, 2006, **18**, 3039–3045.

141. S. Suganuma, K. Nakajima, M. Kitano, S. Hayashi and M. Hara, *ChemSusChem*, 2012, **5**, 1841–1846.

142. O. M. Gazit and A. Katz, *J. Am. Chem. Soc.*, 2013, **135**, 4398–4402.

143. D. Yamaguchi, M. Kitano, S. Suganuma, K. Nakajima, H. Kato and M. Hara, *J. Phys. Chem. C*, 2009, **113**, 3181–3188.

144. P.-W. Chung, A. Charmot, O. M. Gazit and A. Katz, *Langmuir*, 2012, **28**, 15222–15232.

145. Y. Wu, Z. Fu, D. Yin, Q. Xu, F. Liu, C. Lu and L. Mao, *Green Chem.*, 2010, **12**, 696–700.

146. J. Pang, A. Wang, M. Zheng and T. Zhang, *Chem. Commun.*, 2010, **46**, 6935–6937.

147. S. Van de Vyver, L. Peng, J. Geboers, H. Schepers, F. de Clippel, C. J. Gommes, B. Goderis, P. A. Jacobs and B. F. Sels, *Green Chem.*, 2010, **12**, 1560–1563.

148. C. Tagusagawa, A. Takagaki, A. Iguchi, K. Takanabe, J. N. Kondo, K. Ebitani, S. Hayashi, T. Tatsumi and K. Domen, *Angew. Chem., Int. Ed.*, 2010, **49**, 1128–1132.

149. Z. Fan, D. Xin, F. Zhen, Z. Hongyan, T. Xiaofei and J. A. Kozinski, *Petrochem. Technol.*, 2011, 4348.

150. L. Pena, M. Ikenberry, B. Ware, K. L. Hohn, D. Boyle, X. S. Sun and D. Wang, *Biotechnol. Bioprocess Eng.*, 2011, **16**, 1214–1222.

151. I. V. Kozhevnikov, *Chem. Rev.*, 1998, **98**, 171–198.

152. N. Mizuno and M. Misono, *Chem. Rev.*, 1998, **98**, 199–218.

153. T. Okuhara, *Chem. Rev.*, 2002, **102**, 3641–3666.

154. M. Cheng, T. Shi, H. Guan, S. Wang, X. Wang and Z. Jiang, *Appl. Catal., B*, 2011, **107**, 104–109.

155. Y. Jiang, X. Li, X. Wang, L. Meng, H. Wang, G. Peng, X. Wang and X. Mu, *Green Chem.*, 2012, **14**, 2162–2167.

156. A. A. Balandin, N. A. Vasyunina, G. S. Barysheva and S. V. Chepigo, *Bull. Acad. Sci. USSR*, 1957, **6**, 403.

157. A. A. Balandin, N. A. Vasyunina, S. V. Chepigo and G. S. Barysheva, *Proc. Acad. Sci. USSR*, 1959, **128**, 839.

158. V. I. Sharkov, *Angew. Chem., Int. Ed.*, 1963, **2**, 405.

159. A. Fukuoka and P. L. Dhepe, *Angew. Chem., Int. Ed.*, 2006, **45**, 5161–5163.

160. N. Yan, C. Zhao, C. Luo, P. J. Dyson, H. Liu and Y. Kou, *J. Am. Chem. Soc.*, 2006, **128**, 8714–8715.

161. C. Luo, S. Wang and H. Liu, *Angew. Chem., Int. Ed.*, 2007, **46**, 7636–7639.

162. N. Ji, T. Zhang, M. Zheng, A. Wang, H. Wang, X. Wang and J. G. Chen, *Angew. Chem., Int. Ed.*, 2008, **47**, 8510–8513.

163. W. Deng, X. Tan, W. Fang, Q. Zhang and Y. Wang, *Catal. Lett.*, 2009, **133**, 167–174.

164. D. S. Park, D. Yun, T. Y. Kim, J. Baek, Y. S. Yun and J. Yi, *ChemSusChem*, 2013, **6**, 2281–2289.

165. A. Ebringerova, Z. Hromadkova and T. Heinze, Hemicellulose, in *Polysaccharides I*, ed. T. Heinze, Springer Berlin Heidelberg, 2005, ch. 1, vol. 186, pp. 1–67.

166. *Hemicellulose and Hemicellulases*, ed. M. P. Coughlan and G. P. Hazlewood, Portland Press, London and Chapel Hill, 1993.

167. R. M. Rowell, R. Pettersen, J. S. Han, J. S. Rowell and M. A. Tshabalala, Cell Wall Chemistry, in *Handbook of Wood Chemistry and Wood Composites*, ed. R. M. Rowell, CRC Press, Taylor & Francis Group, London, 2005, ch. 1, pp. 35–74.

168. A. K. Chandel, F. A. F. Antunes, P. V. de Arruda, T. S. S. Milessi, S. S. da Silva and M. d. G. de Almeida Felipe, Dilute Acid Hydrolysis of Agro-Residues for the Depolymerization of Hemicellulose: State-of-the-Art, in *D-Xylitol*, ed. S. S. da Silva and A. K. Chandel, Springer Berlin Heidelberg, 2012, ch. 2, pp. 39–61.

169. G. F. Fanta, T. P. Abbott, A. I. Herman, R. C. Burr and W. M. Doane, *Biotechnol. Bioeng.*, 1984, **26**, 1122–1125.

170. Y.-Z. Lai, Chemical Degradation, in *Wood and Cellulosic Chemistry*, ed. D. N. S. Hon and N. Shiraishi, CRC Press, Taylor & Francis Group, London, 2nd edn, 2001, ch. 10, pp. 443–512.
171. B. T. Kusema, C. Xu, P. Maki-Arvela, S. Willfor, B. Holmbom, T. Salmi and D. Y. Murzin, *Int. J. Chem. React. Eng.*, 2010, **8**, 1–16.
172. P. L. Dhepe and R. Sahu, *Green Chem.*, 2010, **12**, 2153–2156.
173. B. T. Kusema, G. Hilmann, P. Maki-Arvela, S. Willfor, B. Holmbom, T. Salmi and D. Y. Murzin, *Catal. Lett.*, 2011, **141**, 408–412.
174. R. Ormsby, J. R. Kastner and J. Miller, *Catal. Today*, 2012, **190**, 89–97.
175. P. D. Cara, M. Pagliaro, A. Elmekawy, D. R. Brown, P. Verschuren, N. R. Shiju and G. Rothenberg, *Catal. Sci. Technol.*, 2013, **3**, 2057–2061.
176. P. Bhaumik and P. L. Dhepe, *ACS Catal.*, 2013, **3**, 2299–2303.
177. P. Bhaumik, T. Kane and P. L. Dhepe, *Catal. Sci. Technol.*, 2014, **4**, 2904–2907.
178. P.-W. Chung, A. Charmot, O. A. Olatunji-Ojo, K. A. Durkin and A. Katz, *ACS Catal.*, 2014, **4**, 302–310.
179. Y. Ogaki, Y. Shinozuka, T. Hara, N. Ichikuni and S. Shimazu, *Catal. Today*, 2011, **164**, 415–418.
180. P. Bhaumik and P. L. Dhepe, *RSC Adv.*, 2014, **4**, 26215–26221.
181. J. C. Speck, Jr., *Adv. Carbohydr. Chem.*, 1958, **13**, 63–103.
182. Y. Z. Lai, *Carbohydr. Res.*, 1973, **28**, 154–157.
183. A. J. Shaw Iii and G. T. Tsao, *Carbohydr. Res.*, 1978, **60**, 327–325.
184. J. M. De Bruijn, A. P. G. Kieboom and H. van Bekkum, *Starch – Stärke*, 1987, **39**, 23–28.
185. S. A. Barker, P. J. Somers and B. W. Hatt, *Fructose*, 1973, DE2229064A1.
186. J. A. Rendleman Jr and J. E. Hodge, *Carbohydr. Res.*, 1979, **75**, 83–99.
187. E. Haack, F. Braun and K. Kohler, *D-Fructose*, 1964, DE1163307.
188. C. Moreau, R. Durand, A. Roux and D. Tichit, *Appl. Catal., A*, 2000, **193**, 257–264.
189. J. Lecomte, A. Finiels and C. Moreau, *Starch – Stärke*, 2002, **54**, 75–79.
190. A. Takagaki, M. Ohara, S. Nishimura and K. Ebitani, *Chem. Commun.*, 2009, 6276–6278.
191. A. Tathod, T. Kane, E. S. Sanil and P. L. Dhepe, *J. Mol. Catal. A: Chem.*, 2014, **388–389**, 90–99.
192. S. Lima, A. S. Dias, Z. Lin, P. Brandao, P. Ferreira, M. Pillinger, J. Rocha, V. Calvino Casilda and A. A. Valente, *Appl. Catal., A*, 2008, **339**, 21–27.
193. M. Moliner, Y. Roman-Leshkov and M. E. Davis, *Proc. Natl. Acad. Sci.*, 2010, **107**, 6164–6168.
194. Y. Roman-Leshkov, M. Moliner, J. A. Labinger and M. E. Davis, *Angew. Chem., Int. Ed.*, 2010, **49**, 8954–8957.
195. V. Choudhary, A. B. Pinar, S. I. Sandler, D. G. Vlachos and R. F. Lobo, *ACS Catal.*, 2011, **1**, 1724–1728.
196. C. M. Lew, N. Rajabbeigi and M. Tsapatsis, *Microporous Mesoporous Mater.*, 2012, **153**, 55–58.
197. US Energy Information Administration, http://www.eia.gov/cfapps/ipdbproject/iedindex3.cfm?tid = 79&pid = 79&aid = 1&cid = ww,& syid = 2008&eyid = 2012&unit = TBPD.

CHAPTER 2

Aqueous-phase Reforming of Sugar Derivatives: Challenges and Opportunities

T. M. C. HOANG, A. K. K. VIKLA AND K. SESHAN*

University of Twente, Faculty of Science & Technology, Catalytic Processes & Materials, 7500 AE, Enschede, The Netherlands
*Email: k.seshan@utwente.nl

2.1 Aqueous Phase Reforming (APR) Definition

The steam reforming reaction uses water as an oxidant to generate syngas $(CO + H_2)$ from hydrocarbons, *e.g.*, methane. This strongly endothermic reaction is carried out at high temperatures ($>750\,^{\circ}C$) and low pressures (<25 bar) where water is in the gas phase. Aqueous phase reforming is a variant where conditions are maintained to keep water in the liquid phase. The feedstock in this case is an aqueous phase containing dissolved organics, typically in the range of 5–20 wt%. The reaction equation for carbohydrates, waste from a food industry, for *e.g.*, is:

$$C_6H_{12}O_6 + 6H_2O \rightarrow 6CO_2 + 12H_2 \tag{1.1}$$

The conversion of such aqueous bio/organic wastes ($>80\%$ water) into high value products such as hydrogen and syngas (CO/H_2) using conventional reforming processes that operate at lower pressures and high temperatures is energy intensive due also to the need to account for the latent heat of evaporation of water. Dumesic and co-workers tackled this problem by developing the "aqueous phase reforming" (APR) process in which water is kept in the

RSC Green Chemistry No. 44
Biomass Sugars for Non-Fuel Applications
Edited by Dmitry Murzin and Olga Simakova
© The Royal Society of Chemistry 2016
Published by the Royal Society of Chemistry, www.rsc.org

liquid phase by applying elevated pressures.[1–3] The concept was further demonstrated for steam reforming of diluted oxygenate feeds at mild temperatures in pressurized liquid water (225–265 °C, 29–56 bar) over supported metal catalysts. For example, typical temperatures for fundamental APR studies of model components (*e.g.* ethylene glycol, methanol and sorbitol) are usually performed in the temperature range of 200–265 °C.[2,4] The reforming of higher concentrated feed streams or more complex oxygenated hydrocarbons, which are of more commercial relevance, require higher temperatures to obtain the reaction rates that are necessary for industrial application.[5]

Generation of chemicals from renewable bio-based feedstocks often involve a hydrogenation step. In order to make these conversions completely green, it is essential that the required hydrogen for this is also made available from bio feedstocks. This chapter addresses the issues involved in the production of hydrogen *via* APR of byproduct waste streams. APR has also been used sparingly, to refer to reactions occurring in an aqueous phase to generate valuable aromatic molecules, *e.g.* from lignin *via* depolymerisation. Hydrogen generated in the presence of catalysts under higher pressures and temperatures facilitate this reaction and hence the use of the term APR. This will be discussed later in the chapter.

2.2   Phase Diagram—Water

The phase diagram of water given in Figure 2.1 shows the pressures required to keep hot water in the liquid phase at elevated temperatures. APR conditions can thus be divided into sub- and supercritical regimes. The transition from sub- to supercritical water occurs at 374 °C and 221 bar. A dramatic change in the properties of water occurs when water becomes supercritical and these offer some advantages for the reforming of biomass derived waste streams into gaseous products.[5–8] The density beyond the supercritical point is around $\pm 100 \, \text{kg m}^{-3}$ and a further increase in temperature does not affect the density significantly anymore. The properties of liquid water and compressed gas converge around the supercritical point, resulting in complete mixture of both phases and the removal of the liquid/gas phase boundary. The latter is very beneficial for fast rates of heat and mass transfer,[5] the latter being very relevant for bulky organic molecules.

Justifiably, APR is also termed "gasification in hot compressed water", which includes also supercritical conditions where water is not a liquid and where feed molecules in liquid or solid phase are converted into gases.

2.3   Background—Hydrogen, Steam Reforming and Water Gas Shift

2.3.1   Current status

Currently, syngas is mostly produced by steam reforming of natural gas and other fossil feedstocks. Hydrogen yields are maximised by a subsequent

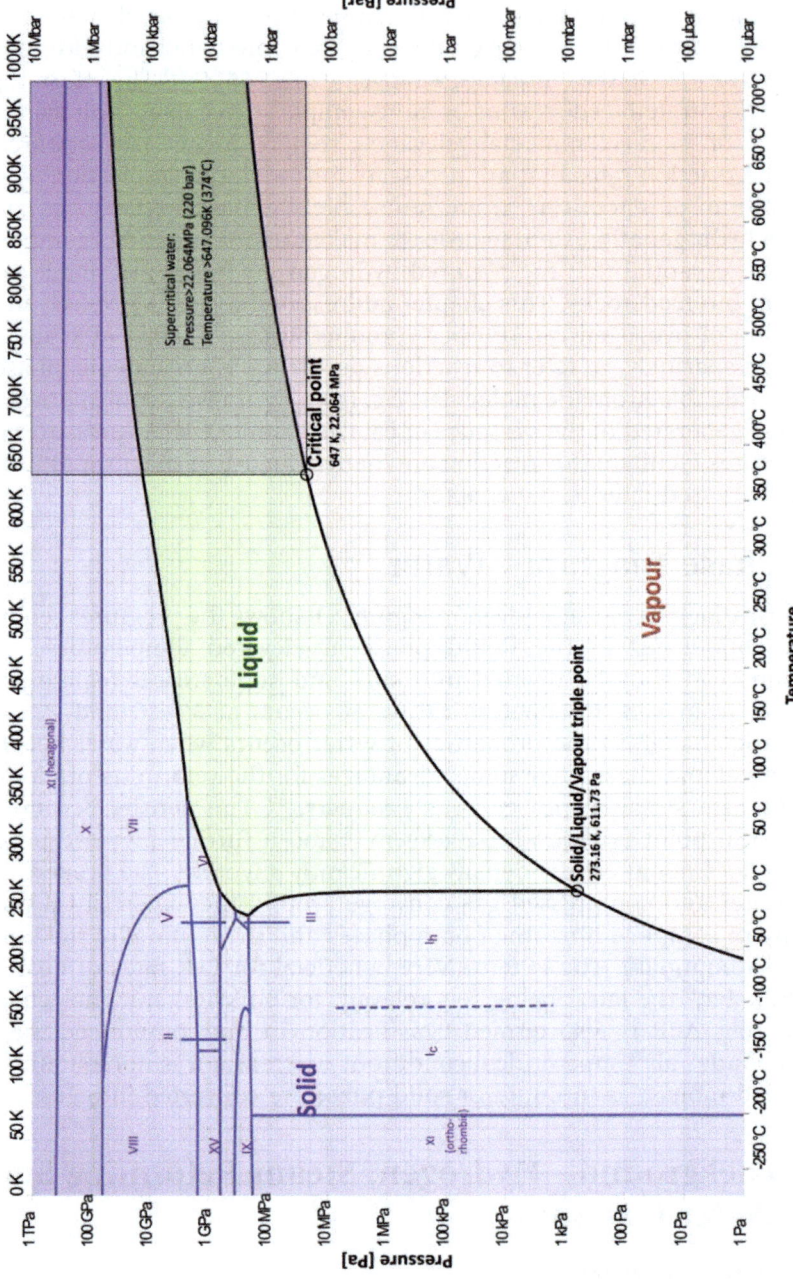

Figure 2.1   Phase diagram of water including triple point and the critical point of water.

water gas shift reaction, discussed later. Demand for hydrogen is ever increasing and predicted to reach 11 billion metric tons by 2016 in the USA alone (Figure 2.2).[9] Currently, hydrogen is widely used in ammonia production for the fertilizer industry, oil refining and the food industry.

Steam reforming of natural gas is the most proven and used technology for the production of hydrogen.[10-12] The first industrial steam reformer was installed as early as 1930.[12] Most of the commercial activities were in the United States where natural gas was abundantly available. Later on, discovery of natural gas reserves in Europe allowed the use of natural gas as a feedstock for hydrogen production. Steam reforming is a strongly endothermic reaction (eqn 2.2). In order to ensure a high methane conversion, it is therefore required to operate at elevated temperatures.

$$CH_4 + H_2O \leftrightarrow CO + 3H_2 \quad (\Delta H \sim 206.3 \text{ kJ mol}^{-1}) \tag{2.2}$$

Additionally, the steam reforming reaction is reversible. Therefore, according to Le Chatelier's principle, low pressures and relatively high steam to carbon ratios will shift the equilibrium to the right, increasing methane conversion, as illustrated in Figure 2.3.[11,13] Methane, particularly, is extremely difficult to activate, as the hydrogen–carbon bonds[6] are strong (435 kJ mol^{-1}) and therefore cleavage of the corresponding hydrogen–carbon bonds requires very high temperatures. The presence of a catalyst, however, allows for milder conditions as it enhances the hydrogen carbon bond rupture. The metals of group VIII of the periodic table are active for steam reforming and Ni particularly appears to be the most cost-effective.[12,13]

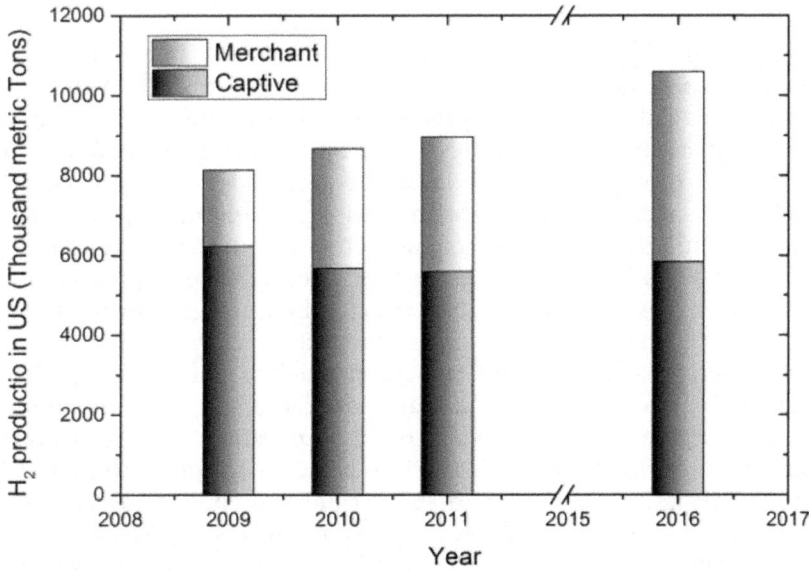

Figure 2.2 Hydrogen production between 2009 and 2011 and the expected production for 2016 (after Joseck[9]).

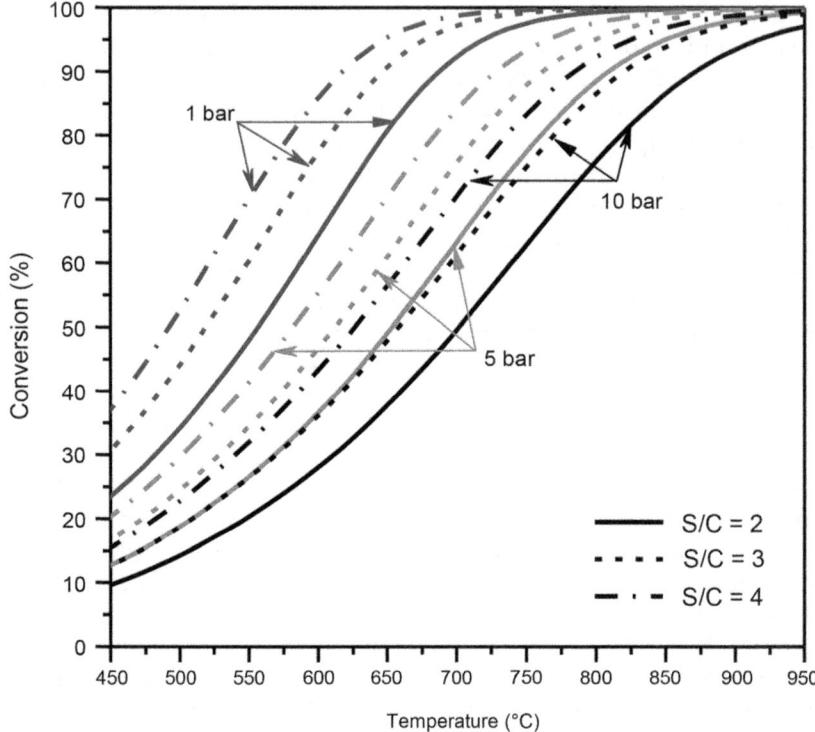

Figure 2.3 Equilibrium conversions of methane at temperatures from 450–950 °C with steam to carbon ratios (S–C) of 2 : 4 and pressuers of 1, 5 and 10 bar.

Industry typically operates this process at around 800 °C, 15–40 bar and a steam to carbon ratio of 2:3 over Ni based catalysts.[14,15] Although the presence of a catalyst makes the process technically more feasible compared to non-catalytic conditions, the elevated temperatures applied often result in large energy consumption; efficiency of the process is typically around 70%. The relatively large amounts of steam used in the reforming process ($\sim$ 3 as against 1 required by reaction stoichiometry) also decrease the overall process efficiency. The significant heating and cooling steps required to recycle the unconverted steam can explain this. Although high steam to carbon ratios result in high energy consumption, this is essential to overcome catalyst deactivation due to coke formation. Coking, which is a side reaction in many industrial processes, is a major issue and still needs attention. Besides resulting in catalyst deactivation it also causes serious operational problems.[16,17] Particularly, Ni is known to be very susceptible to coking. This makes the design of an active and stable catalyst a challenging task. In this context, extensive research has been carried out by Rostrup-Nielsen[11] and the report discusses coking during steam reforming in detail. Addition of promoters to steam reforming catalysts has been widely reported as a successful tool to improve the catalytic properties in terms of activity and stability.[18,19]

Numerous studies have been reported on the role of potassium[20–22] on the properties of the unmodified catalysts and it is widely accepted that the presence of potassium or other alkali/alkaline earth metals improves resistance to coking by assisting in the formation of OH groups on oxide supports, which enhance coke gasification. Formation of an OH group is accepted as the intermediate in the breakdown/activation of water and helps in the oxidation.[23] Some other authors have reported the beneficial influence of La as a promoter in the improvement of catalyst stability for reforming reactions.[24] In the case of dry reforming, La is suggested to react with CO_2 forming a new crystalline phase ($La_2O_2CO_3$), which provides oxygen from its structure and in this way, contributes to coke removal.[25,26] Additionally, La is known to stabilize metal particles and oxides by being present at the grain boundaries and preventing sintering.[27,28]

In the beginning of the last century, naphtha was the most economic feedstock in Europe, in contrast to the natural gas available in the United States. Steam reforming of naphtha, light alkanes such as ethane and propane associated with natural gas has also become a major industrial route for hydrogen production (eqn 2.3). In this case, the temperatures required for the process are lower than those applied in the steam reforming of methane. This is due to the bond dissociation energies of higher hydrocarbons being lower than those of methane and therefore easier to activate. For the same reason, formation of carbon deposits is more severe compared to methane and is one of the main issues. Multiple studies have been performed to elucidate the mechanism of coke formation/growth and to circumvent this bottleneck.[29,30] A separate catalyst (pre-reforming) is used for the process, which operates at much lower temperatures (<450 °C).[31]

$$C_nH_m + nH_2O \rightarrow nCO + \left(n + \frac{m}{2}\right)H_2 \quad (n > 1; \text{ endothermic reaction}) \quad (2.3)$$

2.3.2 Reaction Mechanism for Steam Reforming of Methane and Higher Hydrocarbons

A Langmuir–Hinshelwood type mechanism has been proposed for the steam reforming of methane over supported Ni catalysts.[12,32] It is commonly agreed, for example in the case of methane, that decomposition of methane on the nickel surface *via* C–H rupture is the first step. Subsequently, the carbonaceous species (CH_x, $1 \leq x \leq 3$) formed on the surface react with steam or surface oxygen species generated by activation of water[30] to produce syngas. Similarly, conversion of higher hydrocarbons takes place by irreversible adsorption to the nickel surface on a dual site, with subsequent breakage of terminal C–C bonds one by one until, eventually, the hydrocarbon is converted into C1 components.[27]

As for water activation, Rostrup-Nielsen[11] suggested that water is adsorbed and split on the oxide support, *e.g.* alkali modified aluminas, hydroxylating their surface. OH groups thus formed react with C1 species formed from hydrocarbons to form syngas. In this case, both metal Ni and support oxide provide catalytic sites to activate methane and water, respectively, in a Langmuir–Hinshelwood type mechanism involving non-competing sites. Most importantly, these catalytic sites are on the metal and the support oxide, thus providing for bi-functional catalysts. Additionally, it has been reported that nickel surfaces are able to dissociate water *via* nickel oxidation equilibrium (eqn 2.4).[11] Thus, the use of nickel-based catalysts provides additional active sites for water activation.

$$Ni + H_2O \rightarrow NiO + H_2 \quad (\Delta H = 2.12 \text{ kJ mol}^{-1}) \tag{2.4}$$

and allows for higher steam reforming activities. Similar to the steam reforming of methane, both metal and oxide supports also participate in the steam reforming of higher hydrocarbons. Such a Langmuir–Hinshelwood type mechanism was proposed by Praharso *et al.*[32] to describe steam reforming of iso-octane over a nickel-based catalyst.

We have also shown that a bi-functional mechanism is operational for WGS on Pt/ZrO$_2$; *i.e.*, CO activation takes place on Pt and steam activation on ZrO$_2$. The reaction occurs at the Pt periphery in close proximity of ZrO$_2$. Thus catalysts with identical Pt particle sizes give different intrinsic activities (TOF) indicating the role of support[33] (see Figure 2.4). In mechanistic sequence, activation of water is usually achieved *via* formation of hydroxyl groups on oxide supports. Hydroxyl groups then react with the "C" residue

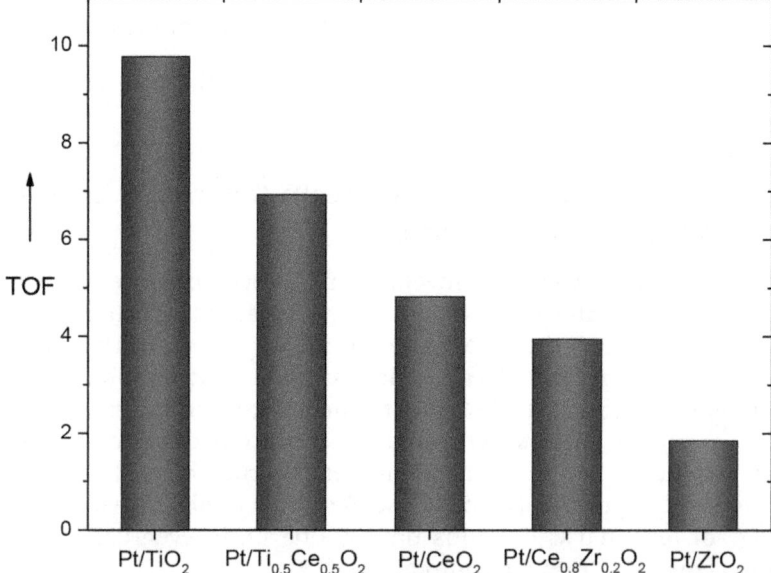

Figure 2.4 Turn over frequency of CO over Pt supported oxides. Conditions: dispersion of Pt $\sim 60 \pm 5\%$, $P_{CO} = 60$ mbar, $P_{steam} = 150$ mbar, $P = 2$ bar, $T = 300\,^\circ C$, GHSV $= 2.1 \times 10^6$ h^{-1} (adapted from Azzam *et al.*[33]).

on the Pt metal to complete SR/WGS reactions.[23,34,35] In ceria based supports, water activation also occurs *via* a redox involving the cerium cation. This bi-functional mechanism requires the active metal sites, such as Pt, to be situated in the close proximity of the support OH groups.

$$CO + H_2O \rightarrow CO_2 + H_2 \quad (-41.1 \text{ kJ mol}^{-1}) \quad (2.5)$$

The main objective of the WGS reaction (eqn 2.5) when coupled with steam reforming is to maximize hydrogen production or adjust the H_2–CO ratio needed for the end application.[27,48] It is a well-known reversible, exothermic reaction. Therefore CO conversions are favoured at low temperatures. In order to overcome this thermodynamic limitation and thus increase conversions, the WGS of the outlet gas from the steam reformer is carried out in multiple stages. The first step (high-temperature shift, HTS), which operates at 300–450 °C, is catalysed by Fe–Cr based catalysts. This rather low activity catalyst is also resistant to poisons present in syngas, such as chlorine and sulfur, and it acts as a poison trap. Additionally, this catalyst is resistant to an adiabatic temperature increase. Since Fe–Cr catalysts have low activity, higher temperatures are required to get favourable kinetics. Thermodynamics dictates that the equilibrium towards hydrogen production is unfavourable at high temperatures and thus a high temperature shift results in low CO conversions. In order to reach high conversions, a second step (low-temperature shift, LTS), which occurs at lower temperatures (180–230 °C) over a very active catalyst (Cu–Zn based catalyst), is used.[49] The sensitivity of this catalyst to sintering and sulfur and chlorine poisoning does not make this catalyst suitable to be used by itself in a single step or in the high-temperature shift. This two stages approach is not desirable for mobile applications because of its technical complexity. In this respect, extensive research is reported on the design of robust and active catalysts that can be applied in one single stage WGS reactor. Supported precious metal catalysts, *e.g.* Pt, show promise in this respect.[50–52]

2.4 Steam Reforming/Gasification of Solid Biomass

Most of the research and commercial production activities for gasification of solid lignocellulosic biomass based feedstocks draw their inspiration from coal gasification. Gasification can be carried out to equilibrium conversions above 1100 °C. Modern entrained flow gasifiers are operated at elevated pressures (up to 80 bar) and temperatures >1250 °C, and are designed for coal or oil applications. A comprehensive overview of gasification of dry biomass, reactor types and the problems can be found in excellent reviews on the subject.[36,37] Bio-oil produced by pyrolysis liquefaction is also considered as a more easy to handle feedstock for gasification.[38]

In order to produce a clean tar and methane free syngas at lower process temperatures (<950 °C), various research groups have studied the application of catalysts to biomass gasification. Lowering the temperature of the gasifier below the melting temperature of the ashes reduces equipment costs and allows for gasification at a smaller scale compared to entrained flow gasification. Catalysts are either pre-mixed with the biomass, used (partly or fully) as bed material in fluid bed gasifiers or applied downstream of the gasifier for product gas upgrading.

Cheap disposable catalysts have been used to create a valuable fuel gas rather than to produce actual syngas. Dolomite has gained the most attention because it is very cheap.[39–42] It is applied inside the gasifier to promote direct tar cracking or separately in a bed downstream of the gasifier. Although it can almost fully convert tars, it is more often used as a tar-reducer and a guard material, allowing the usage of more active but also more sensitive catalysts downstream[43] However, dolomite is not able to effectively convert methane and suffers from attrition losses.[42–44] Olivine is much more resistant to attrition than dolomite with a somewhat lower activity for tar destruction.[44,45] Impregnation of the olivine with nickel makes it possible to enhance its activity while maintaining its strength.[46] Alkali metals are most effective when impregnated into the biomass, promoting tar free gas production, especially when potassium carbonate is being used. Catalyst deactivation, catalyst make-up and fluidization problems still need a lot of research before these catalysts can be effectively applied.[39] When, in addition to tars, complete methane conversion is also desired, the high steam (and dry) reforming activity of the catalyst is of vital importance. Ni/Al_2O_3 catalysts have been used in the industry for naphtha and natural gas reforming for many years and since biomass gasification was initially carried out with off the shelf catalysts, it was therefore also logical that they were applied for biomass gasification applications. Caballero *et al.*[43] and Simell *et al.*[47] have been able to effectively eliminate the tars in the biomass derived gas and realize a significant decrease of methane using a crushed and/or as-received commercial catalyst or dedicated monolith beds. For complete tar and methane elimination, using downstream secondary reactors after the gasifier has been successful in creating a clean gas. However, until now, none of the proposed processes have reached commercialization.

2.5 APR of Biomass Streams

Aqueous waste streams are dilute solutions, typically containing <25 wt% organic molecules. For instance, the aqueous fraction of flash pyrolysis oil commonly consists of 80 wt% water and ~20 wt% of a complex mixture of different oxygenates such as aldehydes, ketones, alcohols, acids and sugars.[48] Conventional steam reforming (SR) of aqueous biomass derived streams is economically unfeasible because of their high water content and the energy required to carry out the reactions in the gas phase.[4] Catalytic aqueous phase reforming (APR) is an attractive alternative to produce 'green' hydrogen.[1,49,50] During APR, water is kept in the liquid phase by applying elevated pressures. Based on system enthalpies, an energy of 3.7 MJ is required to condition 1 kg of water from ambient (25 °C, 1 bar) to gas phase

SR conditions (700 °C and 30 bar), while only 2.6 MJ is required to bring 1 kg of water to APR conditions (400 °C and 250 bar). This implies feasibility for APR applications.

Fundamental APR studies are commonly performed with low concentration aqueous solutions of model compounds at temperatures >225 °C and pressures >34 bar.[3,51,52] However, reforming of industrially relevant feed concentrations (up to 20 wt% organics) requires higher temperatures of 310–650 °C (110–350 bar) to achieve commercially interesting reaction rates (*e.g.* residence time in catalytic bed <5 seconds).[5,53] In the future, hydrogen from renewable resources is also projected to partly replace conventional fuels to reduce anthropogenic CO_2 emissions.[54] Hydrogen is expensive and is sold on the market for $4500 per ton. The integration of the bio-oil refinery with hydrogen production from the aqueous fraction is a promising route to obtain completely green chemicals and fuels. Production of hydrogen by eliminating bio-organic waste streams is very interesting from both an economic and environmental viewpoint. Examples given next show how hydrogen is produced from waste by-product streams in bio-based processes. They allow the hydrogen to be used in the main conversion process and make the whole process totally green.

2.5.1   APR of Waste Streams to Feedback Hydrogen

2.5.1.1   *Example 1—Isosorbide from Cellulose*

Isosorbide is a versatile intermediate. It is a potentially useful bio-feedstock, used in pharmaceutical industries, *e.g.,* isosorbide nitrates for angina pectoris (chest pain), chemicals/polymers, *e.g.*, polycarbonates, polyesters[55] and in fuels, *e.g.,* diesel additives such as dimethyl isosorbide, or isosorbide di-nitrate as an ignition improver.[56] It is obtained in a two-stage process involving hydrogenation of glucose, which gives sorbitol. Isosorbide is obtained by double dehydration of sorbitol (Figure 2.5). Hydrogenation of glucose to sorbitol in aqueous media can be achieved at 100–120 °C; 20–40 Bar H_2 for 120–180 min with (5 wt%) Ru/carbon catalysts. The Ru/carbon catalyst is very active in glucose hydrogenation to sorbitol with complete glucose conversion and sorbitol selectivity above 98%. Side products reported are mannitol (below 1%) and gluconic acid (below 0.1%).[57,58] These reactions are carried out in an aqueous phase and the by-products formed during the reaction find themselves in the aqueous phase.

Sorbitol conversion to isosorbide occurs in two stages with H_2SO_4 as a catalyst. The first reaction is performed at 120 °C for 2 h followed by a second reaction at 130 °C for 0.5 h. Total conversion of sorbitol is 97.6% with the following yields: isosorbide: 83.2%, 1,4-sorbitan: 3.3% and 2,5-sugars: 12.5%.[59] In $ZnCl_2$ solution (70% in water), sorbitol is converted to isosorbide with 85% molar based yield. At low temperatures, the reaction rate is low; with an increase in temperature (over 200 °C), the rate improves but extensive amounts of by-products are formed.[60] With a solid catalyst

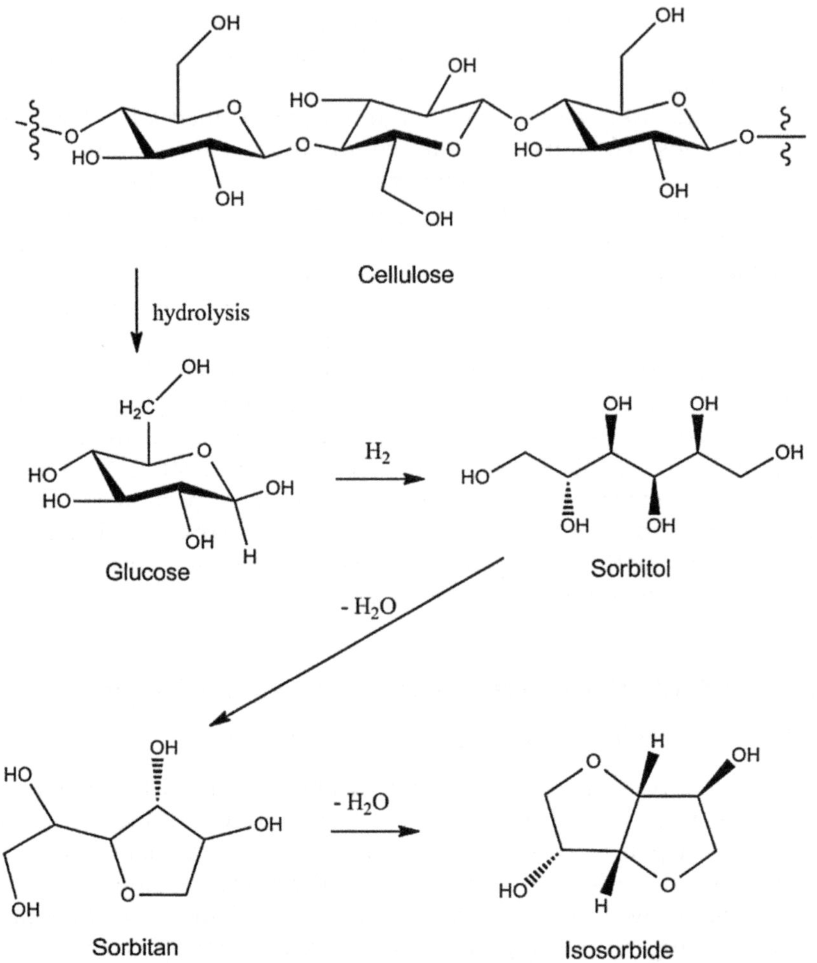

Figure 2.5 Hydrolysis of cellulose to glucose and further hydrogenation lead to formation of sorbitol. Dehydration of sorbitol results in sorbitans and further dehydration to isosorbide.

(silico-tungstic acid), the isosorbide yield is not higher than 40% with sorbitans (mixture of 1,4 or 1,5-sorbitan, and 2,5-sorbitan) as the main intermediate products.[61] Thus the reaction produces a variety of organic molecules dissolved in the aqueous phase resulting in amounts in excess of 12 wt%. The hydrogen required for the reduction of sugars to sorbitol, if obtained from these side products *via* APR, can make the process completely sustainable. For example, 1 mol of sorbitan ($C_6H_{12}O_5$) can yield 13 mol of H_2 *via* SR/WGS. One mole of glucose in comparison requires only 1 mol of hydrogen for conversion to sorbitol. Thus there is enough potential to generate all the required hydrogen for the conversion from the aqueous waste containing by-products.

2.5.1.2   Example 2—Humin Byproducts During Sugar to HMF/Levulinic Acid Conversion

Hydroxymethyl furfural (HMF) and levulinic acid (LA) are addressed as important versatile platform chemicals derived from carbohydrates.[62,63] A wide range of chemicals used as solvents (*e.g.*, GVL, THF, dimethyl THF), monomers (*e.g.*, furan dicarboxylic acid, caprolactam, adipic acid, *etc.*), and commodities (*e.g.*, butene) can be produced from these platform molecules. The pathways to convert lignocellulosic biomass to these platform chemicals include separation of carbohydrates from biomass, hydrolysis of these polysaccharides to oligomer/mono carbohydrates (sugars) and dehydration of these sugar monomers (illustrated in Scheme 2.1). Hydrogenation of LA to GVL can be conducted in a vapour or liquid phase with high conversion and selectivity using especially Ru catalysts (*e.g.*, 5 wt% Ru/C, 15 wt% RuRe (3:4)/C, 5 wt% RuSn (3.6:1)/C). Wright and Palkovits have summarised recent advances in catalyst development in their mini-review.[64] Both hydrolysis and dehydration reactions are conventionally performed in the aqueous phase using acid catalysts such as HCl or H_2SO_4.[65,66] Yield and selectivity of HMF or LA are strongly dependent on process parameters such as sugar substrates, concentrations, reaction temperatures, catalyst *etc.* However, for most of the potential process configurations, the conversion of carbohydrate substrates to useful products is only about 55–65%.[67–70] The rest of the carbohydrates are converted to degradation by-products, called humins (50–66 wt% C, 29–46 wt% O and the rest is H).[67–72] Based on their solubility in the reaction medium, they are further divided into two groups: soluble humins and solid humins. Weingarten *et al.*[68] reported that for LA from cellulose, the selectivity of humins in the aqueous phase is approximately from 22–80 % (on a carbon basis). Thus, the process making LA from biomass can produce large amounts of organic waste in the aqueous phase.

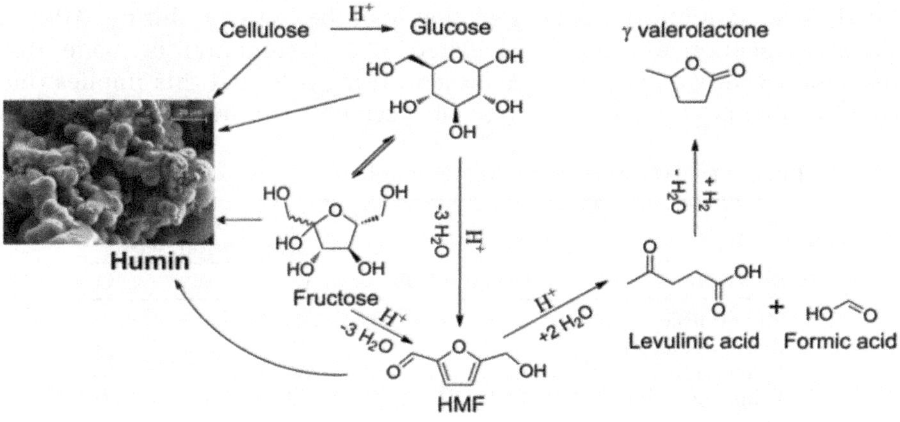

Scheme 2.1   Process scheme for GVL derived from cellulose. SEM image of humin adapted from Hoang *et al.*[74]

Besides, the solid humin (yield of 25–35 wt%)[67–70,72] can also be steam re-formed to H_2.[73] It is estimated that H_2 produced from soluble and solid humin is in surplus to the amount required for hydrogenation of LA to GVL. Thus, they are a potentially renewable source for H_2 production in a bio-refinery and valorisation of humins, carbohydrate-derived wastes, can be achieved.

2.5.2 Evaluation of This Concept in Terms of Hydrogen Economics

Pyrolysis of lignocellulose is a route to liquefaction of solid biomass. Fractionation of the pyrolysis oil in an organic and aqueous phase is the first step in obtaining high value bio-oil. The typical composition of the bio-oil is given in Table 2.1. The next step to upgrade involves de-oxygenation of the organic phase with hydrogen. APR of the aqueous phase of pyrolysis oil is an interesting route to produce the hydrogen required for de-oxygenation of the organic phase.

A concept for the integration of APR in a bio-refinery is evaluated in terms of hydrogen economics as shown in Table 2.1 and Figure 2.6.[48] In this section, the maximum amount of hydrogen that can be obtained theoretically by APR of the aqueous feed is compared with the amount of hydrogen required to upgrade the bio-oil fraction to commercially relevant hydrocarbons, which can be used in a naphtha cracker to obtain olefins, building blocks for chemicals and polymers. It can be seen from Table 2.1 that 25 wt% of the pyrolysis oil is the desired organic phase. The remaining 75 wt% aqueous phase consists of water soluble compounds (oxygenates and H_2O), and is referred to as the aqueous fraction. The oxygenates account for 50 wt% of the pyrolysis oil and these have a weight averaged composition of $C = 46.8$, $H = 6.0$ and $O = 47.2\%$ as shown in Figure 2.6. The oxygenates in the aqueous phase are ideal feedstocks for hydrogen production by APR. Firstly, the amount of hydrogen that can be formed during APR of the aforementioned feed is calculated. The assumption is made that maximum hydrogen yields are achieved during APR and this implies that (i) all hydrogen atoms in the oxygenate recombine to molecular hydrogen

Table 2.1 Typical composition of bio-oil.

Component	Fraction wt%	Average molecular composition (wt%)		
		C	H	O
Water soluble				
Acids, Alcohols	10	36	6	58
Ether-solubles	10	60	6	34
Ether-insolubles	30	46	6	48
H_2O	25	–		
Water insolubles	25	66	7	27

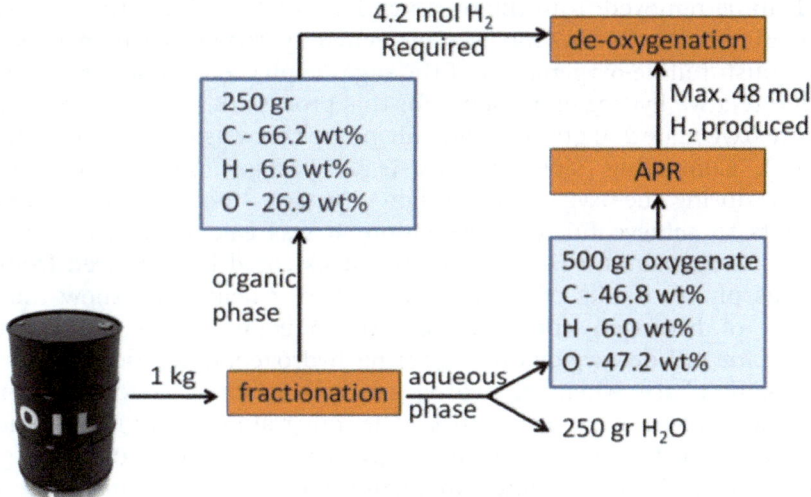

Figure 2.6   Amount of hydrogen produced by APR of the aqueous phase of pyrolysis oil and the amount needed for upgrading bio-oil.

and (ii) carbon undergoes full oxidation by water to form CO_2 and H_2 (WGS). The oxygenates (which represent 50 wt% of the pyrolysis oil) contain 6 wt% hydrogen, indicating that 30 grams (or 15 mol) of H_2 can be produced per kg of initial pyrolysis oil. Furthermore, hydrogen can also be formed by the water gas shift reaction. During WGS, carbon is oxidized by water to form H_2. A single carbon atom can undergo two oxidations by water ($C + H_2O \rightarrow CO + H_2$, and $CO + H_2O \rightarrow CO_2 + H_2$) and therefore, in the optimal case, two hydrogen molecules can be formed per carbon atom. However, the carbon in the oxygenate molecule is already partly oxidized and this should be taken into account when calculating the maximum hydrogen yield possible. It can be calculated for the oxygenates that the molar ratio of oxygen–carbon is 0.77, while fully oxidized carbon has an oxygen–carbon ratio of 2. This indicates that 1.23 molecules of hydrogen can be formed per carbon atom in the case of optimal hydrogen yields. Oxygenates are 46.8 wt% composed out of carbon, indicating that the water-soluble oxygenate fraction (50 wt% of pyrolysis oil) contains ∼19 moles of carbon per kg of pyrolysis oil. This indicates that another 23 moles of H_2 can be produced by complete oxidation of the already partly oxidized carbon. These considerations show that a total of 48 moles of H_2 can be produced maximally from the aqueous fraction of 1 kg of pyrolysis oil. Next, the amount of hydrogen required for upgrading the organic phase of pyrolysis oil is calculated. The pyrolysis oil is composed of 25 wt% organic phase. During the de-oxygenation process, oxygen in the organic compounds is hydrogenated and removed from the molecule in the form of water. Two hydrogen atoms are necessary to remove one oxygen atom in the form of water ($O + 2H \rightarrow H_2O$). The oxygen content in the organic phase is 27 wt%, indicating that 68 grams (or 4.2 mol) of atomic oxygen needs to be removed from the organic phase of 1 kg of pyrolysis oil. These calculations show that 4.2 moles of molecular hydrogen (H_2) are necessary to accomplish full de-oxygenation of the organic phase. Bridgwater[75] reported similar relative hydrogen amounts for this process (8 moles of hydrogen to fully de-oxygenate 250 grams of organic phase that was composed of 50 wt% oxygen). Competing side-reactions (*e.g.* hydrogenation of unsaturated bonds) during de-oxygenation might increase the required hydrogen amounts to achieve full de-oxygenation. It was calculated earlier in this section that maximum H_2 yields of 48 moles could be produced from the aqueous phase of 1 kg of pyrolysis oil. These calculations show that the amount of hydrogen obtained from the aqueous phase is more than enough for full de-oxygenation (ignoring hydrogenation side reactions) of the organic phase, when the hydrogen selectivity for the reforming for the aqueous fraction is at least 10%. The integration of APR to generate hydrogen from the aqueous phase of pyrolysis oil to upgrade the organic phase is graphically presented in Figure 2.6, and shows this to be conceptually feasible.

2.6 Thermodynamics of APR

Dumesic *et al.*[76] reported the thermal dependence of the standard Gibbs free energy for water gas shift and steam reforming of various oxygenate molecules in the liquid and vapour phase as shown in Figure 2.7. A negative value for the standard Gibbs free energy ($\Delta G/RT < 0$) indicates that the process is spontaneous. The Gibbs free energy for the water gas shift reaction in the liquid phase was reported to be negative and temperature independent in the temperature range 27–375 °C. In the case of the vapour phase water gas shift reaction, the reaction becomes less favourable at higher temperatures. The reforming of ethylene glycol, for example, in the liquid phase compared to the vapour phase becomes more favourable above 175 °C. The advantages of aqueous phase reforming are (i) no need for evaporation of the water and (ii) the water gas shift activity and reforming rates are more favoured in the liquid phase than in the vapour phase at temperatures above 175 °C.

2.7 Why Hydrogen is Favoured at APR Conditions

Davda *et al.*[77] have explained why APR conditions favor hydrogen formation. According to them, aqueous-phase reforming of oxygenates to produce H_2 takes place *via* steam reforming and WGS. The lowest partial pressure of CO that can be achieved depends on the thermodynamics of the WGS reaction and the operating conditions, (eqn 2.6).

$$P_{CO} = \frac{P_{CO_2} P_{H_2}}{K_{WGS} P_{H_2O}} \qquad (2.6)$$

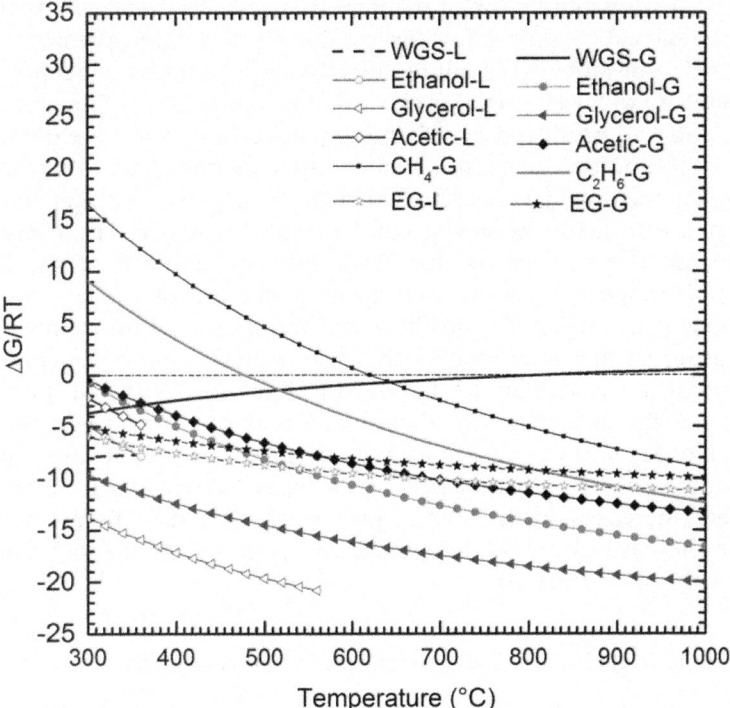

Figure 2.7   Gibbs free energy for different oxygenates at various temperatures including WGS reaction.

K_{WGS} is the equilibrium constant for the vapor-phase WGS and P_j are partial pressures. Since H_2, CO_2, and small amounts of alkanes (primarily CH_4) are produced by aqueous phase reforming,[8] gas bubbles are formed within the liquid-phase flow reactor. The pressure in these bubbles can be approximated to be equal to the system pressure, (eqn 2.7).

$$P_{bubble} \approx P_{system} = P_{H_2O} + \sum_{products} P_i \qquad (2.7)$$

The partial pressures of the reaction products and water vapor are dictated by the feed concentrations, system pressure and temperature, as outlined later. For dilute product concentrations and system pressures above the saturation pressure of water, the bubbles contain water vapor at a pressure equal to its saturation pressure at the reactor temperature, and the remaining pressure is the sum of the partial pressures of the product gases. The extent of vaporization, y, is defined as the percentage of water in the vapor phase relative to the total amount of water flowing into the reactor. In contrast, for systems operated at pressures that are near the saturation pressure of water, all the liquid water may vaporize, and the composition of the bubble is dictated by the stoichiometry of the feed stream. At this condition, the partial pressure of water is below its saturation pressure because the reforming product gases dilute the water vapor. Higher concentrations of oxygenate lead to lower partial pressures of water because of greater dilution from H_2 and CO_2 produced by reforming reactions. As the system pressure is increased, the partial pressure of water vapor increases until it reaches the saturation pressure of water, at which point any further increase in the system pressure leads to partial condensation of water. These arguments indicate that the conditions that favor the lowest levels of CO from reforming of oxygenates are those that lead to the lowest partial pressures of H_2 and CO_2 in the reforming gas bubbles; and these conditions are achieved by operating at system pressures that are near the saturation pressure of water and at low oxygenate feed concentrations. As the system pressure increases and the extent of vaporization decreases below 100%, the partial pressures of H_2 and CO_2 in the bubble increase, thereby leading to higher equilibrium concentrations of CO. Similarly, as the oxygenate concentration in the feed increases, higher partial pressures of H_2 and CO_2 are developed, even for the case of complete vaporization, again leading to higher equilibrium CO concentrations.

2.8   Mechanism of APR/Reaction Routes

Catalytic APR studies with model compounds are usually carried out to simplify the process and a gain fundamental understanding of catalytic reforming. The purpose of these studies has been to identify pathways that lead to hydrogen and prevent the formation of others. Important reactions during reforming are C–H, C–C and C–O bond cleavages. To prevent alkane formation, it is preferred that every carbon atom is connected to one oxygen atom to enable reforming of the molecule to CO and H_2. Ethylene glycol (EG) is often chosen as a model compound to study fundamental catalytic behaviour because it is the smallest molecule (hence avoiding the occurrence of complicated side reactions) with all carbon atoms bonded to oxygen (preventing intrinsic methane formation) where both desired and undesired pathways (C–C, C–O and C–H cleavage) can occur. The reforming of ethylene glycol has been studied intensively by Dumesic and colleagues[3] and a reforming mechanism has been proposed by them as shown in Figure 2.8. EG first undergoes dehydrogenation leaving adsorbed species on the catalyst surface. The formed intermediate compound can further react through two pathways. The desired pathway to form hydrogen involves C–C cleavage, which results in H_2 gas and adsorbed CO. Hydrogen yields are further enhanced by the water gas shift reaction $(CO + H_2O \rightarrow CO_2 + H_2)$. The undesired pathway involves cleavage of the C–O bond leading to such species as alcohols or rearrangement to form acids. These can further undergo sequential de-oxygenation on metal sites to form alkanes. Other pathways leading to undesired products include dehydration of ethylene glycol to produce vinyl alcohol. Sequential hydrogenation of vinyl alcohol results in the

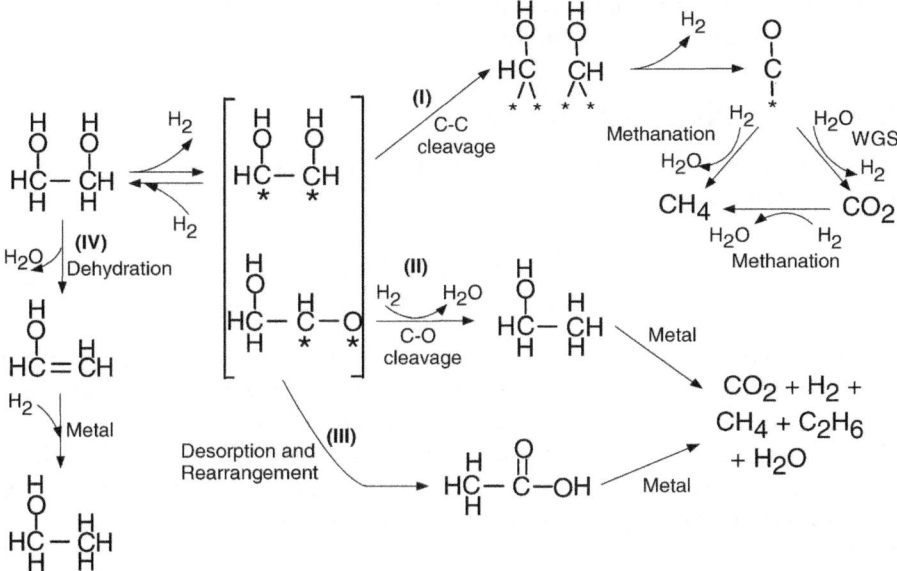

Figure 2.8   Reaction network of ethylene glycol conversion to H_2, CO_2 and alkanes (adapted from Davda *et al.*[4]).

formation of ethanol. Direct hydrogenation of CO_x can also lead to the formation of CH_4 or even higher alkanes through the Fischer–Tropsch sequence.

2.9   Definitions/Calculations for Laboratory Experiments

2.9.1   Carbon to Gas Conversion

Carbon to gas conversion can be determined directly from the GC and volumes of gases formed, however the many assumptions (*e.g.* ideal gas law) and parameters involved in this method compromise the accuracy of this direct approach. Therefore, an indirect approach could be used since it gives better accuracy. In this method, carbon to gas conversion is defined as the percentage of carbon in the feed that is transformed to carbon in gaseous products (CO_x, hydrocarbons) and is calculated according to:

$$X_{C \text{ to Gas}} = \frac{n_{\text{carbon in feed}} - n_{\text{carbon in reactor effluent}}}{n_{\text{carbon in feed}}} \times 100\% \qquad (2.8)$$

Carbon removed from the feed can either go to gas phase, liquid phase products or to coke. This method assumes that the amount of carbon transformed to coke is negligible compared to the amount of carbon in the gas phase. If not, it can be determined *via* thermo-gravimetry. Carbon contents in liquid streams can be determined with a TOC analyzer. Typical errors of $\sim$1% are applicable in TOC analysis, resulting in a 2% error (relatively) in conversion numbers.

2.9.2   Gas Phase Selectivities

Selectivity to carbon containing gas phase products $(i = CO_x, CH_4$ and $C_{2+})$ can be calculated according to:

$$S_i = \frac{n_{\text{Carbon in species } i}}{\sum n_{\text{Carbon in species } i}} \times 100\% \tag{2.9}$$

The selectivity to carbon containing gas phase products indicates the purity of the produced gas and is calculated based on the distribution of carbon in these molecules. Side-reactions leading to carbon containing molecules in the liquid phase are not taken into account for these selectivity calculations. Selectivity to hydrogen is defined as the percentage of the maximal theoretical amount of hydrogen that can formed based on the gasified carbon and is expressed as:

$$S_{H_2} = \frac{n_{H_2 \text{ produced}}}{n_{\text{Carbon in gas phase}}} \times \frac{1}{RR} \times 100\% \tag{2.10}$$

The reforming ratio (RR) is the maximum amount of H_2 (including water gas shift reaction) that can be produced per gasified carbon atom (end products being H_2 and CO_2). The RR for ethylene glycol and acetic acid reforming is 2.5 and 2.0, respectively, as shown below.For ethylene glycol RR is 5/2:

$$C_2H_6O_2 \rightarrow 3H_2 + 2CO \quad \text{(Optimal reforming)} \tag{2.11}$$

$$2CO + 2H_2O \rightarrow 2H_2O + 2CO_2 \quad \text{(WGS)} \tag{2.12}$$

$$\overline{C_2H_6O_2 + 2H_2O \rightarrow 5H_2 + 2CO_2 \quad \text{(Overall)}} \tag{2.13}$$

For acetic acid RR is 4/2:

$$CH_3COOH \rightarrow 2H_2 + 2CO \quad \text{(Optimal reforming)} \tag{2.14}$$

$$2CO + 2H_2O \rightarrow 2H_2 + 2CO_2 \quad \text{(WGS)} \tag{2.12}$$

$$\overline{CH_3COOH + 2H_2O \rightarrow 5H_2 + 2CO_2 \quad \text{(Overall)}} \tag{2.16}$$

2.10   Current Status and Challenges for APR

Many APR studies[1,49,51,78,79] were already undertaken by different research groups to study the reforming of model compounds in subcritical water conditions (175–265 °C and 32–56 bar). These APR conditions are ideal for fundamental reforming studies; however, reaction rates are relatively slow at these low temperatures and therefore studies are limited to feed solutions of low concentrations or long residence times. The ultimate goal should be to develop a commercially and technologically feasible (catalytic) process for the production of hydrogen by APR of industrial biomass derived aqueous streams that contain complex mixtures (15–25 wt%) of oxygenates. Therefore, much higher reaction rates are required compared to low temperature APR to achieve this goal. Achieving high H_2 yields by reforming of highly concentrated feeds is not only a matter of increasing the catalytic efficiency. Thermodynamics predict an increase in alkane formation for reforming reactions with higher feed concentrations.[51] Production of alkanes should be avoided as it competes with the hydrogen yields. Furthermore, low temperature APR is also reported to be subject to mass transfer limitations, which can severely hinder the catalytic reaction for high feed concentrations. As an example, it was shown by Shabaker and colleagues[2] that APR of ethylene glycol (225 °C and 29.3 bar) with a 3.4 wt% Pt/Al_2O_3 (63–125 µm particle size) catalyst was affected by intra-particle mass transfer limitations when an EG feed concentration of 10 wt% was used.

Catalytic supercritical water reforming is promising for achieving high reforming rates and overcoming mass transfer limitations. Supercritical water provides a medium with better heat transfer than is commonly in the gas phase and a higher diffusivity than in the liquid phase.[5,80] Guo *et al.*[5] published a nice overview of catalytic supercritical water reforming studies of biomass derived compounds. High reaction rates for the reforming of high concentrated oxygenated hydrocarbon streams were reported. However, stability and selectivity issues with the studied catalysts are serious drawbacks for industrial exploitation of this process.[5] Catalyst stability issues in hot compressed water are mainly related to sintering of the supported metal particles[5] or instability of conventional metal oxide catalyst supports (*e.g.* Al_2O_3, TiO_2 and ZrO_2).[81,82] Issues with metal oxide supports are already experienced at low temperature APR conditions.[69] Several authors have discussed the hydrothermal stability of catalysts in APR of model feedstocks.[83–86] Higher polyol feeds can enhance the stability of alumina against dehydration, however, under acid feeds the support suffers from severe deactivation. To summarise, the challenge to make sub- or supercritical water reforming of biomass derived waste streams commercially feasible involves the development of catalysts that (i) show high stability in hot compressed water and are stable against different type of impurities and poisons, such as tars, (ii) are able to convert high concentrated feed streams under industrial relevant residence times, and (iii) produce high H_2 yields. A fundamental understanding of the reforming pathways and deactivation mechanisms should help in the development of such catalysts. These issues are addressed later in terms of studies carried out generally by us on model oxygenates.

In general, catalysts for APR should be active for C–C bond cleavage in order to be able to decompose the organic molecules and they should further enhance the WGS reaction to maximize hydrogen yields. Methanation and Fischer–Tropsch are undesired side reactions, as they consume the desired product, H_2. In order to suppress these reactions, catalysts should not be active for C–O bond breaking.[76] Lower alkane selectivities are therefore expected for catalysts with low affinity towards C–O bond dissociation. In 2003, Davda *et al.*[76] reported the relative activities of different metals for C–C and C–O bond dissociation, and for the WGS reaction. They reported high C–C bond breaking activities for Ir, Ru and Ni. Therefore, these metals might be promising catalysts for breaking down oxygenates. Adsorbed oxygenate fractions on the catalyst surface resulting from C–C bond dissociation can undergo dehydrogenation, yielding H_2 and CO.[76] Follow-up reactions, such as methanation and WGS, strongly control the final hydrogen yield. Ir showed very low methanation activity but in contrast to Ru and Ni, almost no WGS activity was reported. Furthermore, Ru and Ni have been reported to have high activities for methane formation.[76]

Catalytic reforming of ethylene glycol (5 and 15 wt%) in supercritical water (450 °C and 250 bar) in the presence of alumina supported mono- and bi-metallic catalysts based on Ir, Pt and Ni has been reported by us earlier.[53] Pt catalyst showed the highest hydrogen yields compared to Ir and Ni. Varying the Pt loading (0.3–1.5 wt%) showed that the intrinsic reforming activities improved with decreasing Pt loadings as shown in Table 2.2. However, a lower Pt loading had a large negative effect on the H_2 selectivity and catalyst stability. It was found that the presence of Ni in a Pt–Ni bimetallic catalyst improved hydrogen yields by suppressing methane formation. Moreover, the presence of Ni also enhanced catalyst stability. Results reported here were obtained at WHSV of 18 h^{-1}. Pt–Ni/Al_2O_3 having a total metal loading of 1.5 wt% (molar ratio Pt–Ni = 1), is identified as a promising catalyst for the reforming of ethylene glycol in supercritical water. The bi-functional reforming mechanism involved with APR over alumina supported Pt catalysts requires water activation. Water activation happens on the alumina support through the formation of hydroxyl groups. Smaller Pt particles were found to give higher intrinsic reforming activities due to a larger interfacial (metal/

Table 2.2 Experimental data for 15 wt% ethylene glycol reforming (WHSV 17.8 h^{-1}) at 450 °C and 250 bar using 0.3, 0.6 and 1.5 wt% Pt supported on γ-alumina (adapted from de Vlieger *et al.*[53]).

	0.3 wt% Pt	0.6 wt% Pt	1.5 wt% Pt
Carbon to gas (%)	48	48	42
Carbon in liquid (%)	58	59	61
Selectivity (%)			
H_2	42	51	80
CO_2	50	60	79
CO	20	14	14
Alkanes	30	26	7

support) area, which enabled better interaction between the reforming fragments on the Pt and the hydroxyl groups on the alumina support. A positive effect of smaller Pt particle size on dehydrogenation and decarboxylation activity has also been reported by the group of Lercher.[87] On the other hand, in our study, smaller Pt particles were found to be responsible for lower hydrogen selectivities due to increased methane formation. Methane can be formed by either dehydration of EG and/or by the hydrogenation of CO_x. The Pt–Ni bimetallic catalyst shows enhanced hydrogen yields by suppressing methane formation. The new Pt–Ni catalyst also shows two times more activity compared to a monometallic Pt catalyst due to supposed lower CO coverage caused by the presence of Ni and an enhanced accessibility to Pt sites for reforming. Moreover, the addition of Ni also enhanced the catalyst lifetime and stable catalytic performances were observed for the reforming of ethylene glycol under supercritical water conditions.

For the reforming of methanol, a selectivity of 8% towards CH_4 was observed for the Pt catalyst. Formation of alkanes during the catalytic reforming of 1 wt% acetic acid, methanol and ethanol along with conversions to the gas phase at 275 °C and 200 bar are shown in Figure 2.9.[86] The selectivity towards alkanes (mainly methane) was found to be the highest ($\pm 47\%$) during acetic acid reforming. In the case of ethanol reforming, the Pt catalyst showed CH_4 selectivity of $\pm 20\%$. The higher amount of methane observed during ethanol reforming is attributed probably to the CH_x fragment formed by C–C bond breaking of ethanol[88] on the Pt surface. CH_x fragments can recombine with adsorbed H atoms on the Pt surface to form methane. A similar sequence is also suggested for acetic acid on Pt based

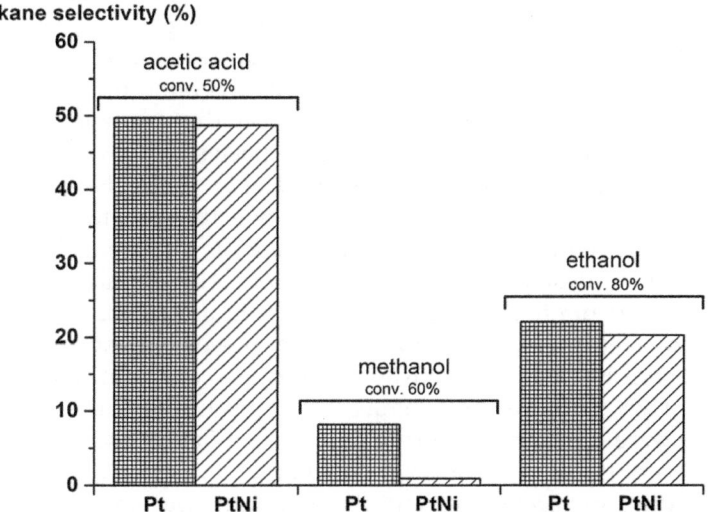

Figure 2.9   Alkane selectivity for catalytic reforming of acetic acid, methanol and ethanol at 275 °C per 200 bar. Conversion levels are shown above the selectivity values (adapted from de Vlieger *et al.*[86]).

catalysts.[89,90] In the case of methanol, C–O bond breaking is necessary to generate these adsorbed CH_x species that lead to the production of methane. High C–C bond breaking activities for Pt during reforming of ethanol and low activity for C–O bond breaking[76] during reforming of methanol can explain the observed differences for methane formation. For methanol reforming with the Pt–Ni catalyst, almost no methane was formed. Methanol was found to be one of the major products during the reforming of EG, which explains why the Pt–Ni/Al_2O_3 catalyst shows almost no methane formation in agreement with earlier studies[3] and our own results.[53] The lower methane production observed during methanol reforming in the presence of Pt–Ni can be explained by the enhanced dehydrogenation activity of the Pt–Ni catalyst,[91] favouring C–H cleavage instead of C–O cleavage in methanol. Promoting the Pt catalyst with Ni did not affect the alkane (methane) selectivity for acetic acid or ethanol reforming as can be expected from the reforming routes. Reforming of these compounds inherently led to CH_3 species on the catalyst surface, which is a precursor for methane formation. In case of methanol reforming, the intermediate CH_3O must undergo C–O cleavage to form CH_3 (and hence methane). Methane formation during methanol reforming can be prevented by avoiding C–O cleavage.

Reforming experiments with 1 wt% methanol (Figure 2.10A) and ethanol (Figure 2.10B) solutions showed stable activity levels for both Pt and Pt–Ni catalysts. Catalysts used for acetic acid reforming deactivated during the reaction (Figure 2.10C). Both catalysts showed similar deactivation rates. The catalysts lost all activity after 3 h on stream with final conversion levels ($\sim 5\%$) being similar to the reforming experiment without a catalyst. Deactivation of the catalyst with acetic acid is commonly observed in reforming reactions.[89] The formation of methane resulting from acetic acid decomposition is reported in the literature to be a possible cause for deactivation of catalysts.[92] It is suggested that C–H bond breaking in methane further results in CH_x species ($1 \leq x \leq 2$) on the catalytic surface. These fragments can further oligomerize to coke species.[89] However, catalysts used for ethanol reforming were found to be stable while also producing high amounts of methane (20%). It is also known that Pt and Ni catalysts are stable for the reforming of methane. Therefore, it is concluded that methane-induced deactivation is unlikely and the dominant deactivation pathway is due to other causes.[53,86] Further, liquid products formed during the APR of acetic acid, methanol and ethanol were investigated to find any role for them in catalyst deactivation. During acetic acid reforming, a conversion of 7% to liquid by-products was observed for the Pt catalyst. By products were identified as formaldehyde and iso-propanol. For ethanol, a conversion of $\pm 6\%$ to liquid by-products was found for the Pt catalyst. The main product was acetic acid, and the remaining was acetaldehyde (<0.5%). The Cannizzaro reaction (disproportionation of an aldehyde to alcohol and carboxylic acid) could be responsible for some of the acetic acid formed

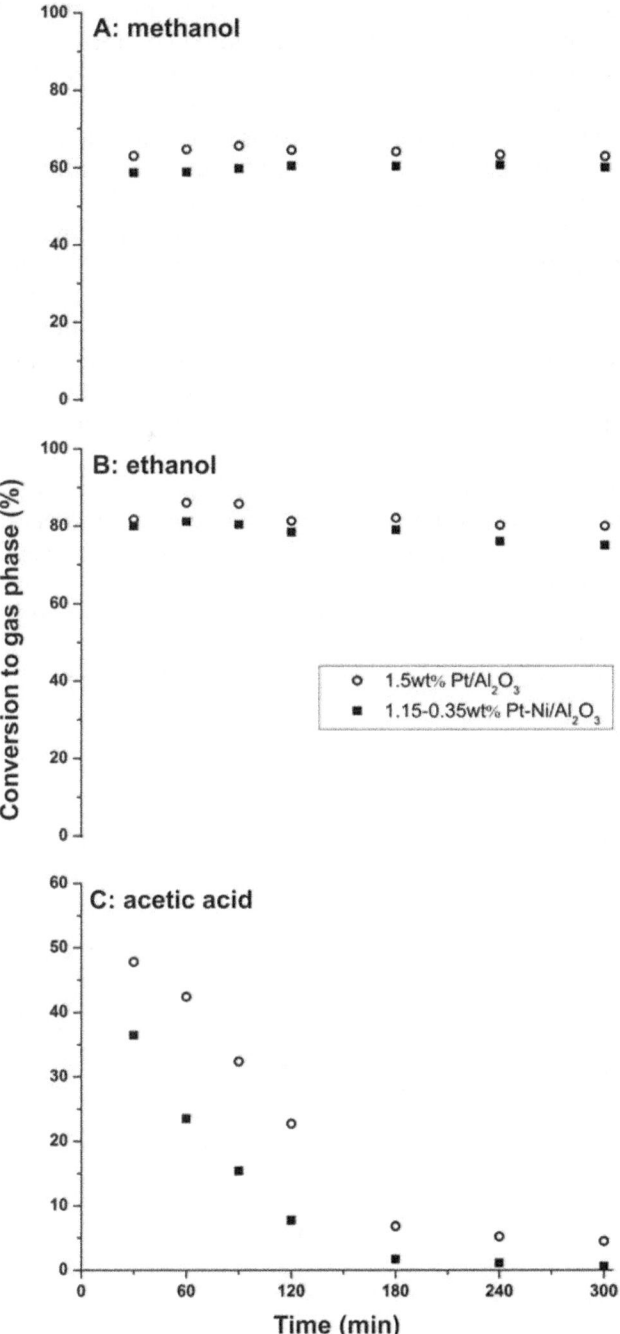

Figure 2.10 Carbon to gas conversion of (A) methanol, (B) ethanol and (C) acetic acid over 1.5 wt% Pt/Al$_2$O$_3$ and 1.15 wt% Pt–Ni/Al$_2$O$_3$.[86]

during ethanol reforming. The Pt and Pt–Ni catalysts studied did not influence the amount of acetic acid and acetaldehyde formed during ethanol
reforming, while during EG reforming, the amount of acetic acid was
strongly dependent on the type of catalyst. This shows that the Cannizzaro-
derived reaction is not a dominant contributor to acetic acid formation
during EG reforming. Aqueous phase reforming experiments with liquid
products formed during the APR of EG were carried out using individual
components (Figure 2.10). Reforming experiments (with acetic acid,
methanol, or ethanol) were conducted at 275 °C and 200 bar in the presence of Pt/Al$_2$O$_3$ or Pt–Ni/Al$_2$O$_3$. Feed concentrations of 1 wt% were used so
as to work in the same concentration range as the liquid by-products
formed during EG reforming experiments. For example, a total conversion
of 15% to methanol, ethanol and acetic acid was observed during the
reforming of 20 wt% EG, resulting in a total concentration of 3 wt% liquid
by-products.

FT-IR spectra of the used catalysts were subtracted from each other as
shown in Figure 2.11 to study differences in chemical groups on the surfaces
of the used Pt catalysts. Subtracting the spectrum of the catalyst that was
exposed to only water from that of aqueous methanol solution resulted in a
more or less flat line, indicating that the surface composition of the active
catalysts (used for methanol and water only) is similar. Subtracting the
spectrum of the catalyst subjected to only water from the spent deactivated

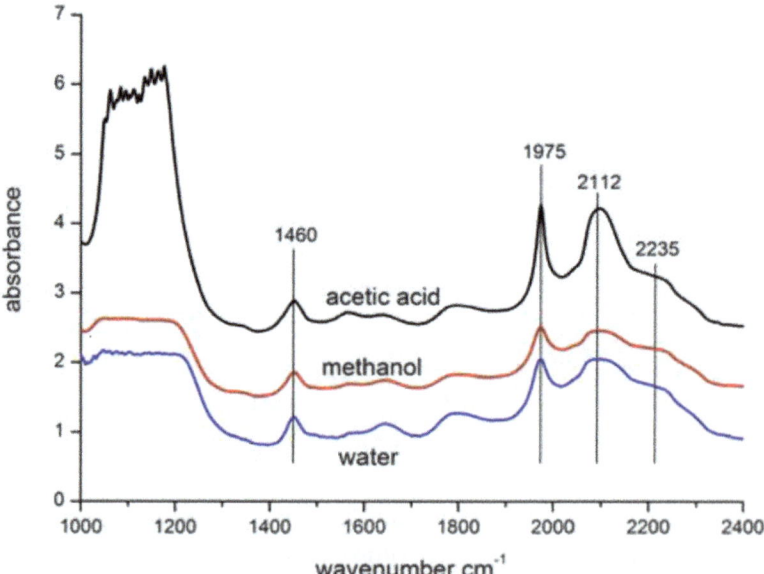

Figure 2.11   FT-IR spectra of 1.5 wt% Pt/Al$_2$O$_3$ catalyst used for APR of acetic acid,
methanol and only water (275 °C and 200 bar).[86]

acetic acid catalyst resulted in three strong bands; one lattice vibration in the region 1000–1200 cm^{-1} and two sharp OH related bands located at 1975 and 2112 cm^{-1}. Also in the FT-IR spectra, the acetic acid treated sample showed the highest O–H intensity (Figure 2.11). These results indicate the presence of a highly hydroxylated type of boehmite in the acetic acid deactivated catalyst compared to the active catalyst used in methanol APR.

Acetic acid was shown to be responsible for the deactivation of Pt and Pt–Ni catalysts by hydroxylation of the Al$_2$O$_3$ surface. Re-deposition of the dissolved alumina on the catalyst leads to the blocking of catalytic Pt sites (see Figure 2.12) and hence deactivation of the catalyst.[86] The increased dehydrogenation activity of the Pt–Ni catalyst was found to suppress the formation of acetic acid during ethylene glycol reforming and thereby increasing the H$_2$. Stability issues of catalyst support materials in supercritical water (SCW) are a major setback for these reactions and stall the further development and industrial exploitation of this technique. The development of stable catalytic support materials for reactions in SCW is therefore of much importance. Carbon nanotubes (CNT) are widely recognized for their significant physical and chemical stability, high heat conductivity and open structure. These properties are already explored for different applications. We have shown that CNT to be a promising stable catalyst support material for reactions in SCW.[93] The efficiency of Pt/CNT as a catalyst for the production of hydrogen by re-forming of ethylene glycol and acetic acid in SCW was studied and illustrated the applicability of CNT as a catalyst support in SCW (Figure 2.13).

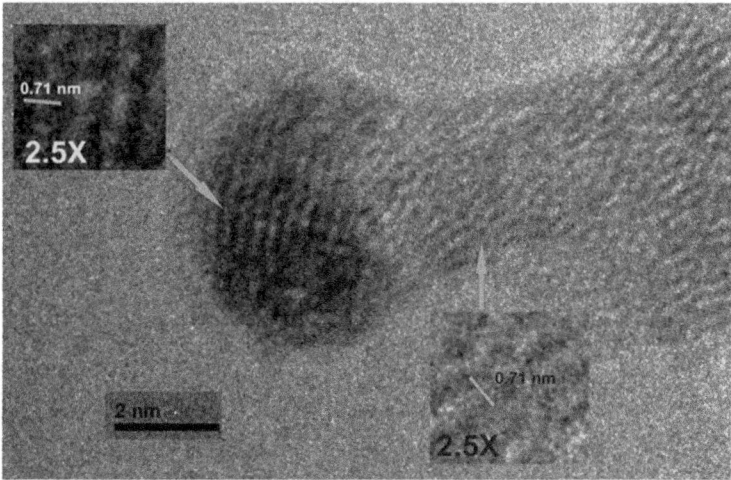

Figure 2.12   TEM image of Pt/Al$_2$O$_3$ catalyst spent for APR of 1 wt% acetic acid at 275 °C per 200 bar. Figure shows coverage of Pt particle with Al$_2$O$_3$ (adapted from de Vlieger *et al.*[86]).

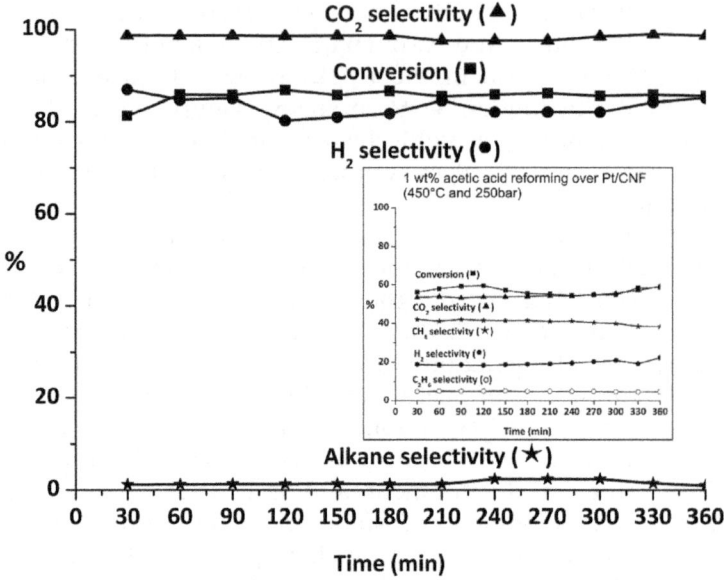

Figure 2.13 Conversion and gas phase selectivites of 5 wt% ethylene glycol reforming in super-critical water (450 °C and 250 bar) using Pt-CNT
catalyst. In the inset conversion and gas phase selectivity of 1 wt%
acetic acid reforming in super-critical water (450 °C and 250 bar) using
Pt-CNT catalyst (adapted from de Vlieger *et al.*[93]).

2.11 APR for Green Chemicals

Lignin comprises 10–40 wt% of lignocellulose.[65,94] Currently, lignin is produced in the paper industry as the by-product of the pulping process where it
is mainly used for combustion to generate heat, steam and electricity. As it is
constructed from methoxylated phenyl propane units, lignin has been
addressed as a potential resource for bio-derived aromatics.[95–97] The solu-
bilisation and conversion of various types of lignin (*e.g.*, kraft lignin, Alcell
organosolv lignin, soda lignin) under mild liquid phase reforming (LPR)
conditions (*i.e.*, $T \sim 225$ °C, $P \sim 25$–58 bar) were explored by Weckhuysen
et al.[98] Among these lignin feedstocks investigated in their study, the solubility followed the order Alcell organosolv > soda > kraft. About 93 wt% of
lignin could be dissolved in compressed water at 225 °C, while only 38 wt%
of kraft lignin dissolved under the same conditions. However, the average
molecular mass (M_w) of the depolymerised lignins was still above 2000 Da
and the decrease of M_w varied from 22–57%.[96,98] Co-solvents (*e.g.*, ethanol,
formic acid and phenol *etc.*) and co-catalysts such as NaOH or H_2SO_4 were
reported to help improve the monomer yields due to enhancement of lignin
solubility and prevent re-condensation of lignin fragments.[96,98–101] In addition, the co-solvents are also partially reformed under the liquid phased
reforming conditions, producing H_2 which in turn contributes to the
hydrogenolysis and hydro-de-oxygenation of lignin. The conceptual process
for lignin valorization *via* LPR was later proposed and is illustrated in
Figure 2.14.[99] The products (Figure 2.15) include oil type soluble aromatics,

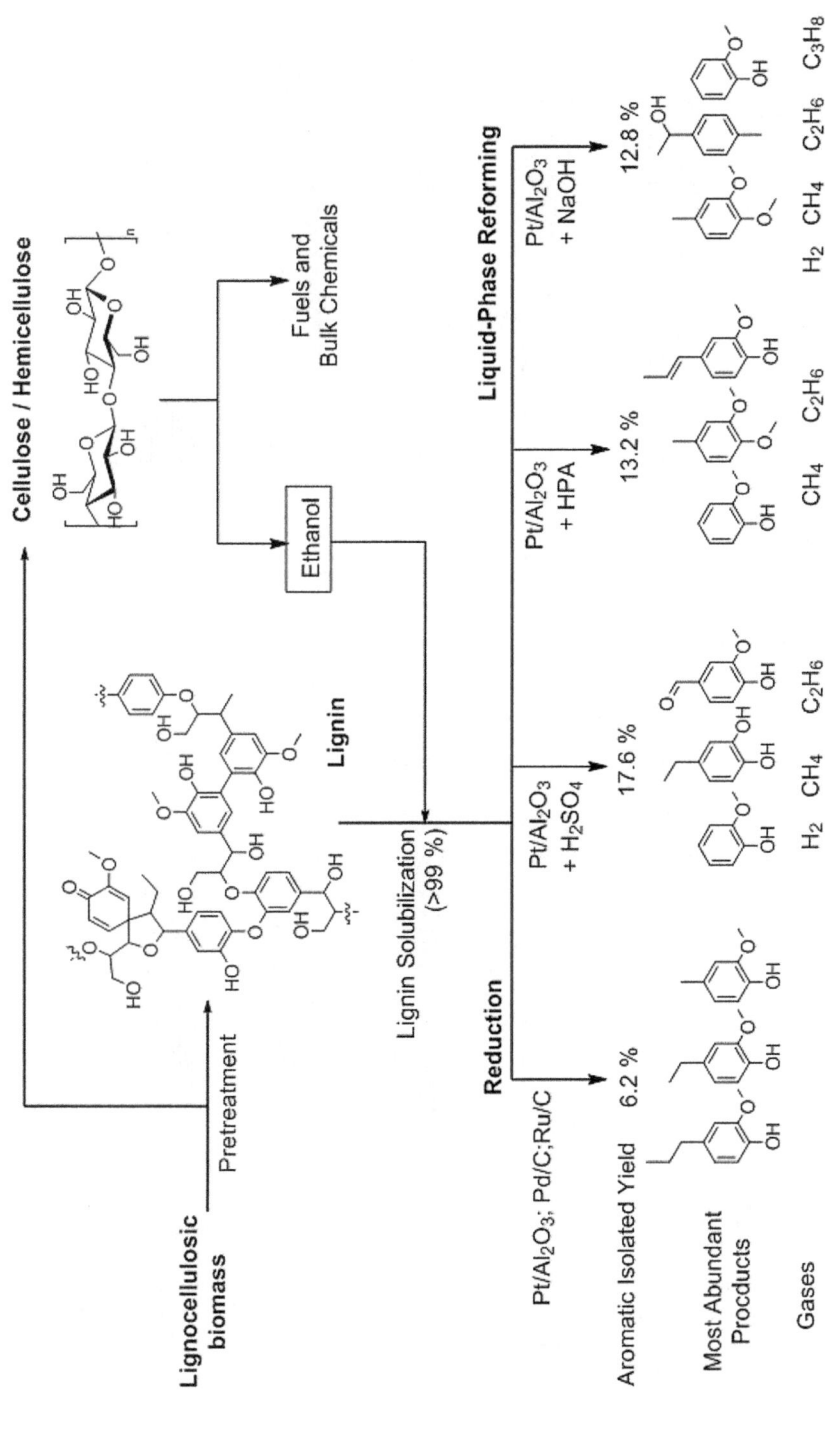

Figure 2.14   Valorisation of lignin *via* liquid phase reforming produces aromatic chemicals, gas and valuable solvent (adapted from Zakzeski *et al.*[99]).

(a) 623 K

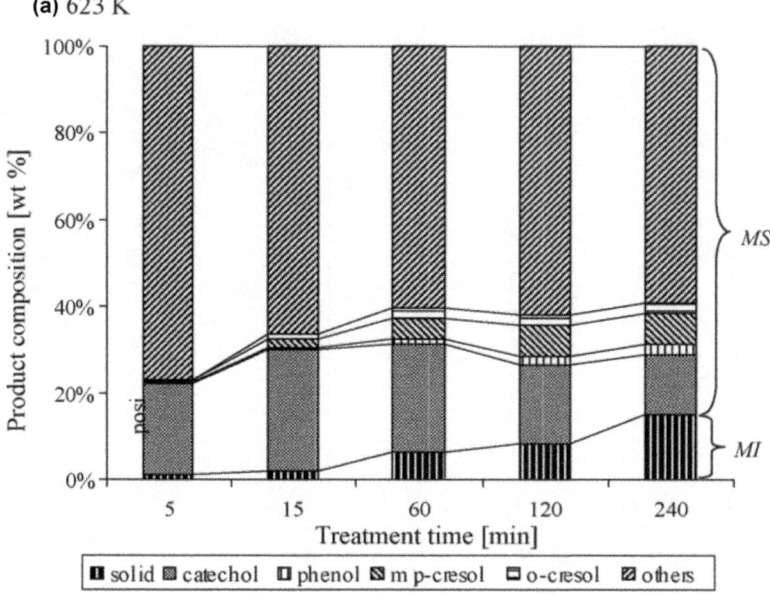

(b) 673 K

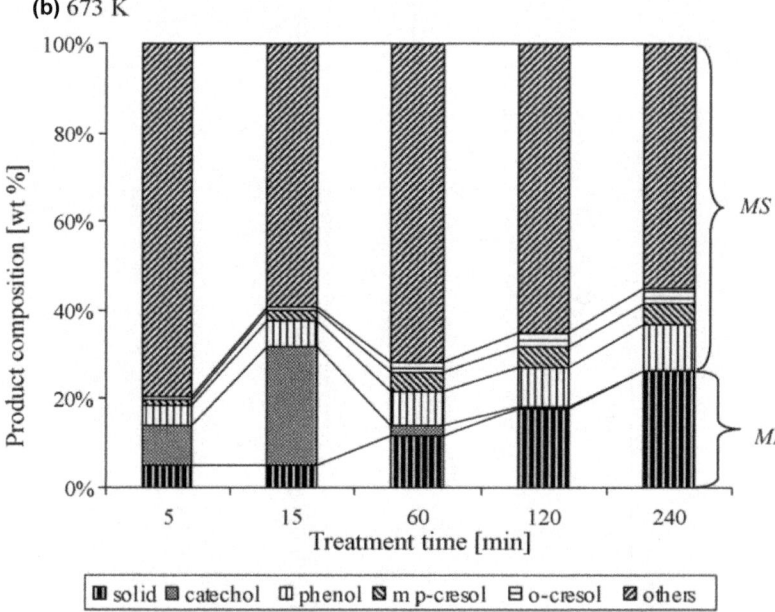

Figure 2.15   Product distribution after treatment of lignin in near and supercritical water at 30 MPa and 350 °C (a) and 400 °C (b), respectively. MS – methanol soluble products, MI – methanol insoluble products (adapted from Wahyudiono *et al.*[97]).

insoluble char and gases (*e.g.*, CO_2, alkane, H_2 *etc.*).[97] At subcritical and supercritical water conditions, the catechol yield peaked at short retention times then it gradually decreased while the yields of phenol and cresols increased along time on stream.[97,102] Condensation or polymerisation reactions were also observed, resulting in higher mass molecules and further insoluble products.[96,97,101] At subcritical conditions, the yield of solid products varied from ~10 to ~60 wt% depending on the temperatures, catalyst used and retention time.[97,101–103] The solid char can also be used as a feedstock for steam reforming for hydrogen. Various noble or transition metal supported catalysts have been tested for the reaction (*e.g.*, Pt/Al_2O_3, Ru/C, Pd/C). Although the Al_2O_3 support is notorious for phase transition to boehmite under APR conditions, its stability was enhanced with the presence of lignin.[85] The authors attributed this stability improvement to the interactions of oxygen functionalities of lignin with the support.

2.12   Challenges for the Future

As outlined in this chapter, the feasibility of APR has been shown to be possible with model molecules, for example sorbitol.[84,104] For instance, Virent has patented a process to produce liquid fuels from sugar derivatives by incorporating *in situ* hydrogen production by APR.[105] Sorbitol transformations to fuels with industrial respects is discussed thoroughly in a recent review.[106] The conceptual process is still at the research stage in laboratories. This is because studies have been carried out with simplistic molecules such as methanol, ethanol, polyols (glycols, glycerol), acetic acid *etc.*; in reality, aqueous waste streams contain complex mixtures of organic molecules. However, studies on model systems have provided a good understanding of requirements for efficient catalyst design. Catalysts are mandatory for the development of an efficient APR process. Research work carried out in various groups has provided this knowledge and this is a positive and good start.

However, a variety of scientific and technological questions have yet to be answered. In the case of catalysts, achieving long-term stability is of paramount importance. The severity of conditions used in APR is problematic due to sintering and loss of the active surface area of supports and metal phases, as well as leaching. The other major problem to be solved for APR is the tendency of the organic molecules to be very reactive and undergo a variety of reactions. Some of these also result in secondary oligomerization/condensation reactions, which lead to severe coking and subsequent blockage of catalytic sites and catalyst deactivation. Bio-oxygenates also have a severe tendency to form homogeneous char in the reaction medium and heterogeneous char on the catalyst. Efficient catalysts should minimise heterogeneous char as well as have the ability to gasify homogeneous char. However, the problem of homogeneous char, in our opinion, cannot be solved by catalyst development alone. A smart reactor and feed injection nozzle into the reactor system are critical. Developments in feed systems for turbines have faced similar issues and such knowledge can be applied. In conclusion, aqueous phase reforming is a promising concept with potential for commercial development.

Acknowledgements

The authors thank the funding ACTS of the Dutch Academy of sciences, European Union's Seventh Framework Programme for research, technological development and demonstration within the project under grant agreement No 310490 (www.susfuelcat.eu). We also acknowledge the CatchBio program, support from the Smart Mix programme of the Netherlands Ministry of Economic Affairs, Agriculture and Innovation and the Netherlands Ministry of Education, Culture and Science, and University of Twente, The Netherlands. We are grateful for discussions with various colleagues working in the area at different times; they have been reported extensively in the reference section.

References

1. R. D. Cortright, R. R. Davda and J. A. Dumesic, *Nature*, 2002, **418**, 964–967.
2. J. W. Shabaker, R. R. Davda, G. W. Huber, R. D. Cortright and J. A. Dumesic, *J. Catal.*, 2003, **215**, 344–352.
3. G. W. Huber, J. W. Shabaker, S. T. Evans and J. A. Dumesic, *Appl. Catal., B*, 2006, **62**, 226–235.
4. R. R. Davda, J. W. Shabaker, G. W. Huber, R. D. Cortright and J. A. Dumesic, *Appl. Catal., B*, 2005, **56**, 171–186.
5. Y. Guo, S. Z. Wang, D. H. Xu, Y. M. Gong, H. H. Ma and X. Y. Tang, *Renewable Sustainable Energy Rev.*, 2010, **14**, 334–343.
6. Y. Matsumura and T. Nakamori, *Appl. Catal., A*, 2004, **258**, 107–114.
7. A. Yamaguchi, N. Hiyoshi, O. Sato, K. K. Bando, M. Osada and M. Shirai, *Catal. Today*, 2009, **146**, 192–195.
8. M. H. Waldner, F. Krumeich and F. Vogel, *J. Supercrit. Fluids*, 2007, **43**, 91–105.
9. F. Joseck, Current U.S. Hydrogen Production, in: D.o. Energy (Ed.), Hydrogen and Fuel Cells Program Record, number 12014 ed., US Departement of Energy 2012.
10. R. M. Navarro, M. A. Pena and J. L. Fierro, *Chem. Rev.*, 2007, **107**, 3952–3991.
11. J. R. Rostrup-Nielsen, *Catal.: Sci. Technol.*, 1984, **5**, 1–117.
12. F. Joensen and J. R. Rostrup-Nielsen, *J. Power Sources*, 2002, **105**, 195–201.
13. D. J. Moon, *Catal. Surv. Asia*, 2008, **12**, 188–202.
14. K. Urasaki, Y. Sekine, S. Kawabe, E. Kikuchi and M. Matsukata, *Appl. Catal. A*, 2005, **286**, 23–29.

15. N. Martín, M. Viniegra, R. Zarate, G. Espinosa and N. Batina, *Catal. Today*, 2005, **107-108**, 719–725.
16. D. L. Trimm, *Catal. Today*, 1999, **49**, 3–10.
17. F. Arena, F. Frusteri and A. Parmaliana, *Appl. Catal., A*, 1999, **176**, 189–199.
18. R. Martínez, E. Romero, C. Guimon and R. Bilbao, *Appl. Catal., A*, 2004, **274**, 139–149.
19. U. Olsbye, O. Moen, A. Slagtern and I. M. Dahl, *Appl. Catal., A*, 2002, **228**, 289–303.
20. J. W. Snoeck, G. F. Froment and M. Fowles, *Ind. Eng. Chem. Res.*, 2002, **41**, 3548–3556.
21. M. A. Henderson, *Surf. Sci. Rep.*, 2002, **46**, 1–308.
22. Z. Zhang and X. E. Verykios, *Appl. Catal., A*, 1996, **138**, 109–133.
23. K. Takanabe, K. I. Aika, K. Seshan and L. Lefferts, *J. Catal.*, 2004, **227**, 101–108.
24. A. Gołebiowski, K. Stołecki, U. Prokop, A. Kuśmierowska, T. Borowiecki, A. Denis and C. Sikorska, *React. Kinet. Catal. Lett.*, 2004, **82**, 179–189.
25. X. E. Verykios, *Int. J. Hydrogen Energy*, 2003, **28**, 1045–1063.
26. N. S. Figoli, P. C. Largentiere, A. Arcoya and X. L. Seoane, *J. Catal.*, 1995, **155**, 95–105.
27. H. Schaper, E. B. M. Doesburg and L. L. Van Reijen, *Appl. Catal.*, 1983, **7**, 211–220.
28. E. Nikolla, J. W. Schwank and S. Linic, *Catal. Today*, 2008, **136**, 243–248.
29. E. Nikolla, A. Holewinski, J. Schwank and S. Linic, *J. Am. Chem. Soc.*, 2006, **128**, 11354–11355.
30. H. Shu-Ren, *Int. J. Hydrogen Energy*, 1998, **23**, 315–319.
31. J. R. H. Ross, M. C. F. Steel and A. Zeini-Isfahani, *J. Catal.*, 1978, **52**, 280–290.
32. A. A. Praharso, D. L. Adesina, Trimm and N. W. Cant, *Chem. Eng. J.*, 2004, **99**, 131–136.
33. K. G. Azzam, I. V. Babich, K. Seshan and L. Lefferts, *J. Catal.*, 2007, **251**, 163–171.
34. S. R. A. Kersten, W. P. M. Van Swaaij, L. Lefferts, K. Seshan, in *Catalysis for Renewables*, ed. G. Centi and R. A. Van Santen, Wiley-VCH Verlag GmbH & Co. KGaA, Weinheim, 2007, pp. 119–146.
35. B. Matas Güell, I. V. Babich, L. Lefferts and K. Seshan, *Appl. Catal., B*, 2011, **106**, 280–286.
36. C. Higman and M. v. d. Burgt, *Gasification*, Gulf Professional Pub./Elsevier Science, Amsterdam, Boston, 2nd edn, 2008.
37. K. Maniatis, in *Progress in Thermochemical Biomass Conversion*, ed. A. V. Bridgwater, Blackwell Science, Oxford, Malden, MA, 2001.
38. J. A. Medrano, M. Oliva, J. Ruiz, L. García and J. Arauzo, *J. Anal. Appl. Pyrolysis*, 2009, **85**, 214–225.
39. D. Sutton, B. Kelleher and J. R. H. Ross, *Fuel Process. Technol.*, 2001, **73**, 155–173.

40. J. Corella, A. Orío and P. Aznar, *Ind. Eng. Chem. Res.*, 1998, **37**, 4617–4624.

41. A. Olivares, M. P. Aznar, M. A. Caballero, J. Gil, E. Francés and J. Corella, *Ind. Eng. Chem. Res.*, 1997, **36**, 5220–5226.

42. J. Corella, M.-P. Aznar, J. Gil and M. A. Caballero, *Energy Fuels*, 1999, **13**, 1122–1127.

43. M. A. Caballero, J. Corella, M. P. Aznar and J. Gil, *Ind. Eng. Chem. Res.*, 2000, **39**, 1143–1154.

44. S. Rapagnà, N. Jand, A. Kiennemann and P. U. Foscolo, *Biomass Bioenergy*, 2000, **19**, 187–197.

45. J. Corella, J. M. Toledo and R. Padilla, *Energy Fuels*, 2004, **18**, 713–720.

46. C. Courson, L. Udron, D. Świerczyński, C. Petit and A. Kiennemann, *Catal. Today*, 2002, **76**, 75–86.

47. P. Simell, E. Kurkela, P. Ståhlberg and J. Hepola, *Catal. Today*, 1996, **27**, 55–62.

48. D. J. M. d. Vlieger, Design of efficient catalysts for gasification of biomass-derived waste streams in hot compressed water, Towards industrial applicability, PhD Thesis, Universiteit Twente, 2013.

49. R. L. Manfro, A. F. Da Costa, N. F. P. Ribeiro and M. M. V. M. Souza, *Fuel Process. Technol.*, 2011, **92**, 330–335.

50. A. J. Byrd, K. K. Pant and R. B. Gupta, *Fuel*, 2008, **87**, 2956–2960.

51. A. V. Tokarev, A. V. Kirilin, E. V. Murzina, K. Eränen, L. M. Kustov, D. Y. Murzin and J. P. Mikkola, *Int. J. Hydrogen Energy*, 2010, **35**, 12642–12649.

52. A. Ciftci, B. Peng, A. Jentys, J. A. Lercher and E. J. M. Hensen, *Appl. Catal., A*, 2012, **431–432**, 113–119.

53. D. J. M. de Vlieger, A. G. Chakinala, L. Lefferts, S. R. A. Kersten, K. Seshan and D. W. F. Brilman, *Appl. Catal., B*, 2012, **111–112**, 536–544.

54. A. Tanksale, J. N. Beltramini and G. M. Lu, *Renewable Sustainable Energ Rev.*, 2010, **14**, 166–182.

55. M. Rose and R. Palkovits, *ChemSusChem*, 2012, **5**, 167–176.

56. R. M. d. Almeida, C. R. K. Rabello, Diesel cycle fuel compositions containing dianhydrohexitols and related products, PETROLEO BRASILEIRO S.A.-PETROBRAS, Rio de Janeiro, BR, US, 2010.

57. K. Van Gorp, E. Boerman, C. V. Cavenaghi and P. H. Berben, *Catal. Today*, 1999, **52**, 349–361.

58. A. Aho, S. Roggan, O. A. Simakova, T. Salmi and D. Y. Murzin, *Catal. Today*, 2015, **241**, 195–199.

59. J. E. Holladay, J. Hu, Y. Wang and T. A. Werpy, Two-stage dehydration of sugars 2007.

60. J. Li, A. Spina, J. A. Moulijn and M. Makkee, *Catal. Sci. Technol.*, 2013, **3**, 1540–1546.

61. J. U. Oltmanns, S. Palkovits and R. Palkovits, *Appl. Catal., A*, 2013, **456**, 168–173.

62. T. Werpy, G. Petersen, Top Value Added Chemicals from Biomass: Volume 1—Results of Screening for Potential Candidates from Sugars and Synthesis Gas, National Renewable Energy Laboratory, 2004.

63. J. J. Bozell and G. R. Petersen, *Green Chem.*, 2010, **12**, 539.
64. W. R. H. Wright and R. Palkovits, *ChemSusChem*, 2012, **5**, 1657–1667.
65. Y. Sun and J. Cheng, *Bioresour. Technol.*, 2002, **83**, 1–11.
66. G. W. Huber and J. A. Dumesic, *Catal. Today*, 2006, **111**, 119–132.
67. S. M. Sen, D. M. Alonso, S. G. Wettstein, E. I. Gurbuz, C. A. Henao, J. A. Dumesic and C. T. Maravelias, *Energy Environ. Sci.*, 2012, **5**, 9690–9697.
68. R. Weingarten, W. C. Conner and G. W. Huber, *Energy Environ. Sci.*, 2012, **5**, 7559–7574.
69. S. W. Fitzpatrick, *U. S. Pat.* US5608105, 1997.
70. D. M. Alonso, S. G. Wettstein, J. Q. Bond, T. W. Root and J. A. Dumesic, *ChemSusChem*, 2011, **4**, 1078–1081.
71. B. Girisuta, L. P. B. M. Janssen and H. J. Heeres, *Green Chem.*, 2006, **8**, 701.
72. I. van Zandvoort, Y. Wang, C. B. Rasrendra, E. R. van Eck, P. C. Bruijnincx, H. J. Heeres and B. M. Weckhuysen, *ChemSusChem*, 2013, **6**, 1745–1758.
73. T. M. C. Hoang, L. Lefferts and K. Seshan, *ChemSusChem*, 2013, **6**, 1651–1658.
74. T. M. C. Hoang, E. R. H. van Eck, W. P. Bula, J. G. E. Gardeniers, L. Lefferts and K. Seshan, *Green Chem.*, 2015, **17**, 959–972.
75. A. V. Bridgwater, *Catalysis Today*, 1996, **29**, 285–295.
76. R. R. Davda, J. W. Shabaker, G. W. Huber, R. D. Cortright and J. A. Dumesic, *Appl. Catal., B*, 2003, **43**, 13–26.
77. R. R. Davda and J. A. Dumesic, *Angew. Chem.,- Int. Ed.*, 2003, **42**, 4068–4071.
78. F. Z. Xie, X. W. Chu, H. R. Hu, M. H. Qiao, S. R. Yan, Y. L. Zhu, H. Y. He, K. N. Fan, H. X. Li, B. N. Zong and X. X. Zhang, *J. Catal.*, 2006, **241**, 211–220.
79. A. O. Menezes, M. T. Rodrigues, A. Zimmaro, L. E. P. Borges and M. A. Fraga, *Renewable Energy*, 2011, **36**, 595–599.
80. P. E. Savage, *Catal. Today*, 2000, **62**, 167–173.
81. A. J. Byrd and R. B. Gupta, *Appl. Catal., A*, 2010, **381**, 177–182.
82. P. Azadi and R. Farnood, *Int. J. Hydrogen Energy*, 2011, **36**, 9529–9541.
83. R. M. Ravenelle, J. R. Copeland, A. H. Van Pelt, J. C. Crittenden and C. Sievers, *Top. Catal.*, 2012, **55**, 162–174.
84. A. V. Kirilin, A. V. Tokarev, L. M. Kustov, T. Salmi, J. P. Mikkola and D. Y. Murzin, *Appl. Catal., A*, 2012, **435–436**, 172–180.
85. A. L. Jongerius, J. R. Copeland, G. S. Foo, J. P. Hofmann, P. C. A. Bruijnincx, C. Sievers and B. M. Weckhuysen, *ACS Catal.*, 2013, **3**, 464–473.
86. D. J. M. de Vlieger, B. L. Mojet, L. Lefferts and K. Seshan, *J. Catal.*, 2012, **292**, 239–245.
87. A. Wawrzetz, B. Peng, A. Hrabar, A. Jentys, A. A. Lemonidou and J. A. Lercher, *J. Catal.*, 2010, **269**, 411–420.

88. A. Erdohelyi, J. Raskó, T. Kecskés, M. Tóth, M. Dömök and K. Baán, *Catal. Today*, 2006, **116**, 367–376.

89. K. Takanabe, K. Aika, K. Seshan and L. Lefferts, *Chem. Eng. J.*, 2006, **120**, 133–137.

90. B. Matas Güell, I. Babich, K. Seshan and L. Lefferts, *J. Catal.*, 2008, **257**, 229–231.

91. O. Skoplyak, M. A. Barteau and J. G. Chen, *Surf. Sci.*, 2008, **602**, 3578–3587.

92. M. M. V. M. Souza and M. Schmal, *Appl. Catal., A*, 2005, **281**, 19–24.

93. D. J. M. de Vlieger, D. B. Thakur, L. Lefferts and K. Seshan, *ChemCatChem*, 2012, **4**, 2068–2074.

94. N. Mosier, C. Wyman, B. Dale, R. Elander, Y. Y. Lee, M. Holtzapple and M. Ladisch, *Bioresour. Technol.*, 2005, **96**, 673–686.

95. J. Zakzeski, P. C. A. Bruijnincx, A. L. Jongerius and B. M. Weckhuysen, *Chem. Rev.*, 2010, **110**, 552–3599.

96. A. L. Jongerius, P. C. A. Bruijnincx and B. M. Weckhuysen, *Green Chem.*, 2013, **15**, 3049–3056.

97. Wahyudiono, M. Sasaki and M. Goto, *Chemical Engineering and Processing: Process Intensification*, 2008, **47**, 1609–1619.

98. J. Zakzeski and B. M. Weckhuysen, *ChemSusChem*, 2011, **4**, 369–378.

99. J. Zakzeski, A. L. Jongerius, P. C. A. Bruijnincx and B. M. Weckhuysen, *ChemSusChem*, 2012, **5**, 1602–1609.

100. M. Saisu, T. Sato, M. Watanabe, T. Adschiri and K. Arai, *Energy and Fuels*, 2003, **17**, 922–928.

101. J. A. Onwudili and P. T. Williams, *Green Chem.*, 2014, **16**, 4740–4748.

102. H. Pińkowska, P. Wolak and A. Złocińska, *Chem. Eng. J.*, 2012, **187**, 410–414.

103. Wahyudiono, T. Kanetake, M. Sasaki and M. Goto, *Chem. Eng. Technol.*, 2007, **30**, 1113–1122.

104. M. F. Neira D'Angelo, V. Ordomsky, J. Van Der Schaaf, J. C. Schouten and T. A. Nijhuis, *Int. J. Hydrogen Energy*, 2014, **39**, 18069–18076.

105. http://www.virent.com/technology/bioforming/.

106. L. Vilcocq, A. Cabiac, C. Especel, E. Guillon and D. Duprez, *Oil Gas Sci. Technol.*, 2013, **68**, 841–860.

CHAPTER 3

Catalytic Hydrogenation of Sugars

DMITRY YU MURZIN,* ANGELA DUQUE, KALLE ARVE, VICTOR SIFONTES, ATTE AHO, KARI ERÄNEN AND TAPIO SALMI

Åbo Akademi University, Biskopsgatan 8, 20500, Åbo, Finland
*Email: dmurzin@abo.fi

3.1 Introduction

Over the last decade, a strong dependence on fossil feedstock has resulted in an understanding of a need to move towards renewable raw materials. In this sense, biomass has been profiled as the key of a fossil-free industry: renewable, energy rich and CO_2-neutral. Biomass is an abundant and rich source of molecules, which can be further refined to yield diverse chemicals, food ingredients and fuels.

The current state at which biomass energetic and chemical capacities are being exploited still needs to be developed further. Moreover, the key for a successful transition towards an economy based on renewables depends on diversification,[1] implying that the focus should not only be on fuels but also on chemicals.

The so-called first generation biofuels rely mostly on sugar monomers derived from edible crops for the production of bioethanol. While environmentally friendly, competition for cultivable land and high production costs hinder, at the present time, a broader implementation of this technology.

RSC Green Chemistry No. 44
Biomass Sugars for Non-Fuel Applications
Edited by Dmitry Murzin and Olga Simakova
© The Royal Society of Chemistry 2016
Published by the Royal Society of Chemistry, www.rsc.org

The second generation of biofuels focuses on waste streams stemming from the paper and agricultural industries, among other sources. This implies a less straightforward task of refining and adding value to complex molecules such as cellulose, hemicelluloses, and lignin. The driving force of this technology lies in its very complexity: long chains of highly functionalized molecules. From here, many processes have been envisioned, with, for example, the platform based on levulinic acid derived from glucose.[2,3] Other platforms based on carbohydrates for the production of specialty and fine chemicals have already been proposed (Figure 3.1): one example being polyesters, polyamides and polyurethane derived from sorbitol, which in turn is obtained from glucose, the building block of starch and cellulose.[4]

In this sense, carbohydrates constitute three quarters of the available renewable biomass.[2] Simpler carbohydrates, such as mono- and disaccharides, are obtained by catalytic hydrolysis of cellulose, hemicelluloses and starch, and their derivatization into useful components, for instance by hydrogenation of sugars to sugar alcohols, is of high importance. Sugar alcohols, traditionally used as low-caloric sweeteners, have experienced an explosive growth in interest, not only due to their potential as platform molecules as mentioned, but also because of their role in aqueous-phase reforming, which has been envisioned as a route to renewable hydrogen or alkanes starting from various sugar alcohols.[5–12] However, not only should the naturally occurring carbohydrates become an important part of the industrial feedstock but also, in order to guarantee a smooth transition to a biomass-based industry, the products derived thereof should be developed having the alternative raw materials in mind, as opposed to seeking routes to reproduce exactly the same chemicals obtained from other technologies.

In agreement with this strategy, naturally occurring polysaccharides, such as cellulose and hemicelluloses, are the ideal starting molecules for the

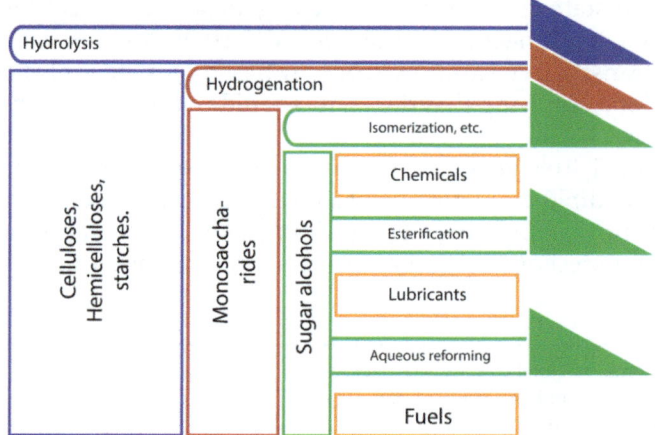

Figure 3.1 Biorefinery scheme with the platform chemical concept.

derivatization to simpler mono- and disaccharides which, in turn, act as substrates for the production of value added molecules.

The term *carbohydrate* was in the beginning used only to refer to molecules made up of carbon, hydrogen and oxygen in the proportions $C_n(H_2O)_n$. However, due to the functional richness of such compounds, which range from the fundamental monosaccharides to highly substituted molecules, the term has lost its specificity. On the other hand, the word *sugar* is mainly used to describe monosaccharides and simpler oligosaccharides, such as glucose and sucrose (table sugar).[13]

Finally, *monosaccharide* refers to the most basic molecules that constitute the building blocks of more complex carbohydrates. Generally, a monosaccharide contains several hydroxyl groups and a carbonyl function. The terms *aldose* and *ketose* refer to monosaccharides whose carbonyl function is an aldehyde or a ketone, respectively.

Besides the number of carbons, the number of hydroxyl groups and the type of carbonyl function present in monosaccharides and, more generally, in carbohydrates, other factors should also be considered when studying these compounds.[13,14] Due to the presence of hydroxyl and carbonyl groups in the same carbon chain, intramolecular hemiacetal formation takes place. The result of these internal reactions is the co-existence of a particular carbohydrate molecule in five-membered rings (furanoses) or six-membered rings (pyranoses), with other cyclic configurations also being possible but uncommon as the strong torsion inside the molecule renders it less stable. Finally, the ring closure at the planar carbonyl group can take place from either side, giving rise to α and β anomers. Besides the cyclic forms, minor fractions of the molecules exist in open-chain forms, in which the carbonyl group is presumed to be active.

Some examples of sugars and their most common tautomeric forms are presented in Figure 3.2. Sugar alcohols, known also as polyols or alditols, are compounds obtained through the reduction of the carbonyl group present in a sugar molecule either by means of chemical agents (such as sodium borohydride) or by molecular hydrogen in the presence of homogeneous or heterogeneous catalysts. The use of molecular hydrogen and heterogeneous catalysis is preferred from the viewpoint of green process technology, since the formation of stoichiometric co-products is avoided and the separation of a heterogeneous catalyst from the reaction mixture is rather straightforward.

Sugar alcohols are versatile molecules, which have found a wide variety of applications. Higher molecular weight polyols (chains longer than four carbon atoms) could be used for manufacturing polyesters, alkyd resins and polyurethanes.[15] They have also found utilization as intermediates in the production of pharmaceuticals and starting molecules for the synthesis of ligands.[16-23] However, their most commonly known use as sweeteners is connected to the fact that polyols possess a sweet taste, contain fewer calories, and produce a lower glycemic response.[24] As proof of the evolving importance of this topic, the global production of sorbitol has increased

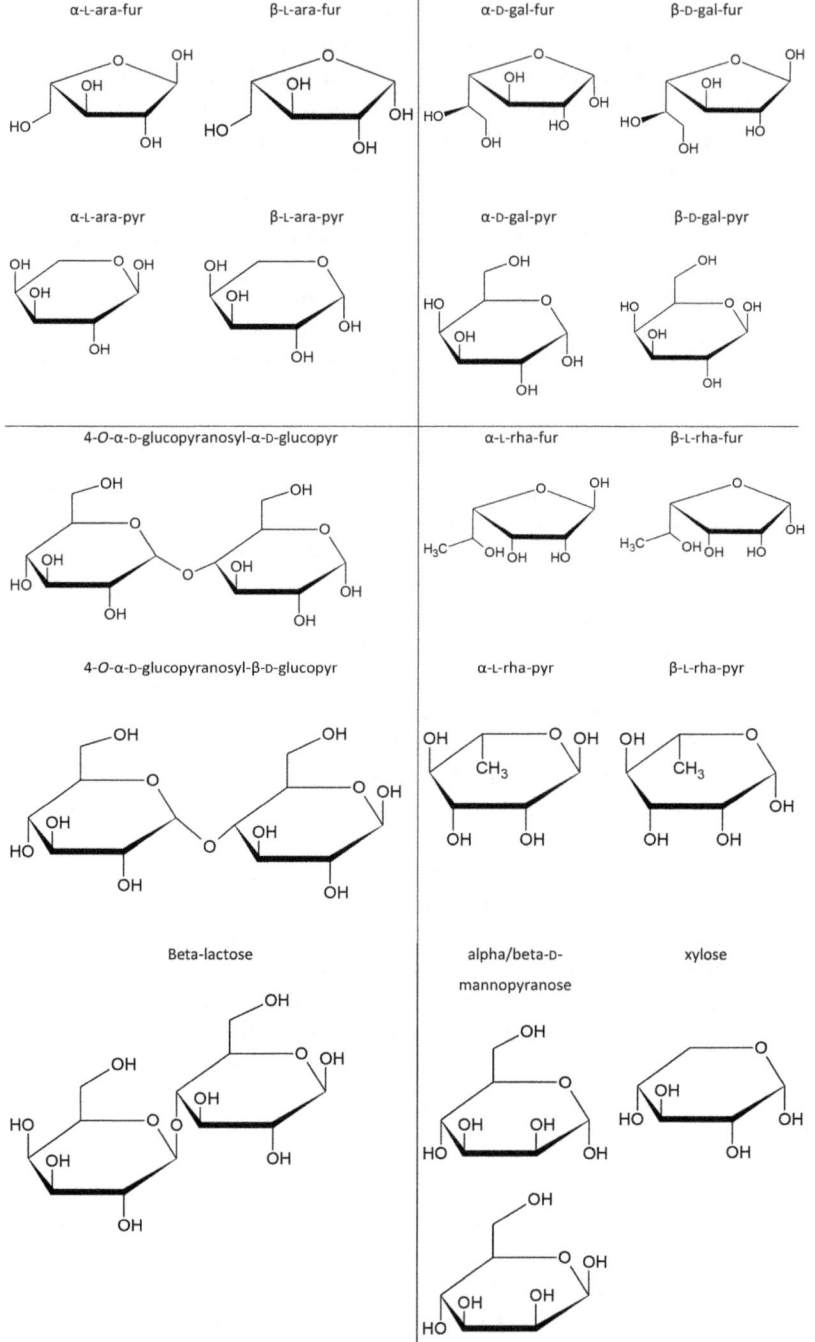

Figure 3.2 Sugar molecules: ʟ-arabinose, ᴅ-galactose, ᴅ-maltose, ʟ-rhamnose, lactose, mannose, xylose. Pyr = pyranose, fur = furanose, ara = arabinose, gal = galactose, rha = rhamnose.

from 700 000 t/a in 2007[15] to over 1 600 000 t/a in 2011 and a demand of above 2 million t/a in 2018.[25]

The best known special sugar alcohols (alditols) are xylitol and sorbitol, but even other sugar alcohols, such as lactitol, maltitol and mannitol have recently gained more importance and more of the market.

The most common solid catalysts used for sugar hydrogenation to sugar alcohols on an industrial scale are based on nickel, such as sponge nickel (Raney nickel), but recently, the use of ruthenium has benefited from a growing interest, since Ru affords good activity and excellent selectivity. Thus, it is expected that Ru will replace Ni as a sugar hydrogenation catalyst, particularly because of the toxic properties of Ni.

Although sugar hydrogenation has been well known for many decades, one of the problems encountered is deactivation and instability of the corresponding nickel catalyst due to leaching and formation of harmful by-products. Aldonic acids, such as lactobionic, gluconic and xylonic acids, represent an example of such harmful by-products formed in the hydrogenation of aldoses to alditols. Deactivation and instability of the catalyst can also lead to problems such as recovery and regeneration of the solid material.

Despite the formation of side products, selectivity towards sugar alcohols is nevertheless high, both on sponge Ni and Ru catalysts, exceeding 95% under optimal conditions. Such data have been reported for the hydrogenation of glucose, xylose and lactose.[27–31] For some special cases, such as hydrogenation of fructose to mannitol, the product selectivity is an important issue; the best selectivity to mannitol is limited to about 60–70% on Cu-based catalysts, the by-product being sorbitol.[32] The conventional production technology of sugar alcohols is based mainly on the use of batch-wise operating slurry reactors: finely dispersed, supported or sponge metal catalysts (catalyst particles smaller than 0.1 mm) are immersed in a batch of an aqueous sugar solution, to which hydrogen is continuously added, maintaining constant pressure. Among the reported operating conditions, it can be found that the hydrogen pressure is typically kept at 30–180 bar and the temperature range from 80 to 150 °C.[26,33] The hydrogen concentration in the liquid phase is one of the limiting factors of the process; therefore, alternative solvents with better hydrogen solubility, such as ethanol, have been proposed.[30]

Quite a number of disadvantages with batch-wise operating nickel catalysts call for more efficient catalytic materials, which could be operated in a continuous mode, either in fixed beds or some other arrangements.

In recent years, continuous operation in sugar hydrogenation has been demonstrated, for instance, for glucose hydrogenation in trickle-bed reactors, monoliths and on activated carbon cloths.[15,26,34] Furthermore, it is possible to carry out sugar hydrogenation in loop reactors, where the sugar solution is recirculated through an external loop system.[26] Since sugar molecules are large and the solubility of hydrogen is limited, the diffusion limitations inside porous catalyst layers become severe, as has been demonstrated experimentally and by numerical simulations.[35,36] Therefore, only egg-shell catalyst pellets and structured catalysts can be considered as feasible alternatives to slurry technology.

For example, one could envisage utilization of structured catalysts (monoliths, foams) where the active phase is coated on a support. Close to such system are microreactors, whose application might require, although not necessarily, novel preparation methods.

The aim of this report is to review the recent literature, covering mainly publications from 2000, related to the catalytic hydrogenation of various sugars. The main emphasis is on describing different types of catalysts which were studied. Kinetics and engineering aspects related to structured catalysts are also addressed.

3.2 Glucose to Sorbitol Hydrogenation

3.2.1 Catalyst Screening

Glucose by no doubt is the molecule that has been studied most frequently in the literature. Hydrogenation of glucose (D-glucose or dextrose) to sorbitol can be readily performed in the presence of a suitable catalyst resulting in high yields (Figure 3.3).

Even though sorbitol, a natural polyol soluble in water, has been isolated from red seaweed and many of the fruits of Rosaceae plant family (pears, apples, cherries, prunes, peaches, and apricots), on a large-scale, it has always been manufactured by the catalytic hydrogenation of glucose.

This is an important reaction for the industry as the main product is a versatile chemical intermediate with promising biorefinery applications in the future. It is used as a low calorie sweetener and sugar substitute for diabetics, as a humectant in cosmetic and pharmaceutical products, in paper and tobacco, as well as other applications. It has no cariogenic activity; hence most toothpastes are based on sorbitol. Additionally, it is utilized as a feedstock for the production of L-ascorbic acid (vitamin C) and for synthesis of polymers through diols.[29,37–41]

Until now, most catalysts for sorbitol hydrogenation have been based on nickel as an active metal. Historically, Raney nickel was used because of its competitive price; more recently, supported nickel catalysts were more frequently applied because of their higher activity, which could be improved further by promoters. However, the use of nickel as mentioned earlier has some disadvantages: namely leaching, causing loss of activity and high

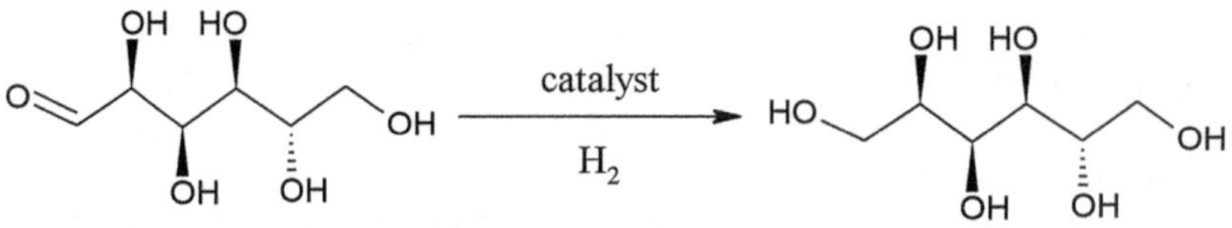

Figure 3.3 Glucose hydrogenation.

nickel content in the sorbitol solution. For food, medical and cosmetic applications, nickel must be removed completely from sorbitol, resulting in high additional costs. Therefore, catalysts based on other active metals were evaluated, including cobalt, platinum, palladium, rhodium and ruthenium. The best activities have been found for supported ruthenium catalysts.[27,42,43] The order of activity has been reported[42,43] as follows: Ru > Ni > Rh > Pd. Additionally, ruthenium is not dissolved under the reaction conditions tested.

Raney nickel is a classical catalyst for hydrogenation processes and it is manufactured from alloys that contain a catalytically active metal, such as Ni, Co or Fe and a soluble metal, *e.g.*, Al, Si or Sn. In addition, a promoter is included to enhance the activity and selectivity of the catalyst. As previously mentioned, commonly used promoters are Ca, Co, Cr, Cu, Fe, Mo and Sn. For instance, a typical Raney Ni catalyst consists originally of a Ni–Al alloy and a promoter. Aluminium is dissolved from the material in an alkaline solution and a porous structure is obtained, as illustrated in Figure 3.4. While Raney nickel has excellent catalytic properties, its major drawback is deactivation[44] and utilization of skeletal catalysis in fixed bed applications presents additional challenges.

Li *et al.*[42] used nickel–boron amorphous alloy catalysts promoted with chromium, molybdenum and tungsten and found the highest activities for catalysts promoted with tungsten. Li *et al.*[45] tested a Raney nickel catalyst promoted with phosphorus. Addition of Mo and Cr promoters was found to be beneficial for the catalyst activity and stability,[37,46–49] whereas the catalyst promoted with Fe and Sn deactivated very rapidly.[37] It was also reported that

Figure 3.4 SEM image of a Raney nickel catalyst.

Pt, Pd, Ru and Rh can have a promoting effect on Ni, enhancing activity by about 20–30%.[27,42,43,50] Raney-type Ni/Ru was considered the most promising catalyst.

Fouilloux[51] claimed that even though Ni was responsible for the hydrogenation capacity, the residual aluminium (in combination with promoters) played a key role in the performance of Raney-type Ni catalysts, because it was able to act as an electron donor to nickel, which rendered the d-band of nickel less electron deficient and thus influenced the adsorption of the reacting species.

Hoffer *et al.*[46] studied the influence of residual Al, the effect of Mo and a combination of Cr and Fe promoters on the performance of Raney nickel catalyst in a three-phase slurry reactor at 40 bar and 120 °C. The promoters were segregated at the surface of the catalysts and had a beneficial effect on the reaction rate, essentially due to an increased surface area and stability of the active phase while an enhanced interaction of glucose with Ni seemed to be secondary. The major drawback was the catalyst deactivation due to the formation of D-gluconic acid that led to a severe loss of Ni.

The main reasons for deactivation of Raney-type catalysts are considered to be loss of the active Ni surface by sintering, leaching of Ni and promoters into the acidic and chelating reaction mixtures, and poisoning of the active Ni surface by organic species generated by side reactions.[37,47,52,53]

Due to these disadvantages of nickel, other catalysts have been investigated. van Gorp *et al.*[54] used ruthenium supported on carbon for batch hydrogenation of glucose, concluding that a ruthenium catalyst was a suitable alternative to skeletal nickel. Despite the higher price of Ru, the possibility of regeneration allows usage of a more expensive metal.

Continuous hydrogenation of glucose with ruthenium catalysts in a trickle bed reactor was performed by Arena[38] and Gallezot *et al.*[41] The former author focused on the deactivation mechanism in the case of ruthenium on alumina catalysts. He found physical changes in the support material and suggested poisoning by gluconic acid, iron and sulfur as the causes for deactivation. Gallezot *et al.*[41] applied ruthenium on carbon supports. A catalyst with a ruthenium loading of 1.6 wt% had a conversion of 98.6% after 52–69 h time on stream (TOS). Conversion dropped to 94.4% after 312 h TOS, indicating a remarkably low deactivation of the catalyst.

Claus and co-workers[55] studied this reaction in batch and continuous reactors using supported nickel and ruthenium catalysts, claiming that Ru is superior to Ni in terms of specific activity and lifetime. Conversion of glucose and selectivity to sorbitol were comparable with a ruthenium loading of 1 wt% to that of nickel in a commercial nickel catalyst. Regarding catalyst stability, significant leaching of Ni (66.8%) was observed, while no leaching of ruthenium was detected. The higher costs of ruthenium were counterbalanced by a much lower metal content and minor downtime due to a prolonged lifetime of the catalyst. The main deactivation mechanism for the ruthenium catalyst was poisoning by metals leached out from the reactor material.

Nickel catalysts supported on ZrO_2, TiO_2 and ZrO_2/TiO_2 mixtures were investigated,[56] being more active, selective and stable than Ni/SiO_2 catalysts in the hydrogenation of glucose to sorbitol.

Selective hydrogenation of glucose to sorbitol over HY zeolite supported ruthenium nanoparticles catalysts was investigated,[57] while the same reaction over Ru nanoparticles embedded in mesoporous hyper-crosslinked polystyrene was performed.[58] In the latter case, two routes of the hydrogenation reaction were revealed. The first route includes the interaction of glucose with the spilled-over hydrogen supplied by the catalyst; the second one includes a classical interaction of the sorbed substrate with hydrogen from the reaction medium. A mathematical description of the reaction kinetics was obtained.

3.2.2 Engineering Aspects

Selective hydrogenation of glucose to sorbitol over a commercial Ru/C catalyst was studied,[59,60] both experimentally and with the aid of detailed mathematical modeling. The experiments were conducted in a laboratory scale trickle bed and in a semibatch stirred tank reactor. Sorbitol was obtained in a packed-bed reactor as the main product, typically with $\sim 90\%$ selectivity within the studied temperature range (90–130 °C), while the side product was mannitol.

Glucose was hydrogenated to sorbitol over carbon supported ruthenium catalysts in a continuously operating parallel multiphase reactor set-up.[59,60] Different types of Ru/carbon catalysts and reaction conditions were screened. Long-term stability testing was carried out for the two best performing catalysts. The parallel multiphase reactor set-up consisted of six parallel reactors operating in co-current trickle flow (Figure 3.5).

The slate of catalysts tested[60] consisted of a commercial 0.77% ruthenium on carbon extrudates denoted as Ru/C and different ruthenium catalysts deposited on pristine (commercial, Baytubes C150HP) or nitrogen doped carbon nanotubes (NCNT). More details regarding catalyst preparation and reactor set-up are given in ref. 60. Ru/NCNT agglomerates were sieved into 125–250 µm size fractions, the Ru/C extrudates were sieved into this size fraction after crushing, therefore internal diffusion resistance could not be completely ruled out.

Different reaction conditions were investigated,[60] namely the reaction temperature, glucose concentration and flow rate as well as hydrogen flow. As expected, different glucose conversions were achieved at different flow rates. However, the sorbitol productivity (moles of sorbitol formed per gram of ruthenium 6and minute) was comparable with minor differences at high and low liquid flows, which can be attributed to channeling or spraying. Selectivity to sorbitol was in the range 97–99.4%.

Figure 3.6 shows the influence of the reaction temperature on the sorbitol productivity over Ru/NCNT catalyst at different flow rates. The apparent activation energy was calculated as 45.7 kJ mol^{-1}. Crezee *et al.*[29] reported

Figure 3.5 Parallel multiphase reactor system.[59,60]

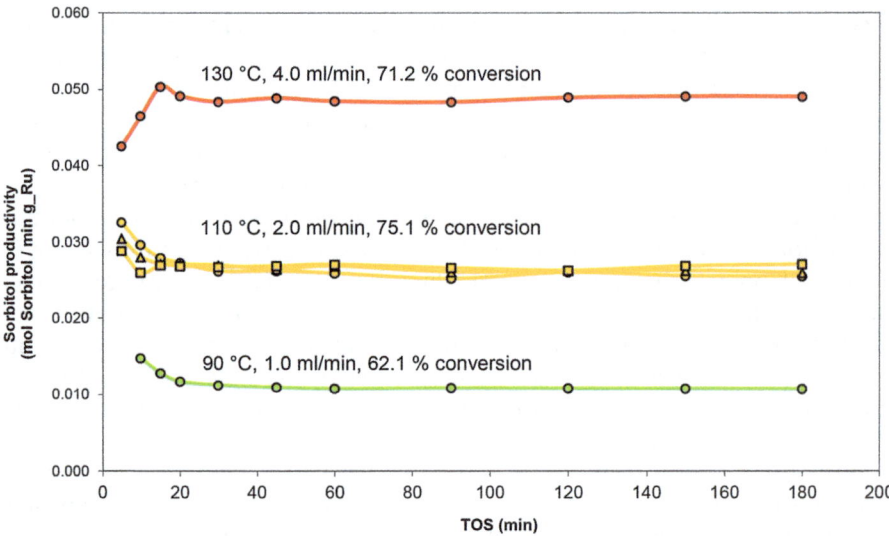

Figure 3.6 Temperature dependence of sorbitol productivity over Ru/NCNT at different temperatures and flow rates using 0.1 mol L^{-1} aqueous glucose solution and 20 bar hydrogen.

slightly higher activation energy, 55 kJ mol^{-1}, for glucose hydrogenation over carbon supported ruthenium catalysts in a semi-batch reactor. Hydrogenation experiments were performed[60] in a semi-batch reactor using the same

Ru/NCNT catalyst and same temperatures as in Figure 3.6. The catalyst was sieved into a fraction smaller than 63 μm in order to calculate kinetic parameters without mass transfer limitations. The activation energy was calculated as 62.1 kJ mol^{-1}. The difference between the activation energies in a continuous and a semi-batch reactor[60] in the former case might be due to the existence of internal diffusion resistance.

The results for catalyst screening performed over the commercial Ru/C, ruthenium on nitrogen doped carbon nanotubes (Ru/NCNT), ruthenium on carbon nanotubes (Ru/CNT) and over a catalyst prepared by a colloidal method (Ru/NCNT-Col) are presented in Table 3.1.

The activity of Ru/NCNT-Col in the sorbitol production is significantly increased by thermal treatment under hydrogen. Surprisingly, the effect of an analogous argon treatment is much less pronounced. A reason for the lower activity of the argon treated sample could be that the removal of PVP, covering the ruthenium particles, was not as good as in the case of hydrogen treatment. The catalysts prepared through a conventional deposition–precipitation method (Ru/NCNT and Ru/CNT), as well as the commercial Ru/C catalyst, are able to hydrogenate glucose more efficiently than the catalyst prepared using the colloidal method.

The selectivity to sorbitol can be considered high for all catalysts, except untreated Ru/NCNT-Col, which gave the lowest selectivity of 93.6%.

Productivity of catalysts based on Ru was calculated[60] and compared with the work of Gallezot *et al.*[41] The corresponding value in ref. 41 was slightly lower than in ref. 60 for most of the catalysts. Note that the reaction conditions in ref. 41 and 60 were, however, different.

Long-term stability testing was performed[60] employing Ru/NCNT and a commercial Ru/C catalyst. It was found that the Ru/NCNT was able to selectively hydrogenate glucose to sorbitol over a period of more than 100 h without any pronounced deactivation, while the commercial catalyst underwent substantial deactivation within this period. The increased stability of the NCNT supported Ru catalyst could be traced back to stronger support–Ru interactions as compared to the Ru/C catalyst. No ruthenium leaching was observed for the Ru/NCNT catalyst, while

Table 3.1 Mass balance, glucose conversion and selectivity to sorbitol, mannitol and fructose over the different catalysts screened at 130 °C, 20 bar, 0.2 mol L^{-1} aqueous glucose solution and 2.0 mL min^{-1} flow rate.

Catalyst	Mass balance	Conversion glucose	Selectivity to		
			Sorbitol	Mannitol	Fructose
2.2 wt% Ru/NCNT	97.3%	91.4%	98.2%	1.6%	0.2%
0.77 wt% Ru/C	99.5%	41.5%	98.7%	0.7%	0.5%
3.1 wt% Ru/CNT	92%	65.0%	97.4%	2.2%	0.4%
6.3 wt% Ru/NCNT-Col (H$_2$)	97.9%	31.2%	97.9%	1.5%	0.6%
6.3 wt% Ru/NCNT-Col (Ar)	95.8%	24.1%	97.9%	1.1%	1.0%
6.3 wt% Ru/NCNT-Col	99.99%	8.3%	93.6%	1.0%	5.4%

leaching from the commercial activated carbon catalyst was found to be significant.

Ruthenium on carbon nanotubes has also been utilized in the hydrogenation of glucose by Pan et al.[61] The authors reported that, among the tested supports, multi-wall carbon nanotubes (MWCT) showed the highest glucose conversion and turnover frequency, with sorbitol being the only reaction product.

Hofmann and Bill[62] studied the hydrogenation of glucose in a slurry batch reactor in the temperature range of 80–130 °C using Raney nickel as a catalyst. The authors found a first order dependence of the reaction rate in both glucose and hydrogen. From the Arrhenius plot, it was observed that the overall activation energy decreased from 82 to 13 kJ mol^{-1} with increasing temperature upon a change in the rate-determining process from kinetics to diffusion limitations.

Bizhanov et al.[50] studied the hydrogenation of glucose in the presence of Raney Ni–Pt and Ni–Rh catalysts. It was shown that for Ni–Pt and Ni–Rh, there was a first order dependence on hydrogen pressure up to 60–80 bar. At higher pressures, zero order behavior with respect to hydrogen was observed. From the Arrhenius plot in the temperature range of 80–130 °C, it was found that the apparent activation energy for the hydrogenation over Ni–Pt (0.1 wt% Pt) was 50–54 kJ mol^{-1}, while for Ni–Rh (0.1 wt% Rh), values of 38–42 kJ mol^{-1} were found.

Brahme and Doraiswamy[63] conducted kinetic experiments using Raney Ni as a catalyst at temperatures varying from 77 to 146 °C and pressures of 4.4–21 bar. From the Arrhenius plot, it was clearly seen that there was a shift in the controlling mechanism from a chemical control in the lower temperature region to mass transfer control in the higher temperature region. Activation energies in the two temperature regions were estimated to be 44 kJ mol^{-1} between 77 and 100 °C and 6 kJ mol^{-1} between 100 and 146 °C. Kinetic modeling was therefore undertaken for temperatures below 100 °C. A mechanism was proposed comprising a reaction step between molecularly adsorbed hydrogen and glucose from the liquid phase (Eley–Rideal mechanism) with desorption of the product (sorbitol) controlling the overall reaction.

Wisniak and Simon[26] hydrogenated glucose in the range 85–130 °C, 3.4–68.9 bar, and a catalyst loading of 0.5–9 wt% with Raney nickel and 5 wt% Ru/C as catalysts. In all cases, the reaction showed a first order dependence with respect to glucose. Ru/C appeared to be the most effective catalyst. External diffusion was supposed to be eliminated since the rate was independent of the steering speed. Furthermore, an almost linear relation of the rate constant with catalyst concentration was found, which points to the absence of gas–liquid mass transfer. For glucose hydrogenation at 41 bar, an activation energy of 18.8 kJ mol^{-1} with 3% Raney nickel and 70.3 kJ mol^{-1} with 3% ruthenium was calculated. A low value of the activation energy for Raney nickel would normally suggest the presence of diffusion effects; however, most probably gas–liquid mass transfer and external diffusion were eliminated, although internal diffusion could still be present. From the behavior of the reaction rate with hydrogen pressure, some differences depending on the catalyst were noticed. With Raney nickel at 85 and 100 °C, the increase in the reaction rate with increase in hydrogen pressure is minor, while at 130 °C, there is a sudden increase in the reaction rate at pressures above 41 bar. The authors suggested that the reaction occurred between glucose from the fluid phase and atomically chemisorbed hydrogen. No mechanism was suggested for hydrogenation at 130 °C and pressures above 41 bar, since the reaction was too fast and probably not controlled by chemical steps in line with the low activation energy observed. For hydrogenation of glucose on ruthenium, two types of reaction models were proposed depending on the hydrogenation pressure. At pressures below 35 bar, an Eley–Rideal model was suggested with the controlling step being the surface reaction between atomically chemisorbed hydrogen and incoming from the fluid phase glucose. At higher pressures, a Langmuir–Hinshelwood model was proposed, in which the surface reaction between atomically chemisorbed hydrogen and adsorbed glucose was the controlling step.

Turek *et al.*[64] suggested another model for the hydrogenation of glucose to sorbitol on a nickel-silica catalyst at 80–150 °C and 16–100 bar. In this model, molecularly adsorbed hydrogen reacts with adsorbed glucose, whereby the adsorption of hydrogen and glucose takes place on different sites.

Déchamp *et al.*[39] studied the kinetics in a high-pressure trickle-bed reactor in concurrent downflow mode. The hydrogenation reactions were performed on silica-alumina supported nickel catalysts of the following composition: 48.4% Ni; 5.15% Al; 8.46% Si in the form of cylindrical extrudates (6.3 mm × 1.6 mm). The operating conditions ranged from 70 to 130 °C in the pressure range of 40–120 bar. Reaction rates were measured with the catalyst extrudates as such or after crushing and sieving them into 0.5–0.8 mm particles. For both sizes, a three-fold increase in the catalyst mass produced only a two-fold increase in the initial rate of sorbitol formation, which indicates that the reaction could be partly limited by gas–liquid mass transfer. Moreover, for the same catalyst loading, the crushing of pellets strongly increased the reaction rate of glucose hydrogenation, which implies that the reaction over the extrudates was also severely limited by internal diffusion. In the case of the crushed extrudates, the authors[39] estimated an effectiveness factor of $\eta = 0.9$ with the approach of Weisz and Hicks. Subsequent kinetic experiments were, therefore, conducted with the crushed catalyst. From the Arrhenius plot of the initial reaction rates measured between 70 and 130 °C and hydrogen pressure lower than 80 bar, an activation energy of 67 kJ mol^{-1} was obtained. This value, much larger than the activation energy of mass transfer limited process (12–21 kJ mol^{-1}), indicates that the reaction rate was controlled by the reaction kinetics on the metal surface. The kinetic data obtained from measurements of the initial reaction rates as a function of temperature, pressure and mass of catalyst were described by a Langmuir–Hinshelwood rate expression based on a model where the reaction between glucose and molecular hydrogen, both adsorbed on the nickel surface, is the rate-determining step.

Tukac[65] also studied this reaction in a high-pressure trickle-bed reactor. The reaction was performed on kieselguhr supported nickel catalysts (12% NiO, 2% Cr_2O_3) with a pellet size of 2–3 mm. The operating conditions ranged from 115 to 165 °C in the pressure range of 5–100 bar. The reaction was found to be first order with respect to glucose and the order in hydrogen was 0.65. The apparent activation energy was equal to 24–49 kJ mol^{-1}, depending on the initial glucose concentration, density and viscosity of the solution. The author stated that the reaction rate was probably influenced by external mass transfer limitations.

Creeze *et al.*[29] investigated the kinetics of this reaction in a semi-batch slurry autoclave operating at 100–130 °C and 40–75 bar hydrogen pressure using Ru/C catalyst. The glucose concentration was varied between 0.56 and 1.39 mol L^{-1} leading to a shift in the order towards glucose with the concentration; at low glucose concentrations (up to *ca.* 0.3 mol L^{-1}) the reaction showed a first order dependence, while at higher concentrations it changed to zero order behavior. A first order dependence with respect to hydrogen was obtained. The kinetic data were modeled using three plausible rate expressions based on Langmuir–Hinshelwood–Hougen–Watson (LHHW) kinetics, assuming that the surface reaction was rate-determining. These three models described the data satisfactorily and the authors claimed that further statistical discrimination between these models was not possible.

Glucose hydrogenation/hydrogenolysis reactions on noble metal (Ru, Pt)/ activated carbon supported catalysts were performed.[66] XPS measurements showed the existence of different kinds of metallic and oxidic surface species, in different proportions, as a function of the reduction method and the amount and type of metal. The authors emphasized the importance of $Ru(O)^{\delta+}$ species for catalysis.

Despite a large number of publications on hydrogenation over ruthenium catalysts, almost no reports on the effect of ruthenium particle size could be found in the literature. Several ruthenium on carbon catalysts with different metal dispersion were studied.[67] The ruthenium average particle size of the catalysts ranged from 1.2 to 10 nm. The catalysts tested at 120 °C showed different behavior in terms of activity. Full conversion was achieved in most cases, although some catalysts displayed lower activity. The highest activities were observed for catalysts supported on active carbon and nitrogen functionalized carbon nanotubes (NCNT) bearing a metal cluster size close to 3 nm. High selectivity, up to 96.1%, at full conversion was achieved with most of the catalysts.

The structure sensitivity of glucose hydrogenation is illustrated in Figure 3.7, which also includes a comparison of the experimental and calculated values. A clear maximum in Figure 3.7 indicates that glucose hydrogenation is a structure sensitive reaction and that an optimal catalyst should contain ruthenium particles around 3 nm in size.

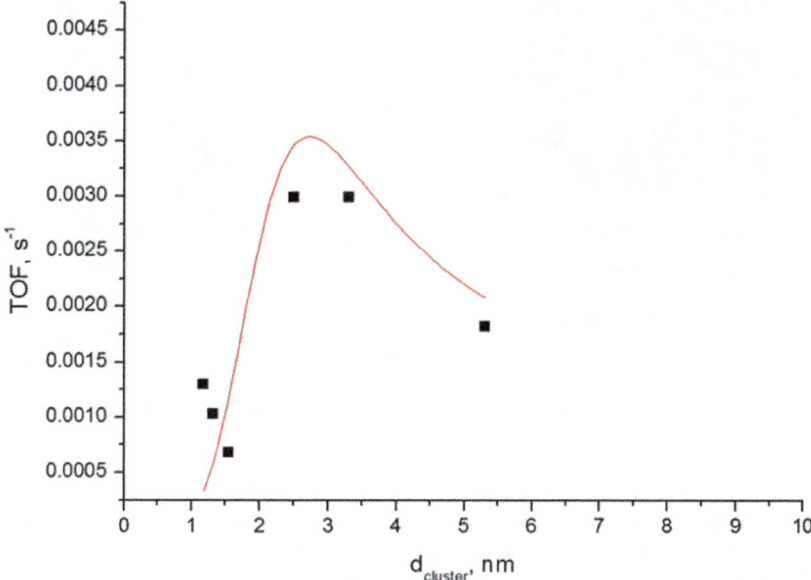

Figure 3.7 Dependence of initial glucose conversion rates on the particle size of ruthenium. Hydrogenation of glucose was performed in semi-batch mode at 120 °C and 19 bar hydrogen over several carbon (active carbon and CNT) supported Ru catalysts.[67]

Hydrogenation of sugars over ruthenium is often described by zero order kinetics in hydrogen at high pressures, resulting in the following expression:

$$\text{TOF} = \frac{ke^{\alpha\chi_G/d_{\text{cluster}}} K_G C_G}{(1 + K_G e^{\chi_G/d_{\text{cluster}}} C_G)} \tag{3.1}$$

where α is the Polanyi parameter for the surface reaction between adsorbed sugar and hydrogen and parameter χ accounts for differences in the Gibbs energy of glucose adsorption on edges and terraces.

Preliminary calculations showed that the value of the Polanyi parameter is very close to 0.5, a typical value for a large number of heterogeneous catalytic reactions, therefore, this value was fixed. Calculations performed[67] indicated a difference in the Gibbs energy of adsorption on edges and terraces for glucose of *ca.* 41 kJ mol^{-1}.

3.3 Galactose Hydrogenation

3.3.1 Engineering Aspects

Ruthenium supported on alumina, carbon cloth (Figure 3.8 left), and alumina washcoated metal monolith (Figure 3.8 right) catalysts were prepared and tested for their activity in galactose to galactitol hydrogenation in a temperature range of 105–135 °C and 20, 40 or 60 bar H$_2$.[68] The surface

Figure 3.8 Carbon cloth and pre-washcoated (alumina) metal monolith.

area of the carbon cloth support is considerably higher (1680 m^2 g^{-1}) than that of alumina (220 m^2 g^{-1}). The pore volumes of both supports are close to each other—0.55 cm^3 g^{-1} for the alumina and 0.6 cm^3 g^{-1} for the carbon cloth. The prepared catalysts were characterized by SEM imaging and their ruthenium content was estimated by SEM-EDX. The metal content in the precursor solution for Ru/C was adjusted either to 5 wt% or 2.5 wt%, which resulted in approximately 3.8 wt% and 2.0 wt% ruthenium content on the final catalyst. For the impregnation of the metal monolith, the ruthenium content in the precursor solution was adjusted to 2.5 wt%, resulting in approximately 1.9 wt% ruthenium content on the support.

The first experimental data for Ru/Al$_2$O$_3$ were obtained in a semi-batch reactor[68] at temperatures ranging from 105 to 135 °C and hydrogen pressures from 30 to 60 bar. First order kinetic models were assumed both for galactitol and by-product formation and a non-linear regression was performed to obtain the numerical values for the activation energies as well as for the pre-exponential factors. Based on the experimental data, it can be concluded that at the lowest experimental temperature (105 °C), the effect of hydrogen pressure on the galactose conversion plays a key role. However, at higher temperatures, the effect of hydrogen pressure was found to be less important. In addition, the results from the parameter estimation showed that the overall fit of the model based on the first order kinetics (galactose to galactitol to by-product) was good. A comparison of calculations with the experimental data is given in Figure 3.9. The values of activation energy for hydrogenation of galactose to galactitol and further transformation of the latter to by-products were, respectively, 54 and 51 kJ mol^{-1}.

Activity, selectivity and durability experiments over the Ru/carbon cloth and the Ru/Al$_2$O$_3$ washcoated metal monolith catalyst were performed in a pressurized fixed bed reactor under 20, 40 or 60 bar and compared with batch-wise hydrogenation of galactose. Considerable galactitol formation over the prepared catalysts was observed, whereas deactivation was not significant. The results from galactose to galactitol hydrogenation as

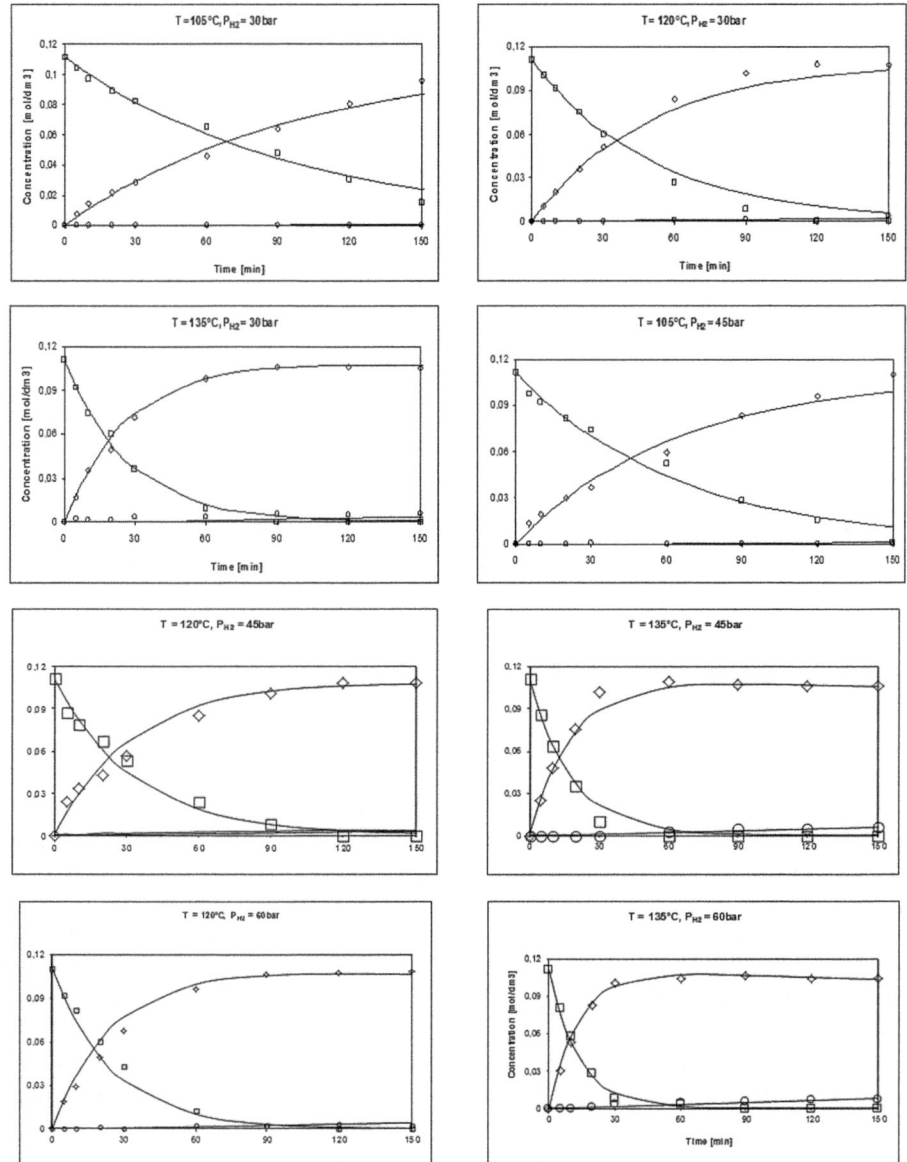

Figure 3.9 Hydrogenation of galactose over Ru/alumina catalyst.[68]

a function of time over the 3.8 wt% Ru/carbon cloth catalyst using 40 mL min^{-1} make-up hydrogen flow in the fixed bed reactor using the recycling mode are shown in Figure 3.10. The effect of hydrogen pressure was minor indicating a zero order reaction in hydrogen.

Ru on carbon cloth was shown[68] to be a stable and active catalyst for sugar hydrogenation to sugar alcohols in a continuous fixed bed reactor system

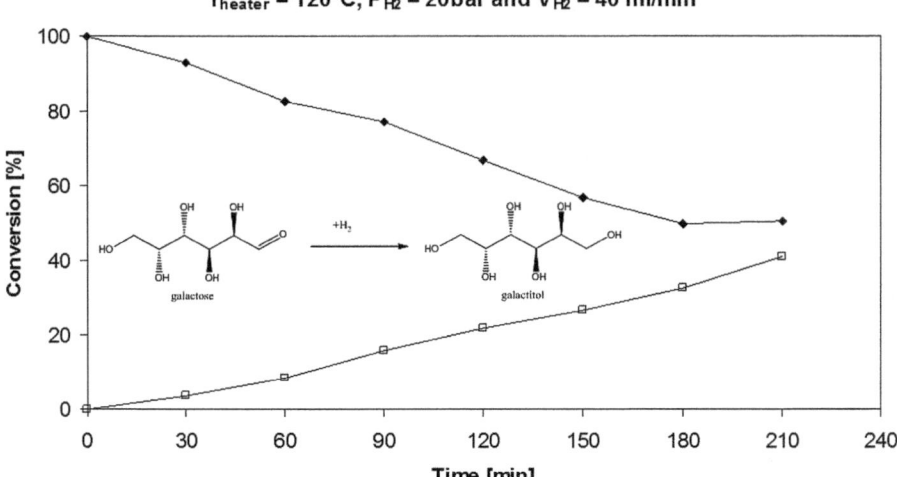

Figure 3.10 Galactose (♦) to galactitol (□) hydrogenation over Ru/C. Heater temperature 120 °C, pressure 20 and make-up hydrogen flow 40 mL min^{-1}.

with recycling. An alternative to carbon cloth catalysts would be the application of structured reactors such as monoliths and/or microreactors. Washcoating of microplates with alumina was considered[68] to be more straightforward than washcoating with carbon. Ruthenium supported on alumina was also shown[68] as stable and active for batch-wise galactose hydrogenation. However, the metal monolith did not result in considerable conversion in sugar hydrogenation, which might be due to an inefficient flow pattern.

Based on the semi-batch reactor experiments with Ru/alumina catalyst, it was expected[68] that Ru/Al$_2$O$_3$ catalyst wash-coated on the microplates would also result in high and stable galactose to galactitol conversion under relevant reaction conditions.

Therefore, microreactor plates (IMM, Mainz, Germany) were washcoated[68] with Ru/Al$_2$O$_3$ and tested in a reactor system built in-house for their activity in galactose to galactitol hydrogenation. However, when Ru/Al$_2$O$_3$ catalyst was implemented on the microplates, the activity was no longer reproducible. It was found that the washcoat was not stable on the microplates but was being rapidly removed from them.

Based on the experimental results (Figure 3.11), the detected rate of the galactose conversion to galactitol was approximately 1.09×10^{-3} mol g$_{cat}$ h^{-1}. Compared to the galactose conversion rate in a batch reactor under similar conditions, the rate in the microreactor was approximately one order of magnitude lower.

A comparison of the opened tested microreactor and an unused microreactor is shown in Figure 3.12.

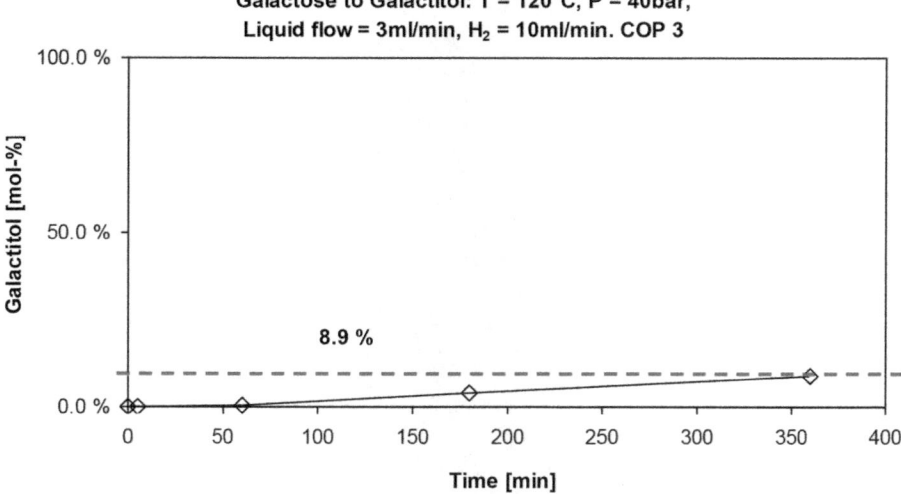

Figure 3.11 Galactose to galactitol conversion over a coated microplate.

Figure 3.12 (a) Opened microreactor, which was tested for its activity in galactose to galactitol hydrogenation. Operational conditions: $T = 105\,^{\circ}C$, $p = 60$ bar, gas to liquid ratios varying from 1 to 20 (mL min^{-1}). Sugar solution: galactose 0.23 mol L^{-1}. (b) Opened unused microreactor.

As can be seen in Figure 3.12a, the washcoat had been almost totally removed from the lower plate, and, in the upper plate, there is an estimated washcoat loss of 30%. Moreover, it was demonstrated[68] that there is a very inhomogeneous distribution of gas and liquid in the channels (Figure 3.13).

Figure 3.13   Flow pattern in microreactor channels, liquid (lighter color) and gas (darker color).

3.4   Maltose Hydrogenation

3.4.1   Catalyst Screening

Maltose is a low cost, large-scale product of the starch-molasses industry, with several million tons of maltose produced annually worldwide.[69] Maltose hydrogenation is commonly performed batch-wise in stirred tank reactors at temperatures ranging from 60 up to 150 °C, hydrogen pressures of 30–80 bar with sponge nickel or transition metals supported on oxides.[70] Maltose hydrogenation is characterized by the formation of several side products including glucose and sorbitol along with formation of the target product—maltitol (Figure 3.14).[71]

Commonly used sponge Ni catalysts have high selectivity up to 96–98%. However, the main problem of this type of catalyst is similar to reactions with other sugars, namely Ni leaching and fast catalyst deactivation. Hydrogenation over ruthenium supported on MgO, SiO_2, Al_2O_3 and TiO_2 led to increased maltose hydrolysis rates and excessive sorbitol formation, therefore selectivity decreased to 20–80%.[69,72] Ru supported on different types of carbon showed high activity in maltose hydrogenation with the selectivity up to 96–97%.[31]

3.4.2   Engineering Aspects

The kinetics of maltose hydrogenation using Ru-containing nanoparticles (NPs) formed in the pores of hyper-crosslinked polystyrene (HPS) was

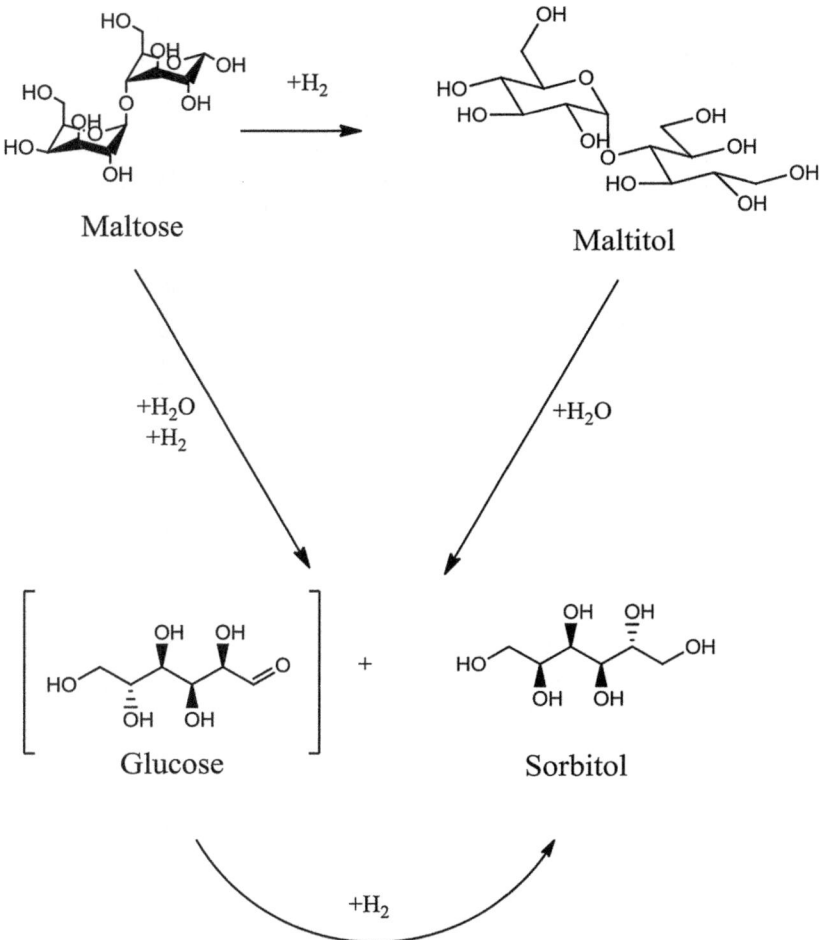

Figure 3.14 Scheme of possible reactions in maltose hydrogenation.

studied[73] at 110–140 °C and 20–70 bar. HPS is a rigid polymer with high mechanical and chemical stability that can be a good matrix for metal nanoparticle synthesis as it has a high surface area and both small and large mesoporous. Small mesopores are suitable for metal nanoparticle formation and larger mesopores can provide transport of the substrate to the active sites.

The study of maltose concentration influence on the catalytic activity and selectivity was performed at 140 °C, overall pressure 40 bar, catalyst and maltose concentration of 8 g L^{-1} and 0.1–0.4 mol L^{-1} respectively. Figure 3.15, displaying the kinetic curves of maltose transformation and the selectivity for HPS-Ru-3%, shows that maltose hydrogenation is a complex process with the main side product being sorbitol. Selectivity to maltitol is characterised with well-defined maxima.

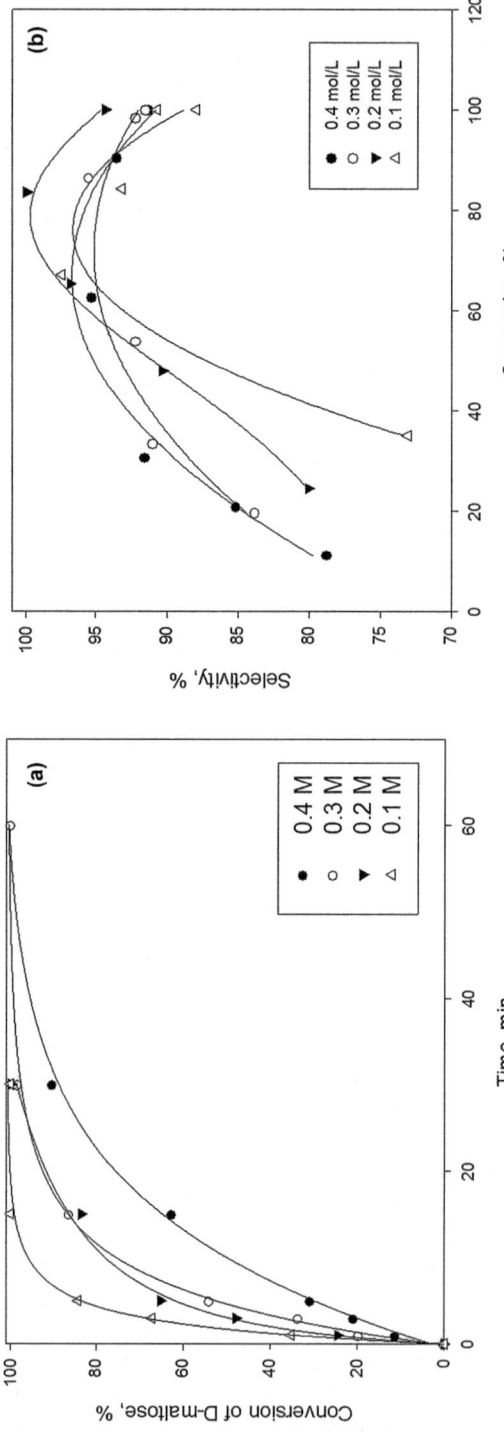

Figure 3.15   Dependence of (a) maltose conversion and (b) selectivity to maltitol for HPS-Ru-3% (reaction conditions: temperature 140 °C, overall system pressure 40 bar, Cc = 8 g L^{-1}). Reprinted with permission from ref. 73. Copyright Elsevier.

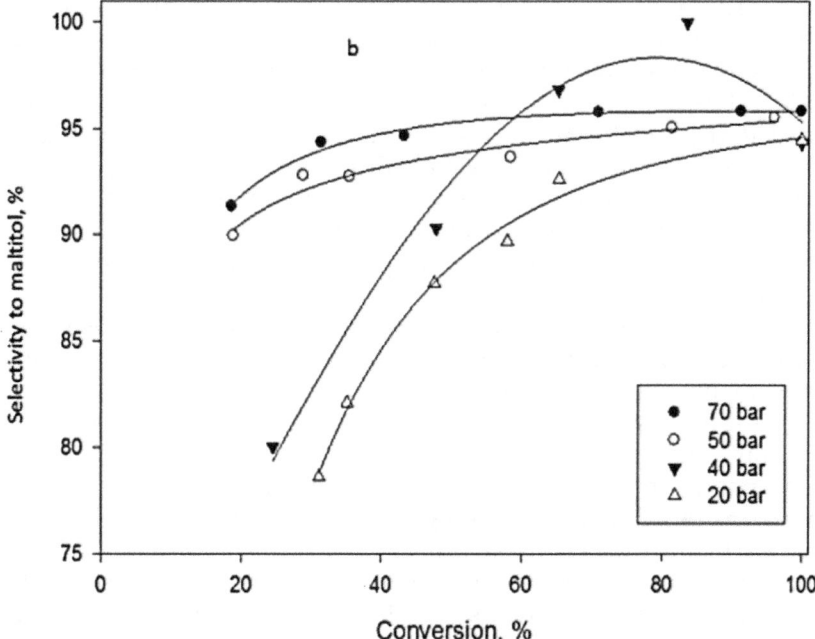

Figure 3.16 Dependence of selectivity to maltitol on conversion for HPS-Ru-3% at different pressures (reaction temperature 140 °C, Cc = 8 g L^{-1}, C0 = 0.4 mol L^{-1}).
Reprinted with permission from ref. 73. Copyright Elsevier.

The increase of hydrogen pressure results initially in the increase of maltose transformation to maltitol, giving thereafter a zero order dependence. Interesting mechanistic observations follow from Figure 3.16. Selectivity to maltitol increased with conversion, which contradicts with the conventional thinking of a consecutive process of maltitol formation from maltose and subsequent hydrogenation to sorbitol. Moreover, selectivity increased with a hydrogen pressure increase.

The apparent activation energy was calculated according to the Arrhenius equation and was found to be 58 kJ mol^{-1}. Maltose, lactose and glucose hydrogenation with HPS-Ru-3%[73] revealed that maltose and lactose hydrogenation rates are comparable while the rate of glucose hydrogenation is substantially higher, which can explain the absence of glucose in the reaction media.

The model developed[73] corresponds to the Langmuir–Hinshelwood mechanism with adsorption of organic compounds and hydrogen of a noncompetitive nature because of large differences in the sizes of molecules.

In order to explain the observed kinetic dependences of selectivity on conversion and hydrogen pressure, a hypothesis was advanced[73] assuming the direct hydrolysis–hydrogenation of maltose to sorbitol and glucose with involvement of two moles of hydrogen and one mole of water per mole of maltose. The formal kinetic expression for this route should have another hydrogen as well as conversion dependence. Tentatively, it can be assumed that this reaction route follows the rate equation with competitive adsorption of reactants on other types of sites, which is in line with the presence of Ru(IV) and Ru(0) species on the catalyst surface. Thus maltose is adsorbed on one type of site, giving hydrogenation of one glucose ring and resulting in maltitol and at the same time it can adsorb in another mode on other types of sites, forming sorbitol and glucose.

The model advanced[73] was able to predict kinetic behavior at different conditions with sufficient accuracy.

Sifontes *et al.*[74] also studied the kinetics of D-maltose hydrogenation as well as some other sugars, such as L-arabinose, D-galactose and L-rhamnose.

Experiments were performed at temperatures ranging from 90 to 130 °C and hydrogen pressures from 40 to 60 bar using an active carbon-supported ruthenium catalyst.

Conversions up to 100% with 100% of selectivity for arabinose and galactose were reported while by-product formation (less than 5%) was observed for maltose and L-rhamnose.

The kinetic model obtained assumed a competitive adsorption of hydrogen and the organic components, and the hydrogenation step on the catalyst surface was taken as the rate limiting, whereas the adsorption and desorption steps were presumed to be rapid quasi-equilibrium steps. An example of the model fit is provided in Figure 3.17.

The hydrogenation behavior of the four selected sugars was similar in terms of the pressure and temperature influence. The effect of the hydrogen pressure was minor, whereas the temperature effect was very profound.

The by-product yields were observed to be dependent on the operating conditions: the more severe the conditions (higher pressures and temperatures), the more diverse and abundant the by-products, amounting to *ca.* 10–12% of the total product yield under the more severe conditions for D-maltose and L-rhamnose, and to *ca.* 5–7% for L-arabinose and D-galactose under the same severe conditions. However, in the majority of conditions, the by-product yields were less than 1%.

3.5   Lactose to Lactitol Hydrogenation

Lactose is a disaccharide, which consists of glucose and galactose moieties. In aqueous solutions at 20 °C, lactose, according to NMR, coexists in two anomeric forms: 62.7% as β-lactose and 37.3% α-lactose.[75] The lactose content of milk originating from different mammals varies between 0 and 9%.[76] The estimated annual worldwide availability of lactose as a by-product from cheese manufacture is several million tons.[76,77] However, only about 400 000 t/a lactose is processed further from cheese whey. Non-processed whey is an environmental problem due to its high biochemical and chemical oxygen demand.[77] A relatively low solubility of lactose in most solvents limits its use in many applications. Another restricting factor is the inability of

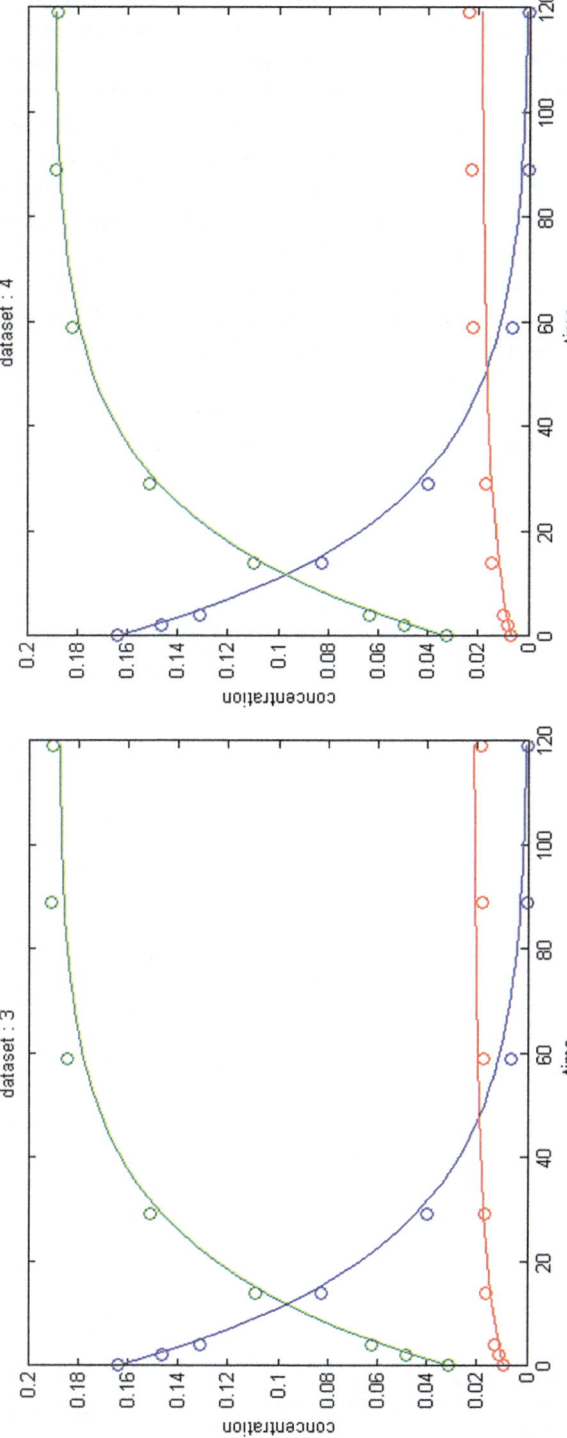

Figure 3.17 Simulation results for ᴅ-maltose hydrogenation at 120 °C and (a) 40 bar and (b) 50 bar. Reprinted with permission from ref. 73. Copyright Elsevier.

lactose intolerant people, with a low level of lactase enzyme in the body, to digest milk sugar.[77] Therefore, the development of value-added products from waste generated during cheese manufacturing processes is important. Lactitol (by hydrogenation), lactulose (by isomerisation) and lactobionic acid (by oxidation in the presence of oxygen) are the industrially most important lactose derivatives. Moreover, the hydrolysis products of D-lactose, D-galactose and D-glucose, can be used as valuable raw materials by the pharmaceutical industry.[78,79] The scheme of lactose hydrogenation in an aqueous environment is shown in Figure 3.18. Due to the reducing atmosphere, lactobionic acid is produced by dehydrogenation of lactose following a mechanism different from the oxidation (oxidative dehydrogenation) of lactose, when the reaction is conducted in the presence of oxygen.

Lactitol is a sugar alcohol, manufactured by reducing the glucose part of disaccharide lactose. It has a clean sweet taste that closely resembles the taste profile of sucrose with only 40 percent of sucrose's sweetening power.

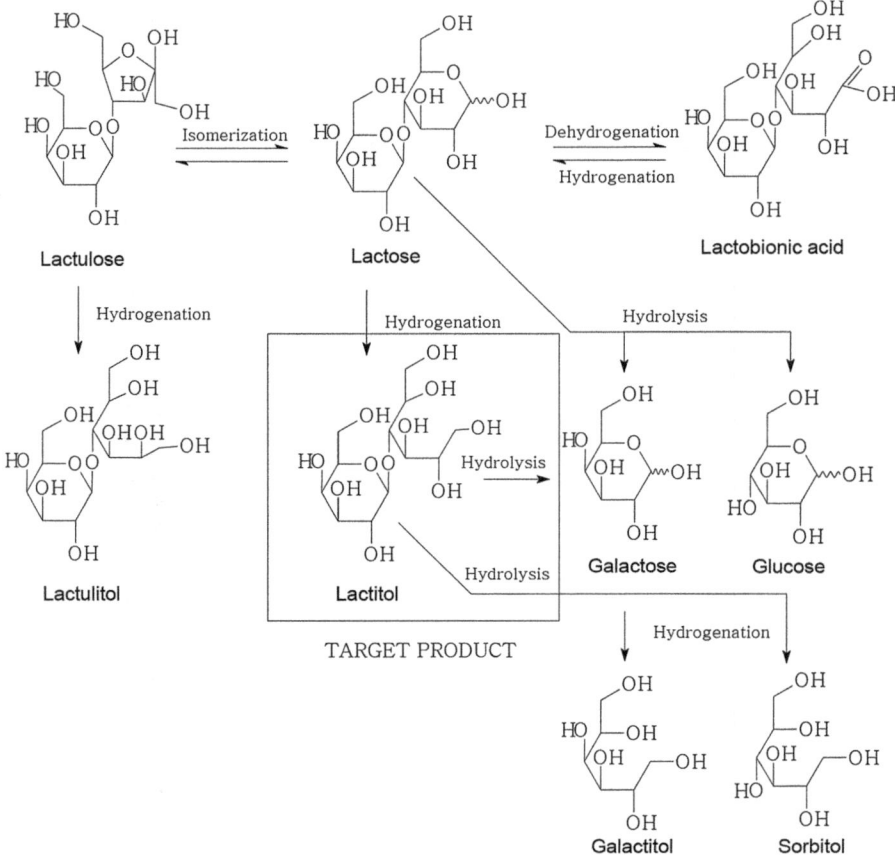

Figure 3.18   Lactose hydrogenation.
            Reprinted with permission from ref. 80. Copyright (2013) American Chemical Society.

This mild sweetness makes it an ideal bulk sweetener to partner with low-calorie sweeteners, such as acesulfame K, aspartame, neotame, saccharin and sucralose. Due to its stability and solubility, and because it tastes similar to sucrose, lactitol can be used in a variety of low-calorie, low-fat and/or sugar-free foods such as ice cream, chocolate, hard and soft candies, baked goods, sugar reduced preserves, chewing gums and sugar substitutes. Unlike lactose, lactitol is not hydrolyzed by lactase. It is neither hydrolyzed nor absorbed in the small intestine. Lactitol is metabolized by bacteria in the large intestine, where it is converted into organic acids, carbon dioxide and a small amount of hydrogen. The organic acids are further metabolized, resulting in a caloric contribution of 2 calories per gram (carbohydrates generally have about 4 calories per gram). As it is metabolized independently of insulin, it is suitable for a diabetic diet.

Only a few studies about D-lactose hydrogenation (Figure 3.18) have been published so far,[28,31,78,80,81] with mainly Ni and Ru catalysts. Kuusisto *et al.*[28] studied the kinetics of lactose hydrogenation over a Mo-promoted sponge nickel catalyst in a batch reactor operating at 20–70 bar and 110 and 130 °C. Selectivity and conversion were investigated under different reaction conditions such as lactose concentration, catalyst loading, hydrogen concentration and reaction temperature. The kinetic data were modeled as well.

The influence of external mass transfer was studied by using various stirring rates. At lower stirring rates (600 rpm) and thus inefficient agitation, a lower conversion and lactitol selectivity indicating external mass-transfer limitations were observed. These values were not affected by changing the impeller rate from 900 to 1800 rpm (Figure 3.19). It was found that inefficient mixing increased the formation of by-products such as lactobionic acid, galactitol, and sorbitol.

The influence of lactose concentration was also investigated, not having, however, any impact either on reaction rate or selectivity. Higher catalyst loadings besides an expected productivity increase led to a lower lactitol selectivity at low lactose conversion levels because of increased formation of lactobionic acid, lactulose, and lactulitol. Low catalyst loadings resulted in elevated lactose hydrolysis and, thus, higher galactitol and sorbitol formation.

It was shown[28] that although the catalyst activity declined, it was possible to regenerate the catalyst by washing with a basic medium, leading to almost complete restoration of the original activity. The dissolved metals at the end of the hydrogenation batches were determined by DCP showing that leaching increased with hydrogen pressure and temperature increase, as well as at low loadings.

Kuusisto *et al.* also studied lactose hydrogenation over several commercial and self-synthesized supported ruthenium catalysts in powder form.[31] Among 5% Ru/C, 5% Ru/Al$_2$O$_3$, 5% Ru/SiO$_2$, 5% Ru/TiO$_2$, 5% Ru/MgO and 3% Ru/hyper-crosslinked polystyrene, the commercial Ru/C catalyst clearly showed the best performance (Figure 3.20).

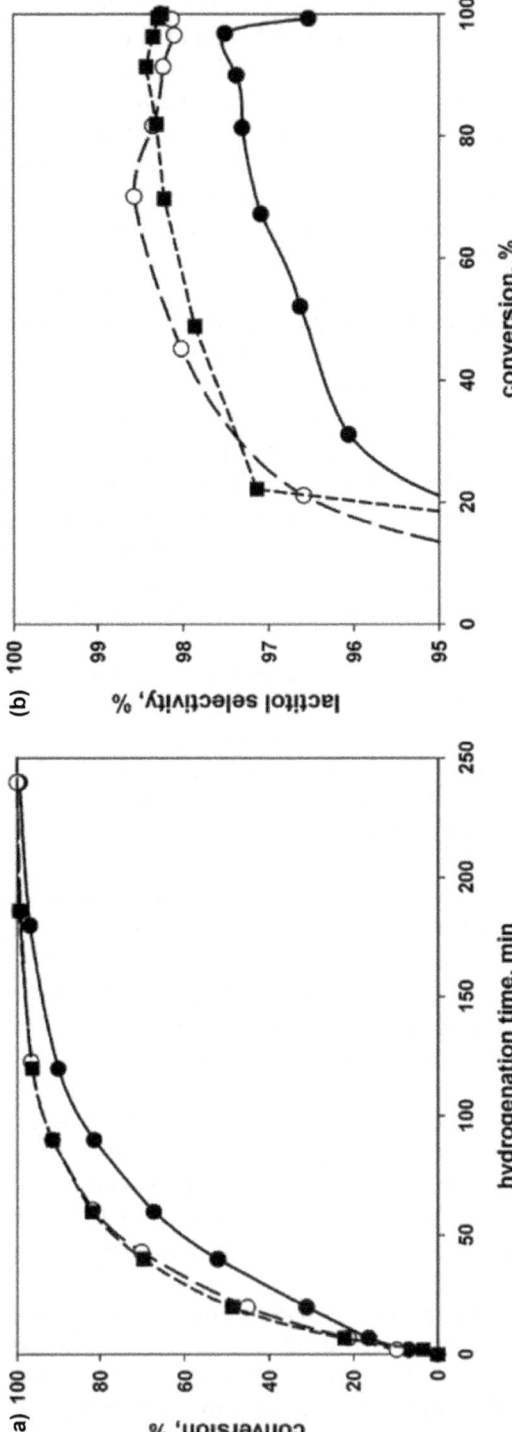

Figure 3.19 Influence of the impeller rate on (a) lactose conversion and (b) lactitol selectivity at 120 °C and 50 bar H_2 over a 5 wt% sponge nickel catalyst. (●) 600 rpm, (○) 900 rpm and (■) 1800 rpm. Reprinted with permission from ref. 28. Copyright (2006) American Chemical Society.

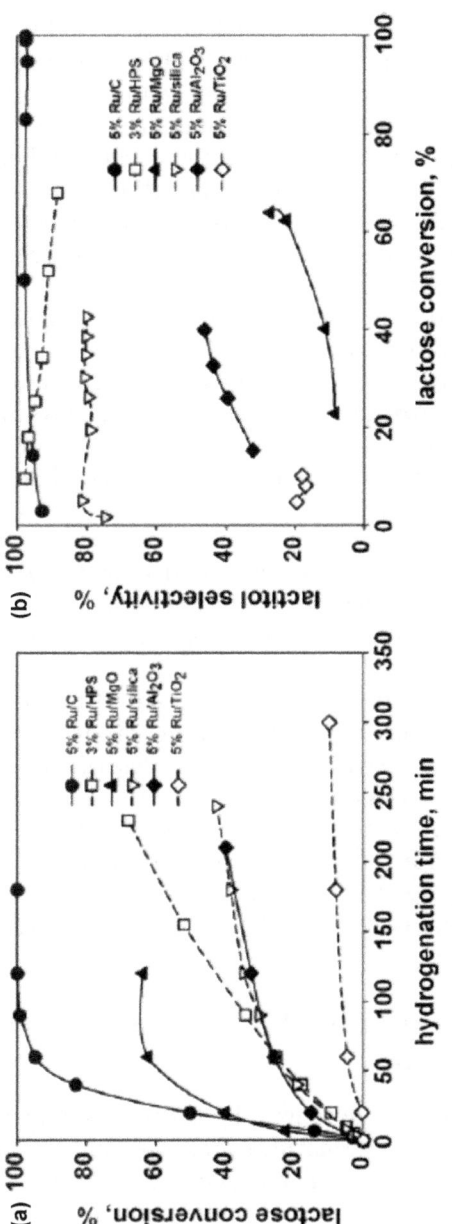

Figure 3.20 Lactose conversion (a) and lactitol selectivity (b) over various supported ruthenium catalysts. $T = 120\,^{\circ}C$, $P = 50$ bar H_2 and catalyst amount 8.96 g L^{-1} in each experiment. Reprinted with permission from ref. 31. Copyright Elsevier.

As with the nickel catalyst, high lactitol yields (>98%) can be achieved with small amounts of by-products. Even though catalyst deactivation was severe during consecutive lactose hydrogenation, deactivation over Ru/C was less prominent than for the sponge nickel. This behavior could be explained by the better ability of Ru/C to hydrogenate lactobionic acid.

Kuusisto *et al.*[31] investigated the kinetics of lactose hydrogenation using Ru/C. The experiments were performed in a semi-batch slurry autoclave reactor at 40–60 bar of hydrogen pressure and 110–130 °C. The main hydrogenation product was lactitol with selectivity varying between 96.5 and 98.5%, while small amounts of lactulose, lactulitol, sorbitol, galactitol and lactobionic acid were detected as by-products. The selectivity improved as the hydrogen pressure increased and the reaction temperature decreased within the studied experimental range.[31]

The kinetic data were modeled based on a Langmuir–Hinshelwood mechanism. The amounts of by-products in the liquid phase and thus also on the catalyst surface were minor and the main reaction turned out to be of first order with respect to lactose. In addition, it is known that the adsorption affinity of sugar alcohols is much less than that of sugar aldehydes. Furthermore, the product desorption step was excluded and the adsorption constants for hydrogen and lactose were presumed to be independent of temperature.

The parameter estimation results indicated that it is not only the main reaction (lactitol formation) that can be described by the model, but the side reactions (formation of by-products) can also be described reasonably well (Figure 3.21).

Several Ru based catalysts with Ru incorporated in the pores of hyper-crosslinked polystyrene modified with amino groups and their catalytic properties in the lactose hydrogenation were described.[80] Kinetic analysis performed[80] included also values of metal dispersion in the calculations.

More recently, Pd/h-BN materials have been synthesized in order to find an appropriate catalyst for selective hydrogenation of lactose into lactitol.[81] The catalysts were prepared by impregnation and characterized by a range of physico–chemical methods. Pd/α-Al$_2$O$_3$ and Pd/γ-Al$_2$O$_3$ were also prepared for comparison.

3.6   Sucrose or Fructose to D-Mannitol and D-Sorbitol Hydrogenation

Sucrose is a disaccharide, which gives, by hydrolysis, glucose and fructose. Hydrogenation of the monosaccharides produces both sorbitol and mannitol.[46] Mannitol has a higher commercial value than sorbitol due to its properties as a sweetener for the food industry, and application in diet and diabetics products.[82] The industrial production of sorbitol and mannitol comes from the catalytic hydrogenation of glucose and/or fructose but higher yields of mannitol are obtained when syrups with high fructose contents or pure fructose are used.[32]

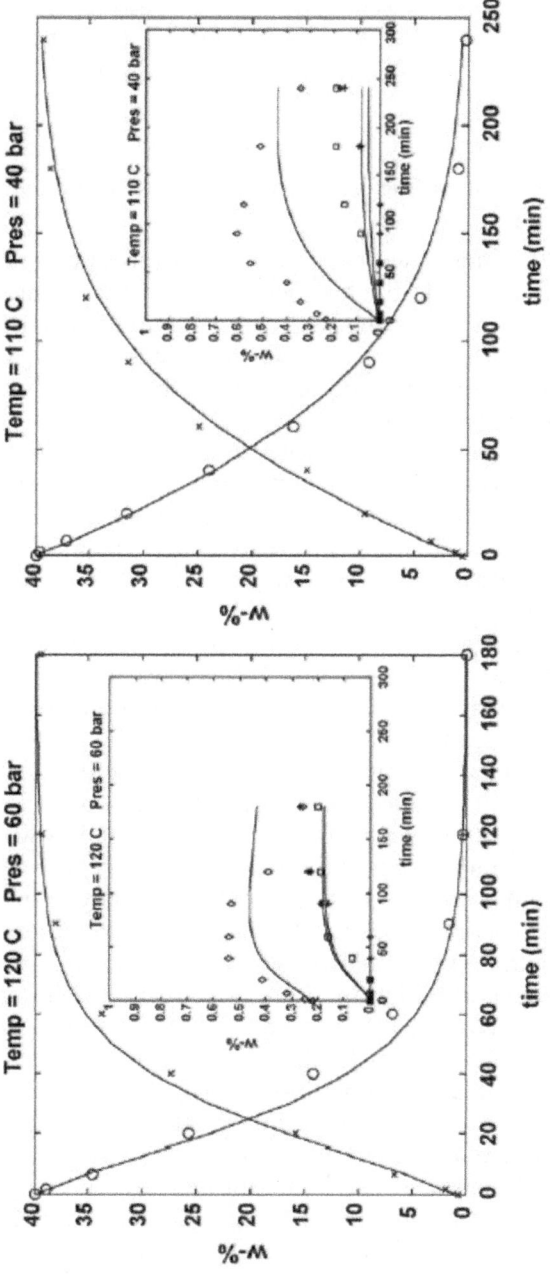

Figure 3.21 Model fit to some lactose hydrogenation experiments: lactose (_), lactitol (×), lactobionic acid (♦), lactulitol (_), galactitol (+) and sorbitol (*).
Reprinted with permission from ref. 31. Copyright Elsevier.

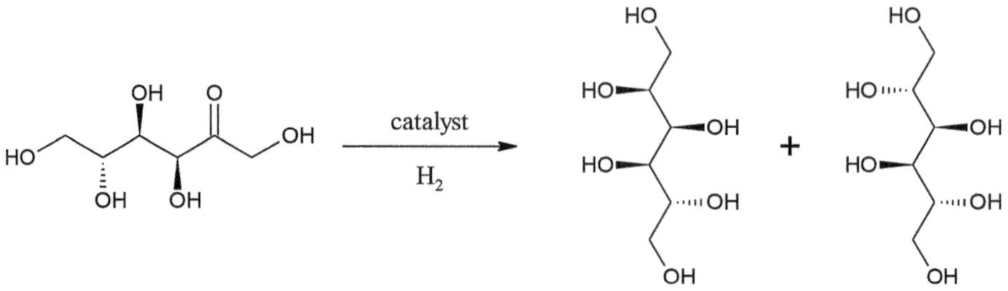

Figure 3.22   Hydrogenation of fructose giving sorbitol and mannitol as products.

Raney nickel is the most used catalyst in this process due to its high activity and low cost[46] but, as in the hydrogenation of other sugars, utilization of ruthenium based catalysts is another way to obtain high activities without leaching under the conditions required for the reaction.[46,83,84] Mannitol can also be produced by chemical and biological methods, which suffer from a low volumetric productivity.[85,86]

Mannitol is a hexavalent sugar alcohol widely distributed in nature, found in olive trees, plane trees, fruits, and vegetables (*e.g.*, strawberry, pumpkin, celery, and onion). Being a low caloric sweetener, non-toxic and non-hygroscopic, it has a low chemical reactivity and is extensively used in the food and pharmaceutical industries.[87] Figure 3.22 displays a fructose hydrogenation reaction.[87]

As in other sugar hydrogenation reactions, the most often studied systems involve nickel catalysts. Hydrogenation of fructose over classical nickel-based catalysts gives mannitol yields between 48 and 50 wt% with sorbitol being the other main product.[32,88]

Heinen *et al.*[82] studied the hydrogenation of different fructose mixtures over Ru/C. The reaction produced a mixture of mannitol and sorbitol in a weight ratio of about 40 : 60 being 100% selective to these two alditols. By changing the initial composition of the fructose mixture, the authors found that there was a competition for adsorption on the catalyst surface between D-fructose and the reaction products. In the same study, the effect of promoters on conversion and selectivity was investigated, suggesting that tin is a good promoter to achieve high mannitol selectivity. Leaching of Ru from the surface was merely 0.03%. It should be kept in mind that the main drawback for Ni-based catalyst is leaching.

Toukoniitty *et al.*[89] tested copper-based catalysts exhibiting higher selectivity to mannitol of about 65%, although such copper-based catalysts were much less active than the nickel-based counterparts.

In the study of Kuusisto *et al.*,[32] the authors determined the reaction orders in catalytic hydrogenation of D-fructose over CuO-ZnO catalyst. Experiments were carried out in a batch reactor operating at 35–65 bar and 90–130 °C. The effect of some reaction parameters, such as stirring rate, reduction temperature, catalyst loading, sugar concentration, type of the solvent and reaction temperature, were tested. Conversion and selectivity values were not affected by changing the stirrer speed from 1200 to 1800 rpm. Elevation of the catalyst reduction temperature from 220 to 300 °C showed higher activity as expected. A second order reaction with respect to hydrogen pressure was determined[32] at 110 °C.

In the kinetic study of Kuusisto *et al.*[32] for hydrogenation of fructose, a competitive adsorption model, in which all the species occupy the same type of sites, was applied. This model predicted the same reaction order for both mannitol and sorbitol, according to the observed product distribution. The adsorption of hydrogen was assumed to be dissociative even though the hydrogen atoms were presumed to be added pair-wise to the carbonyl group. Figure 3.23 shows that the kinetic model was in good correspondence with the kinetic data.

Amorphous alloy catalysts of NiB and CoB types exhibited better activity, selectivity and resistance to sulfur poisoning than in general nickel and cobalt catalysts.[90–93] Bimetallic amorphous CoNiB and polymer-stabilized CoNiB catalysts displayed activity higher than NiB, CoB and Raney nickel, even though the selectivity of mannitol improved slightly.[94]

Liaw *et al.*[94] studied the effect of temperature (ranging from 60 to 100 °C), pressure (from 20 to 60 bar) and initial concentration (from 20 to 50 wt%) on the reaction rate. The authors found that the reaction rate had a maximum at 80 °C and observed a first-order reaction with respect to hydrogen pressure. At 70 °C and 40 bar, the initial rates decreased when increasing the initial concentration of the sugar from 20 wt% to 50 wt%, indicating a negative order dependence on fructose concentration.

Fructose hydrogenation kinetics has been studied, for example, by Maranhão *et al.*,[95] who accomplished kinetic studies with a nickel catalyst supported in carbon using sucrose, glucose and fructose solutions. Based on experimental results, the authors proposed a Langmuir–Hinshelwood model to represent the kinetics of the process, assuming occurrence of acid (metal) and basic (support) sites on the catalyst surface. Saccharide adsorption on acid sites through the C=O bond was suggested, along with dissociative hydrogen adsorption on basic sites with a subsequent reaction.

Heinen[82] also proposed a Langmuir–Hinshelwood model to represent the hydrogenation kinetics over a Ru/C catalyst.

In another work, Barbosa *et al.*[96] evaluated the kinetics of sucrose hydrogenation over ruthenium containing Y zeolites. The Langmuir–Hinshelwood approach was assumed and the reaction parameters were determined, obtaining a pseudo first-order kinetic model where the hydrolysis to glucose and fructose was favored by the zeolite catalyst.

In a later study, Castoldi *et al.*[97] evaluated sucrose hydrogenation kinetics over Ni/Raney and Ru/Al$_2$O$_3$ catalysts using a hybrid autocatalytic kinetic model to simulate the reaction and account for experiment observations. In this model, the authors considered that glucose as well as fructose could generate sorbitol and mannitol and the products could be transformed into isomers in irreversible reactions (Figure 3.24).

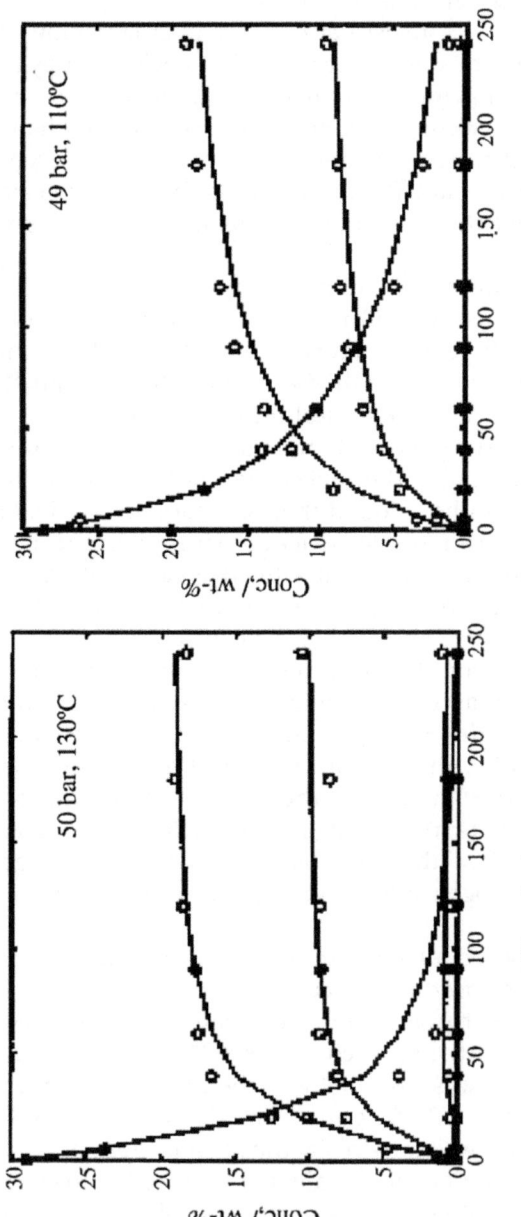

Figure 3.23   Kinetic curves of fructose hydrogenation experiments at different temperature and pressure values (○, experimental values; —, model fit).
Reprinted with permission from ref. 32. Copyright Elsevier.

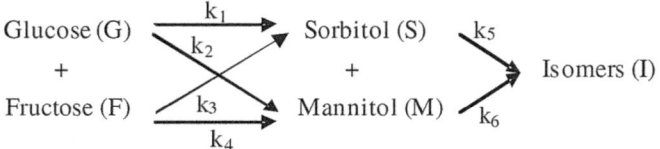

Figure 3.24 Reaction kinetic network considered in the modeling.
Reprinted with permission from ref. 97. Copyright Springer.

The model was able to describe a maximum in sorbitol and mannitol concentrations observed experimentally for Ru/Al_2O_3 (Figure 3.25).

3.7 Hydrogenation of Xylose to Xylitol

3.7.1 Catalyst Screening

Xylose is a hydrolysis product of xylan, hemicellulose, present in, for example, birch and reed. Hydrolysis is performed under acidic conditions to obtain the corresponding monosaccharide, xylose, whose carbonyl group is subsequently hydrogenated, leading to xylitol in water and aqueous alcohols as solvents. Quite recently, xylitol has become a popular sweetening agent because of its high sweetening capacity and anti-caries property.

The main characteristics of xylitol are that it has high solubility in water. Xylitol has sweetness equal to that of table sugar (sucrose), but with 40% fewer calories and no insulin requirements.[98] These properties are the reason that xylitol is even used in pharmaceuticals, cosmetics, synthetic resin and alimentary industries.[68]

Even though a lot of effort has been put towards the microbial production and metabolic engineering of xylitol recently, synthesis from xylose to xylitol is industrially carried out as a catalytic hydrogenation with Raney nickel catalyst.[28,30–32] The traditional approach is based on the analysis of the main components only, although it is known that xylulose, arabinitol, furfural and xylonic acid can appear as by-products (Figure 3.26). Purification of xylose, usually carried out by means of ion-exchange, filtering and crystallization, is essential for the success of the subsequent step, *i.e.* the hydrogenation of xylose to xylitol, since the most commonly used catalyst, Raney nickel, is easily poisoned and deactivated by the impurities in the starting material.[27,30,99,100]

Hydrogenation of xylose is economically feasible at elevated temperatures and pressures: a high enough temperature (80–130 °C) is required to achieve a sufficient reaction rate, however, too high a temperature leads to the formation of undesired by-products. Since the solubility of hydrogen in the aqueous phase is rather low, elevated pressures are needed (generally 40 bar or higher).

Hydrogenation of xylose on nickel is selective and the pH should be maintained at neutral, or moderately acidic. The by-products of the

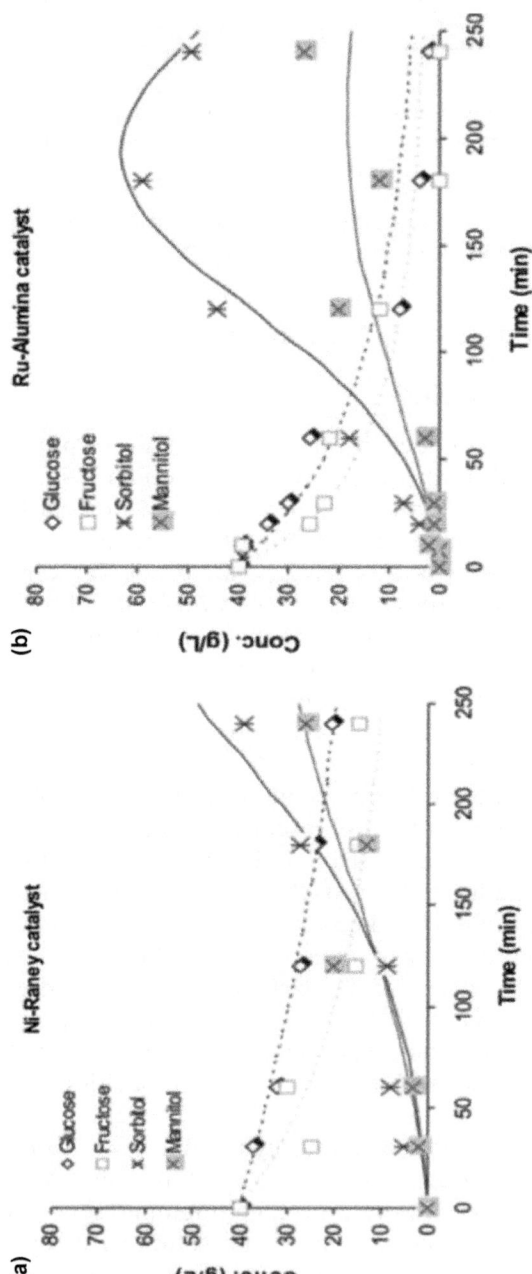

Figure 3.25   Correlation between the system of hybrid kinetic models and the experimental data from the hydrogenation with Ni-Raney (a) and Ru/Al$_2$O$_3$ (b). Reprinted with permission from ref. 97. Copyright Springer.

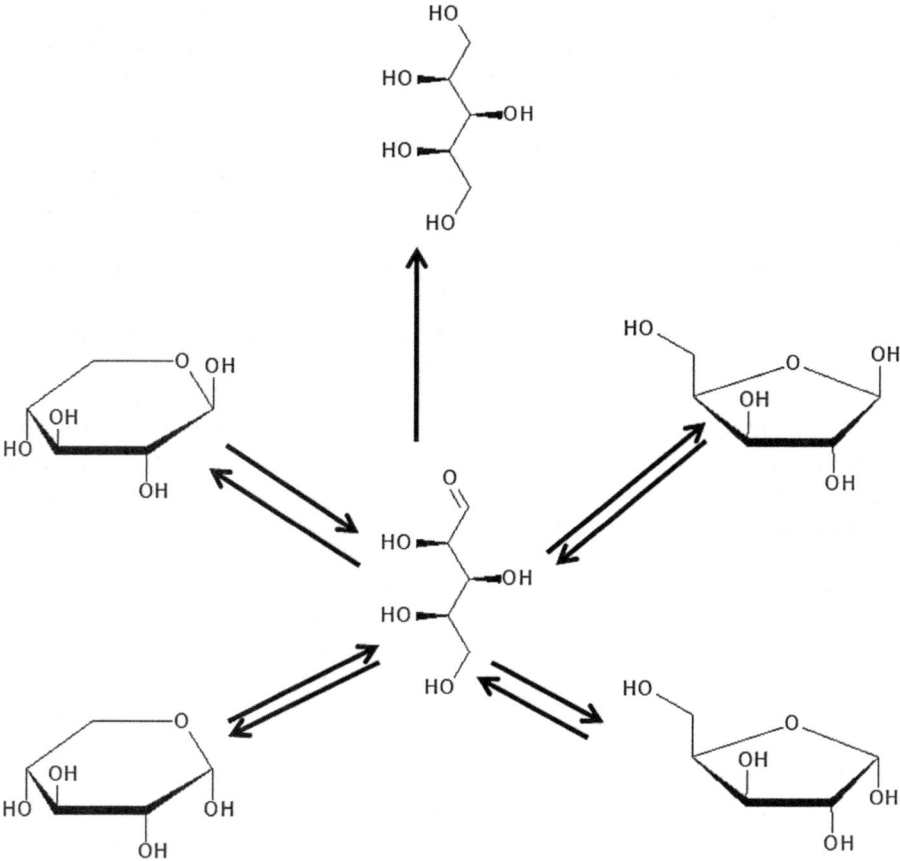

Figure 3.26 Anomeric equilibria of xylose and by-products.

hydrogenation are D-xylulose, D-arabinitol, furfural and D-xylonic acid. Formation of by-products, such as D-xylulose and furfural, is favored by a shortage of hydrogen on the catalyst surface.

This reaction has also been studied with mild conditions: temperatures between 20–70 °C and reaction pressures of 5–300 bar using ruthenium supported on zirconia as a catalyst. The main drawbacks are the by-products, wastes generated and long reaction times.[101]

The solubility of hydrogen in water is quite limited, but the hydrogenation rate can be enhanced by using alternative solvents. Unfortunately, the solubility of xylose in many non-polar solvents, such as ethanol, is very low, restricting the usability of organic solvents. Therefore, it is reasonable to expect that a mixture of water and a non-poisonous organic solvent, with a better solubility of hydrogen, enhances the hydrogenation rate. Furthermore, the reaction system is complicated by the fact that xylose mutarotation takes place upon dissolution in water. Two types of both furanose (five-ring internal ether) and pyranose (six-ring internal ether) as well as very small amounts of acyclic aldehyde are formed, co-existing in equilibria. It is generally believed that the acyclic carbonyl form is the active one in hydrogenation. Nevertheless, the ring forms strongly dominate in aqueous solutions at 30 °C.[30]

Leikin *et al.*[102] hydrogenated xylose with 4–80% Raney nickel and found that pH dropped sharply during the first 30 min, except when the amount of catalyst was 8%. Similar but less pronounced results were obtained when using buffer solutions.

Lately Mikkola *et al.*[103] have shown that the formation of by-products is primarily influenced by such factors as temperature, pH of the reaction media and hydrogen mass transfer. The higher the temperature, the more prominent the formation of by-products. Alkaline conditions facilitate generation of D-xylonic acid *via* the Cannizzaro reaction. It has been observed that xylulose was formed at the beginning of the experiments and, thereafter, hydrogenated to xylitol and D-arabinitol. In the experiments carried out below 90 °C, neither xylulose nor D-arabinitol were formed. The formation of by-products is also suppressed by high hydrogen pressures.

A patent[101] disclosed a process for the hydrogenation of sugars, focusing on xylose, by using a catalyst in which ruthenium is supported on zirconia with a metal dispersion of 10% or higher and in which the chlorine content is less than 100 ppm. The conditions are mild; temperatures between 20–70 °C and reaction pressures of 5–300 bar were used in a fixed-bed tubular reactor achieving a purity of 99.5% or more without a complicated separation process. The major drawbacks are the formation of side-products, generation of waste and long reaction times (up to 100–1000 hours). For the catalyst preparation, ruthenium chloride is preferred in comparison to ruthenium nitrate, ruthenium nitrosyl nitrate, and ruthenium acetylacetonate.

Zeolite Y supported ruthenium nanoparticle catalysts prepared by impregnation were evaluated in the hydrogenation of xylose to xylitol.[104] The reaction conditions were optimized to achieve the maximum conversion of xylose and selectivity to xylitol.

The activity of the Ru catalyst on a new class of NiO modified TiO_2 support, Ru/(NiO-TiO_2), was studied in the liquid phase catalytic hydrogenation of xylose to xylitol.[105] The TiO_2 support was modified by impregnation with a nickel chloride precursor and subsequent oxidation.

3.7.2 Engineering Aspects

Although hydrogenation kinetics of monosaccharides has been studied extensively, only a few investigations have been published concerning the hydrogenation of xylose.

Wisniak *et al.*[43] proposed that for Raney nickel catalyzed hydrogenation, at 100 °C the reaction follows a pseudo-first-order with a surface-reaction-controlling step between atomically adsorbed hydrogen and adsorbed xylose while xylose reacts directly from the liquid phase at higher temperatures.

A later study by Mikkola *et al.*[53] revealed that the reaction order with respect to xylose is lower than unity, which indicates a retarding effect of xylose adsorption on the catalyst surface.

Mikkola *et al.*[99] developed a semi-competitive model for hydrogenation taking into account that large-sized organic molecules occupy more space on the surface than hydrogen. In this model, adsorption and desorption steps were assumed to be rapid, whereas the irreversible hydrogenation steps on the surface were presumed to be rate controlling. Hydrogen adsorption was assumed to be dissociative, even if hydrogen atoms were supposed to be added pair-wise to the organic species.

Experiments[99] revealed that the product distribution was not influenced by the catalyst deactivation. Thus, all of the reactions were affected in a similar way and the catalyst deactivation kinetics was described by a semi-empirical model disappearance, assuming a loss of active sites because of promoter leaching and surface restructuring.

3.8 Hydrogenation of Sugar Mixtures

To study the influence of the sugar molar ratios on the hydrogenation kinetics of both sugars (L-arabinose and D-galactose) in the mixture, several experiments were carried out[106] at different molar ratios of D-galactose to L-arabinose (0.1–10), the average ratio of galactose-to-arabinose units in arabinogalactan (hemicellulose) being about 5 : 1.[107] Typically, very high yields (close to 100%) and selectivity (95–99%) were achieved, the products being arabitol and galactitol. No by-products were detected.

Several experiments were performed[106] to study the effect of hydrogen pressure on the hydrogenation kinetics of mixtures under operating conditions of 40, 50 and 60 bar and 105 °C, and at two different molar ratios. It was found from these experiments that the influence of hydrogen pressure on the reaction rate was insignificant, *i.e.* the kinetics was close to zero order with respect to the hydrogen pressure.

In the case of D-galactose, hydrogenation in the mixture did not exhibit any unexpected behavior: the higher the concentration of galactose in the mixture, the faster the reaction proceeded with no important difference between the limiting case of pure sugar and the higher ratios. The experimental behavior can be explained by a competitive adsorption effect: the presence of a competing sugar retards the hydrogenation rate of the other sugar.

However, L-arabinose exhibited an acceleration of the rate with an increased amount of D-galactose starting from 0.1 and stabilizing at ratio 1 (ratios 2 and 3.6 presented a small variation). The most remarkable observation is that at ratio 5 (*i.e.*, the concentration of D-galactose being 5 times that of L-arabinose), the reaction proceeded even faster than the limiting case of pure arabinose. This effect was very clear as confirmed by Figure 3.27.

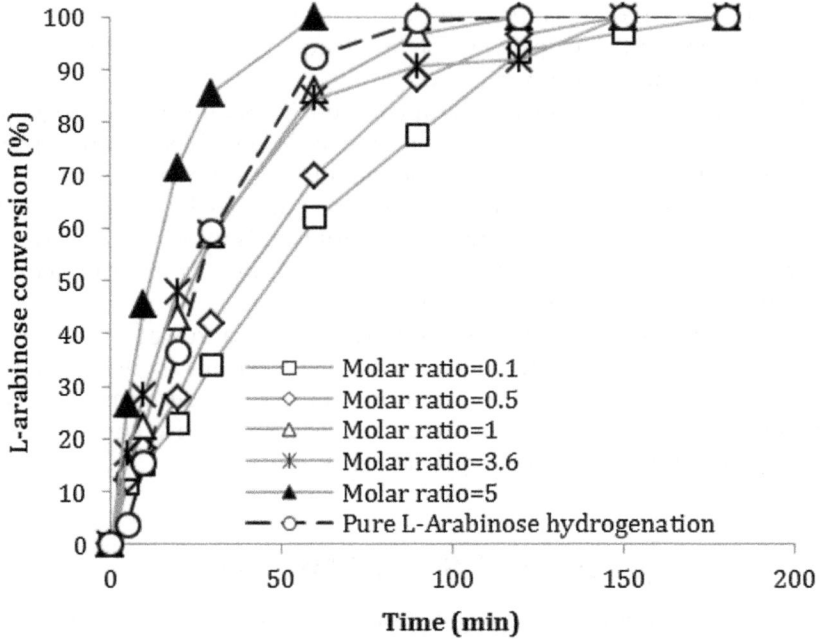

Figure 3.27 Effect of the molar ratio on the hydrogenation of L-arabinose in the presence of L-arabinose at 120 °C and 40 bar.

3.9   Final Remarks

The heterogeneous catalytic hydrogenation of various sugars over a range of catalysts was described. Historically, sponge nickel catalysts were applied mainly, although more recently, ruthenium based catalysts have emerged as a feasible alternative. Various aspects of hydrogenations, such as catalyst selection, reaction kinetics, influence of the sugar structure, structure sensitivity and reactor aspects were covered in this chapter, highlighting the recent advances in this fascinating field of biomass transformations.

References

1. J. Sanders, E. Scott, R. Weusthuis and H. Mooibroek, *Macromol. Biosci.*, 2007, **7**, 105.
2. Top Value Added Chemicals from Biomass Volume I—Results of Screening for Potential Candidates from Sugars and Synthesis Gas, Eds. T. Werpy, G. Petersen, NREL, 2004.
3. S. Murat, C. A. Sen, D. J. Henao, J. A. Braden, J. A. Dumesic and C. T. Maravelias, *Chem. Eng. Sci.*, 2012, **67**, 57.
4. P. Gallezot, *Chem. Soc. Rev.*, 2004, **41**, 1538.
5. R. R. Davda, J. W. Shabaker, G. W. Huber, R. D. Cortright and J. A. Dumesic, *Appl. Catal., B*, 2013, **43**, 13.

6. R. R. Davda, J. W. Shabaker, G. W. Huber, R. D. Cortright and J. A. Dumesic, *Appl. Catal., B*, 2013, **56**, 171.
7. A. V. Kirilin, A. V. Tokarev, E. V. Murzina, L. M. Kustov, J.-P. Mikkola and D. Yu Murzin, *ChemSusChem*, 2010, **3**, 708.
8. A. V. Kirilin, A. V. Tokarev, L. M. Kustov, T. Salmi, J.-P. Mikkola and D. Yu. Murzin, *Appl. Catal. A.*, 2012, **435–436**, 172.
9. Y. T. Kim, J. A. Dumesic and G. W. Huber, *J. Catal.*, 2013, **304**, 72.
10. A. V. Kirilin, A. V. Tokarev, H. Manyar, C. Hardacre, T. Salmi, J.-P. Mikkola and D. Yu. Murzin, *Catal. Today*, 2014, **223**, 97.
11. A. V. Kirilin, B. Hasse, A. V. Tokarev, L. M. Kustov, G. N. Baeva, G. O. Bragina, A. Yu. Stakheev, A.-R. Rautio, T. Salmi, B. J. M. Etzold, J.-P. Mikkola and D. Yu. Murzin, *Catal. Sci. Technol.*, 2014, **4**, 387.
12. A. V. Kirilin, J. Wärnå, A. V. Tokarev and D. Yu. Murzin, *Ind. Eng. Chem. Res*, 2014, **53**, 4580.
13. M. Sinnott, *Carbohydrate Chemistry and Biochemistry: Structure and Mechanism*, 2008, RSC.
14. P. Collins and R. Ferrier, *Monosaccharides: Their Chemistry and Their Roles in Natural Products*, Chichester; New York, Wiley & Sons, 1995, p. 574.
15. A. Corma, S. Iborra and A. Velty, *Chem. Rev.*, 2007, **107**, 2411.
16. M. L. Cunningham and C. B. Walker, *U. S. Pat.* 20040222004.
17. M. Niimi, Y. Hario, Y. Ishii, K. Kataura and K. Kato, *Jp. Pat.* 02042991990.
18. G. Darsow, *EP Pat.* 4235251991.
19. M. Magara, K. Shimazu, M. Fuse, K. Kataura, J. Osada, K. Kato and K. Yoritomi, *Jp. Pat.* 010935971989.
20. J. Kondo, T. Miyamoto and M. Asano, *Jp. Pat.* 53119811978.
21. A. Maisin, A. Lefèvre, A. Germain and M. Wauters, *BE Pat.* 8822791980.
22. V. Benessere, R. Del Litto, A. De Roma and F. Ruffo, *Coord. Chem. Rev.*, 2010, **254**, 390.
23. G. Helmchen and C. Murmann, *EP Pat.* 8858971998.
24. V. B. Duffy and M. Sigman-Grant, *J. Am. Diet. Assoc.*, 2004, **104**, 255.
25. http://www.prnewswire.com/news-releases/global-sorbitol-market-is-expected-to-reach-usd-3-billion-in-2018-transparency-market-research-191646781.html, access 5.6.2015.
26. J. Wisniak and R. Simon, *Ind. Eng. Chem. Prod. Res. Dev.*, 1979, **18**, 50.
27. J. Wisniak, M. Hershkowitz, R. Leibowitz and S. Stein, *Ind. Eng. Chem. Prod. Res. Dev*, 1974, **13**, 75.
28. J. Kuusisto, J.-P. Mikkola, M. Sparv, J. Wärna, H. Heikkilä, R. Perälä, J. Väyrynen and T. Salmi, *Ind. Eng. Chem. Res.*, 2006, **45**, 5900.
29. E. Crezee, B. W. Hoffer, R. J. Berger, M. Makkee, F. Kapteijn and J. A. Moulijn, *Appl. Catal., A*, 2003, **251**, 1.
30. J.-P. Mikkola, T. Salmi and R. Sjöholm, *J. Chem. Technol. Biotechnol*, 2001, **76**, 90.

31. J. Kuusisto, J.-P. Mikkola, M. Sparv, J. Wärnå, H. Karhu and T. Salmi, *Chem. Eng. J.*, 2008, **139**, 69.

32. J. Kuusisto, J.-P. Mikkola, P. P. Casal, H. Karhu, J. Vayrynen and T. Salmi, *Chem. Eng. J.*, 2005, **115**, 93.

33. B. Chen, U. Dingerdissen, J. G. E. Krauter, H. Lansinkrotgerink, K. Mobus, D. J. Ostgard, P. Panster, T. H. Riermeier, S. Seebald, T. Tacke, H. Rotgerink and H. Trauthwein, *Appl. Catal., A*, 2005, **280**, 17.

34. N. Déchamp, A. Gamez, A. Perrard and P. Gallezot, *Catal. Today*, 1995, **24**, 29.

35. T. Salmi, J. Kuusisto, J. Wärnå and J.-P. Mikkola, Chim l'Industria (Milan), 2006, **88**, 90.

36. T. O. Salmi, D. Y. Murzin, J. P. Wärnå, J.-P. Mikkola, J. E. B. Aumo and J. Kuusisto, in *Catalysis of Organic Reactions*, CRC Press, ed. S. Schmidt, 2006, vol. 115, pp. 187–196.

37. P. Gallezot, P. J. Cerino, B. Blanc, G. Flèche and P. Fuertes, *J. Catal.*, 1994, **146**, 93.

38. B. J. Arena, *Appl. Catal., A*, 1992, **87**, 219.

39. N. Déchamp, A. Gamez, A. Perrad and P. Gazellot, *Catal. Today*, 1995, **24**, 29.

40. M. C. Nunes, C. A. Perez and M. Schemal, *Appl. Catal., A*, 2004, **264**, 111.

41. P. Gazellot, N. Nicolaus, G. Flèche, P. Fuertes and A. Perrand, *J. Catal.*, 1998, **180**, 51.

42. H. Li, H. Li and J. F. Deng, *Catal. Today*, 2002, **74**, 53.

43. J. Wisniak, M. Hershkowitz and S. Stein, *Ind. Eng. Chem. Prod. Res. Dev*, 1974, **13**, 232.

44. J.-P. Mikkola, H. Vainio, T. Salmi, R. Sjöholm, T. Ollonqvist and J. Väyrynen, *Appl. Catal,. A*, 2000, **196**, 143.

45. H. Li, W. Wang and J.-F. Deng, *J. Catal.*, 2000, **191**, 257.

46. B. W. Hoffer, E. Crezee, F. Devred, P. R. M. Mooijman, W. G. Sloof, P. J. Kooyman, A. D. van Langeveld, F. Kapteijn and J. A. Moulijn, *Appl. Catal., A*, 2003, **253**, 437.

47. P. J. Cerino, G. Flèche, P. Gallezot and J. P. Salome, in *Heterogeneous Catalysis and Fine Chemicals II*, ed. M. Guisnet, J. Barrault, C. Bouchoule, Elsevier, Amsterdam, 1991, p. 231.

48. R. Albert, A. Strätz and G. Vollheim, *Chem. Eng. Technol.*, 1980, **52**, 582.

49. T. Koscielski, J. M. Bonnier, J. P. Damon and J. Masson, *Appl. Catal.*, 1989, **49**, 91.

50. F. B. Bizhanov, D. V. Sokol'skii, N. I. Popov, N. Y. Malkhina and A. M. Khisametdinov, *Kinet. Catal.*, 1965, **10**, 655.

51. P. Fouilloux, *Appl. Catal*, 1983, **8**, 1.

52. B. Kusserow, S. Schimpf and P. Claus, *Adv. Synth. Catal.*, 2003, **345**, 289.

53. J.-P. Mikkola, H. Vainio, T. Salmi, R. Sjoholm, T. Ollonqvist and J. Vayrynen, *Appl. Catal. A*, 2000, **196**, 143.

54. K. van Gorp, E. Boerman, C. V. Cavenaghi and P. H. Berben, *Catal. Today*, 1999, **52**, 349.

55. B. Kusserow, S. Schimpf and P. Claus, *Adv. Synth. Catal.*, 2003, **345**, 289.
56. R. Geyer, P. Kraak, A. Pachulski and R. Schoedel, *Chem. Ing. Tech.*, 012, **84**, 513.
57. D. K. Mishra, A. A. Dabbawala, J. J. Park, S. H. Jhung and J.-S. Hwang, *Catal. Today*, 2014, **232**, 99.
58. V. N. Sapunov, M. Ye. Grigoryev, E. M. Sulman, M. B. Konyaeva and V. G. Matveeva, *J. Phys. Chem. A*, 2013, **117**, 4073.
59. T. Kilpio, A. Aho, D. Murzin and T. Salmi, *Ind. Eng. Chem. Res.*, 2013, **52**, 7690.
60. A. Aho, S. Roggan, K. Eränen, T. Salmi and D. Yu. Murzin, *Catal. Sci. Technol.*, 2015, **5**, 953.
61. J. Pan, J. Li, C. Wang and Z. Yang, *React. Kinet. Catal. Lett.*, 2007, **90**, 233.
62. H. Hofmann and W. Bill, *Chem. Ing. Tech.*, 1959, **31**, 81.
63. P. H. Brahme and L. K. Doraiswamy, *Ind. Eng. Chem. Process Des. Dev.*, 1976, **15**, 130.
64. F. Turek, R. K. Chakrabarti, R. Lange, R. Geike and W. Flock, *Chem. Eng. Sci.*, 1983, **38**, 275.
65. V. Tukac, *Collect. Czech. Chem. Commun.*, 1997, **62**, 1423.
66. P. A. Lazaridis, S. Karakoulia, A. Delimitis, S. M. Coman, V. I. Parvulescu and K. S. Triantafyllidis, D-glucose hydrogenation/hydrogenolysis reactions on noble metal (Ru, Pt)/activated carbon supported catalysts, *Catal. Today*, DOI: 10.1016/j.cattod.2014.12.006.
67. A. Aho, S. Roggan, O. Simakova, T. Salmi and D. Yu. Murzin, *Catal. Today*, 2015, **241**, 195.
68. K. Arve, V. Sifontes, K. Eränen, T. Salmi and D. Yu. Murzin, unpublished results.
69. H. Li, P. Yang, D. Chu and H. Li, *Appl. Catal., A*, 2007, **325**, 34.
70. Y. Wang, L. Xu, L. Xu, H. Li and H. Li, *Chin. J. Catal.*, 2013, **34**, 1027.
71. I. Toufeili and S. Dziedzic, *Food Chem.*, 1993, **47**, 17.
72. C. Zhao, H.-z. Wang, N. Yan, C.-x. Xiao, X.-d. Mu, P. J. Dyson and Y. Kou, *J. Catal.*, 2007, **250**, 33.
73. E. M. Sulman, M. E. Grigorev, V. Yu. Doluda, J. Wärnå, V. G. Matveeva, T. Salmi and D. Yu. Murzin, Maltose hydrogenation over ruthenium nanoparticles impregnated in hypercrosslinked polystyrene, *Chem. Eng. J.*, in press, DOI: 10.1016/j.cej.2015.04.002
74. V. A. Sifontes, D. Rivero, J. P. Wärnå, J.-P. Mikkola and T. Salmi, *Top. Catal.*, 2010, **53**, 1278.
75. V. H. Holsinger, in *Fundamentals of Dairy Chemistry*, ed. N. B. Wong, Van Nostrand Reinhold Company, New York, 3rd edn, 1988, p. 297.
76. P. Linko, Lactose and lactitol, in *Nutritive Sweeteners*, ed. G. Birch and K. Parker, Applied Science Publishers, New Jersey, 1982, p. 109.
77. M. Hu, M. J. Kurth, Y.-L. Hsieh and J. M. Krochta, *J. Agric. Food Chem.*, 1996, **44**, 3757.

78. A. Abbadi, K. F. Gotlieb, J. B. M. Meiberg and H. van Bekkum, *Green Chem.*, 2003, **5**, 47.

79. H. Berthelsen, K. Eriknauer, K. Bottcher, H. J. S. Christensen, P. Stougaard, O. C. Hansen and F. Jorgensen, WO2003008617, 2003.

80. V. Yu. Doluda, J. Warna, A. Aho, A. V. Bykov, A. I. Sidorov, E. M. Sulman, L. M. Bronstein, T. Salmi and D. Yu. Murzin, *Ind. Eng. Chem. Res.*, 2013, **52**, 14066.

81. N. Meyer, M. Devillers and S. Hermans, *Catal. Today*, 2015, **241B**, 200.

82. A. Heinen, J. Peters and H. van Bekkum, *Carbohydr. Res.*, 2000, **328**, 449.

83. S. Schimpf, C. Louis and P. Claus, *Appl. Catal., A*, 2007, **318**, 45.

84. B. W. Hoffer, E. Crezee, P. R. M. Mooijman, A. D. van Langeveld, F. Kapteijn and J. A. Moulijn, *Catal. Today*, 2003, **79–80**, 35.

85. N. von Weymarn, K. Kiviharju, S. T. Jääskeläinen and M. Leisola, *Biotechnol. Prog*, 2003, **19**, 815.

86. M. Helanto, J. Aarnikunnas, N. von Weymarn, U. Airaksinen, A. Palva and M. Leisola, *J. Biotechnol.*, 2005, **116**, 283.

87. O. Akinterinwa, R. Khankal and P.-C. Cirino, *Curr. Opin. Biotechnol*, 2008, **19**, 461.

88. B.-J. Liaw, C.-H. Chen and Y.-Z. Chen, *Chem. Eng. J.*, 2010, **157**, 140.

89. B. Toukoniitty, J. Kuusisto, J.-P. Mikkola, T. Salmi and D. Yu. Murzin, *Ind. Eng. Chem. Res.*, 2005, **44**, 9370.

90. Y. Z. Chen, B. J. Liaw and S. J. Chiang, *Appl. Catal., A*, 2005, **284**, 97.

91. H. Li, Y. Wu, H. Luo, M. Wang and Y. Xu, *J. Catal.*, 2003, **214**, 15.

92. X. Chen, H. Li, H. Luo and M. Qiao, *Appl. Catal., A*, 2002, **233**, 13.

93. H. Li, X. Chen, M. Wang and Y. Xu, *Appl. Catal., A*, 2002, **225**, 117.

94. B. J. Liaw, S. J. Chiang, C. H. Tsai and Y. Z. Chen, *Appl. Catal., A*, 2005, **284**, 239.

95. L. C. A. Maranhão, F. G. Sales, J. A. F. R. Pereira and C. A. M. Abreu, *React. Kinet. Catal. Lett.*, 2004, **81**, 169.

96. C. M. B. M. Barbosa, E. Falabella, M. J. Mendes, N. M. Lima and C. A. M. Abreu, *React. Kinet. Catal. Lett.*, 1999, **68**, 291.

97. M. C. M. Castoldi, L. D. T. Câmara and D. A. G. Aranda, *React. Kinet. Catal. Lett.*, 2009, **98**, 83.

98. T. B. Granstrom, K. Izumuri and M. Leisola, *Appl. Microbiol. Biotechnol.*, 2007, **74**, 277.

99. J.-P. Mikkola, T. Salmi and R. Sjöholm, *J. Chem. Technol. Biotechnol.*, 1999, **74**, 655.

100. J.-P. Mikkola, R. Sjöholm, T. Salmi and P. Mäki-Arvela, *Catal. Today*, 1999, **48**, 73.

101. B.-S. Kwak, B.-I. Lee, T.-Y. Kim, J.-W. Kim and S.-I. Lee, WO 2006/093364 A1 2006.

102. E. R. Leikin, *Sbor. Tr. Gos. Nauchn.-Issled. Inst. Gidrolizn. i Sul'fitno-Spirt. Prom.*, 1963, **11**, 86, Ref. Zh. Khim. Abstr. No. 24. 1963, 21.

103. J.-P. Mikkola, T. Salmi, A. Villela, H. Vainio, P. Mäki-Arvela, A. Kalantar, T. Ollonqvist, J. Väyrynen and R. Sjöholm, *Braz. J. Chem. Eng.*, 2003, **20**, 263.
104. D. K. Mishra, A. A. Dabbawala and J.-S. Hwang, *J. Mol. Catal. A: Chem*, 2013, **376**, 63.
105. M. Yadav, D. K. Mishra and J.-S. Hwang, *Appl. Catal., A*, 2012, **425-426**, 110.
106. V. Sifontes Herrera, Hydrogenation of L-arabinose, D-galactose, D-maltose and L-rhamnose, PhD thesis, 2012, Åbo Akademi University, Turku.
107. B. T. Kusema, G. Hilpmann, P. Mäki-Arvela, S. Willför, B. Holmbom, T. Salmi and D. Yu. Murzin, *Catal. Lett.*, 2010, **141**, 408.

CHAPTER 4

Advances in Sugar-based Polymers: Xylan and its Derivatives for Surface Modification of Pulp Fibres

BEATRIZ VEGA, OLGA GRIGORAY, JAN GUSTAFSSON AND PEDRO FARDIM*

Laboratory of Fibre and Cellulose Technology, Åbo Akademi University, Porthansgatan 3, FI-20500, Turku, Finland
*Email: pfardim@abo.fi

4.1 Introduction

Hemicelluloses belong to a group of non-cellulosic polysaccharides. Together with cellulose and starch, hemicelluloses are one of the three most common classes of polysaccharides found in nature. Moreover, hemicelluloses constitute one quarter to one third, exceptionally even up to half, of the weight of biomass of annual and perennial plants.[1,2] They can be divided into four different classes based on their chemical structure: (1) xylans, (2) mannans, (3) β-glucans with mixed linkages, and (4) xyloglucans.[2] Unlike cellulose, hemicelluloses have relatively low molecular weight.[3] Depending on the natural origin, polymeric chains of hemicelluloses can be branched and consist of different monomeric units, including glucose, xylose, mannose, galactose, arabinose, fucose, glucuronic acid, and galacturonic acid.[1] In a plant cell wall, hemicelluloses have a less ordered (amorphous)

RSC Green Chemistry No. 44
Biomass Sugars for Non-Fuel Applications
Edited by Dmitry Murzin and Olga Simakova

arrangement compared to cellulose due to their branched structure. Consequently, they are fairly easily prone to depolymerisation or disassembly.

As a result of the extensive resources and possibility of isolation, hemicelluloses have gained an increased academic and industrial research interest for the development of new applications. For instance, hemicelluloses are used as food additives, health and cosmetic products, thickening and strength enhancing additives, adhesives, *etc.*[3–6] The properties of the extracted hemicelluloses can be easily tailored for the desired applications. For example, esterification, etherification, grafting or crosslinking are the most common methods to enhance solubility, strength, thermoplasticity or viscosity of the hemicellulose solution.[7]

Xylan, among other hemicelluloses, are stated to be the most common non-cellulosic polysaccharide, and the second most abundant natural polymer of the plant cell wall after cellulose.[1] Consequently, xylans are now the most studied hemicelluloses to be utilised for the development of new products. In this chapter, the most efficient extraction methods, as well as potential applications of xylan-based materials for fibre-surface engineering, will be presented and discussed.

4.2 Sources and Structures of Xylan

Xylan-type hemicelluloses are found in an extensive list of plants; *e.g.*, wood (especially hardwoods, but also softwoods), agricultural crops (*e.g.*, straw, sugar cane, or corn stalks), grasses, herbs, and seaweed. Preferably, xylans can be obtained as by-products from forestry, food processing and agricultural residues.

Xylans are mostly heteropolymers with a backbone, most often consisting of β-1, 4-linked D-xylopyranose units, which are the pyranose form of pentoses. The backbone can be branched with carbohydrate chains composed of D-xylose, L-arabinose, D- or L-galactose or D-glucose units.[2] The chemical structure of these sugars is presented in this book in the chapter of Murzin *et al.* The primary structure of the different types of xylans is strongly dependent on the natural origin. Moreover, xylans are found with different macromolecular structures even within different locations of the same plant (stem, leaves and seeds).[1,4,8] The average molar mass of xylan polymers depends on the source, the method used for extraction, and also the methods of determination.

Values of 30 000–380 000 g mol^{-1} have been reported.[2] Xylans are usually classified according to their structural differences in degree of substitution and types of side groups as the following:[2,8,9]

1. Linear homoxylans are common in some seaweeds (Scheme 4.1).
2. Glucuronoxylans
   2.1. Methylglucuronoxylans have units substituted with 4-*O*-methyl-α-D-glucopyranosyl uronic acid residues at position 2 (Scheme 4.2). The xylopyranose units are partly acetylated both at positions 2 or

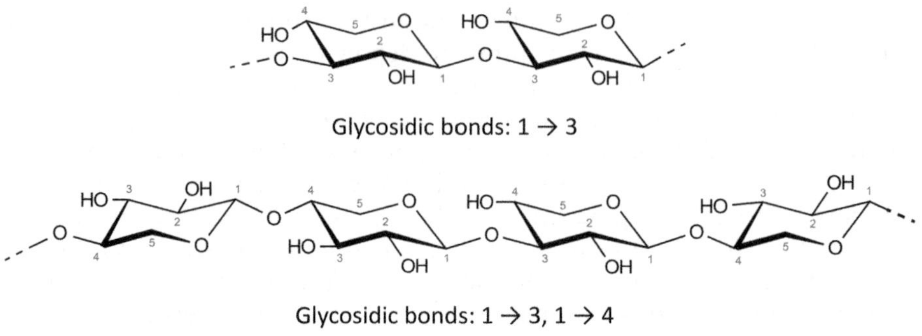

Glycosidic bonds: 1 → 3

Glycosidic bonds: 1 → 3, 1 → 4

Scheme 4.1 Primary structure of two types of homoxylans (adapted from Ebringerová *et al.*[2]).

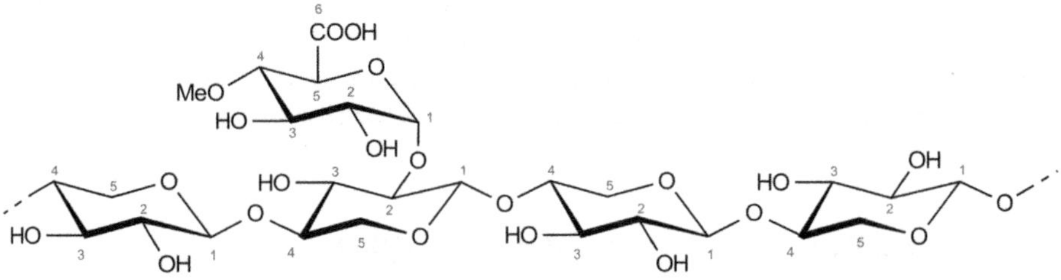

Scheme 4.2 Primary structure of 4-*O*-methyl-D-glucurono-D-xylan (adapted from Ebringerová *et al.*[2]).

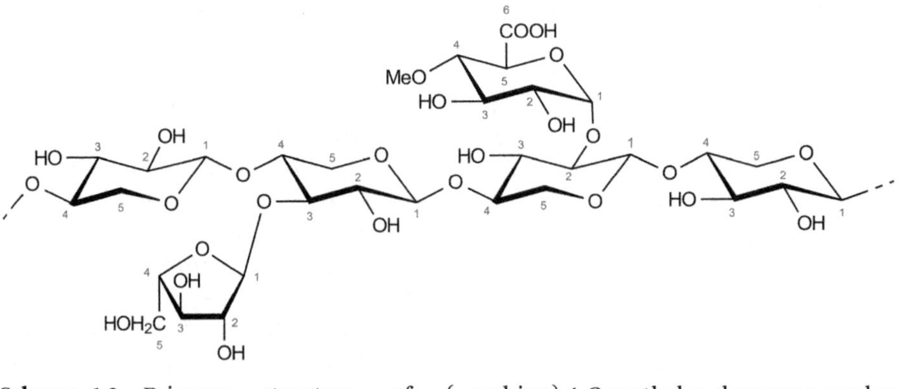

Scheme 4.3 Primary structure of (L-arabino)-4-*O*-methyl-D-glucurono-D-xylan (adapted from Ebringerová *et al.*[2]).

3 (acetyl groups are not shown in Scheme 4.2). Glucuronoxylans are the main hemicellulose component in hardwoods.

2.2. (Arabino)glucuronoxylans have units substituted with 4-*O*-methyl-α-D-glucopyranosyl uronic acid and α-L-arabinofuranosyl at positions 2 and 3, respectively (Scheme 4.3). (Arabino)glucuronoxylans are found for example in softwoods, lignified tissues of grasses, and cereals (Scheme 4.4).

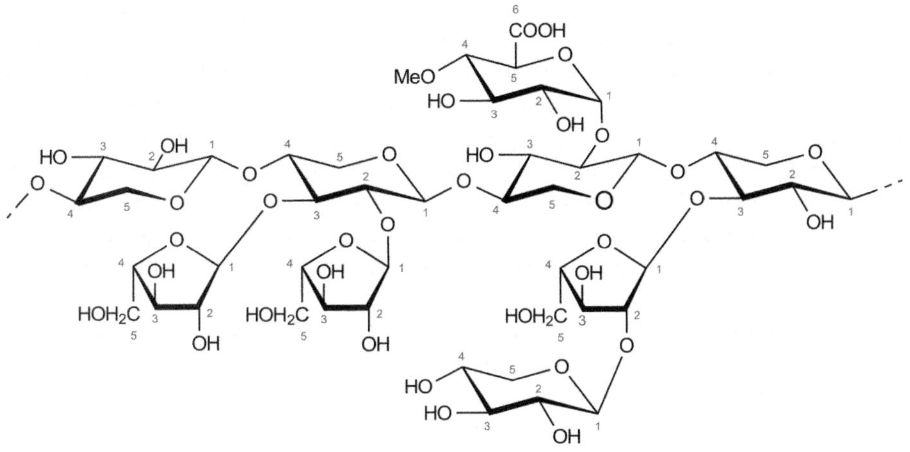

Scheme 4.4 Primary structure of water soluble xylan (adapted from Ebringerová *et al.*[2]).

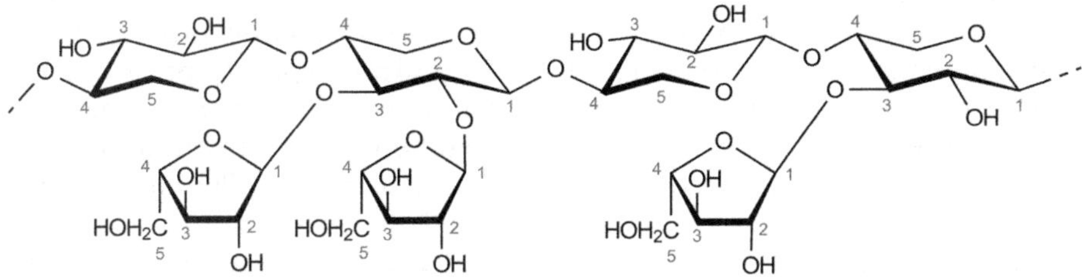

Scheme 4.5 Primary structure of water soluble L-arabino-D-xylan (adapted from Ebringerová *et al.*[2]).

3. Arabinoxylans have xylose units substituted on the backbone with arabinofuranosyl units at position 2 and/or 3 (Scheme 4.5). Moreover, phenolic acids such as ferulic or coumaric acid can be esterified to *O*-5 of some arabinofuranosyl units. Arabinoxylans are commonly found in the starchy endosperm (flour) and the outer layers of cereal grains (bran).
4. Heteroxylans are structurally more complex, with the backbone heavily substituted with various mono- or oligosaccharides side chains. They are present in cereal bran, seeds, gum exudates and mucilage.

4.3 Extraction of Xylan from Biomass and Conversion to Xylan Derivatives

The extraction of high molar mass xylan from biomass requires specific attention. Extensive degradation might occur due to the drastic extraction conditions required for breaking the strong associations between xylans and other cell wall constituents.[2] The choice of biomass type (wood or annual plants) and the optimisation of the extraction method toward sufficiently mild conditions are the keys to obtaining high molar mass xylan with a very low lignin content. On the other hand, the aim for high molar mass hemicelluloses with low contamination level may result in a moderate yield compared to those processes targeting for degradation to sugars.[1,5]

The most common industrial and laboratory scale processes for extraction of xylans are briefly described below. The reader should be aware that the list is not complete. Moreover, the terminology words "pretreatment" and "extraction" are occasionally used interchangeably for the same process in different publications.

Hydrothermal processes use hot water or steam, or a combination of both, to heat the biomass at high pressure. In hot water extraction (HWE) (also known as autohydrolysis), acetic acid from acetyl groups and uronic acids are released from the biomass to the liquid phase, where they act as hydrolysing agents. Due to technical, environmental and material purity advantages, HWE has gained a lot of attention both for extracting polymeric hemicelluloses, and for the bioethanol route.[10-17]

Alkaline processes include the use of sodium, potassium or calcium hydroxide, and also ammonia. The cold caustic extraction process (CCE), which is conducted at room temperature and atmospheric pressure, has proven to have some advantages over alkaline processes that use elevated temperatures.[10-12] The combination of alkaline extraction conditions with ultrasound, or the addition of hydrogen peroxide to the alkaline medium, seem to improve the yield and quality of the isolated xylan.[18]

Although organosolv was originally intended to be used as a pulping method, it should be stated that organosolv pretreatment may also be used to extract hemicelluloses. Rowley et al.[19] extracted xylans from corn stover by dimethyl sulfoxide (DMSO). The effect of a temperature increase from room temperature to 70 °C was investigated with the aim of increasing efficiency without deterioration of the chemical structure. DMSO was chosen as a solvent because it is one of the few solvents that allows the production of a water-soluble xylan by retaining the acetyl groups, which are present in the native state. The results showed that the chemical structure and the yield of xylans extracted from corn stover by DMSO at 70 °C were comparable to the ones extracted at room temperature with alkali (KOH or NaOH), while the extraction time was reduced by 90%.[19]

Immerzeel et al.[20] studied the extraction of water-soluble xylans from wheat bran. Microwave or autoclave heating was conducted as a pretreatment with subsequent extraction in deionised water and isolation by ethanol precipitation. The authors concluded that by using the heat treatment, the yield of water soluble xylan from wheat bran was increased, and the degree of arabinose substitution could be varied. Microwave assisted heating was superior to autoclave heating regarding the amount of extractable xylan. The yield of polymeric xylan showed a peak value of 59% with microwave heating at 185 °C for 10 min. The extraction of xylans from rice straw was investigated by Nie et al.[21] The rice straw required dewaxing by toluene–ethanol as a pretreatment. Different alcohols, all with a concentration of 60%, with addition of acid or alkali as a catalyst, were used to extract xylans from rice straw, targeting high yield and low degradation of the polymeric material. Of the investigated alcohols, the application of ethanol with an alkali catalyst was the most favourable promoting a moderate yield (18% of the total weight of the biomass, corresponding to 55% of the original hemicellulose content) of large-molecular xylan (Mw up to 48 000 g mol^{-1}). On the other hand, lignin was also more efficiently solubilised in the alkaline processes.[21]

Ionic liquids are molten organic salts with a melting point below 100 °C that can dissolve polysaccharides in large amounts. The ionic liquid 1-butyl-3-methylimidazolium chloride has been utilised for the fractionation of bagasse to cellulose, hemicellulose and lignin, where xylan was determined to be the major component of hemicelluloses.[22]

4.3.1 Extraction of Xylan from Wood Chips

As discussed before, several strategies for the extraction of xylans from biomass have been developed. In this section, efforts will be focused on addressing the extraction of high molar mass xylan from hardwood, using the so-called pressurised hot-water extraction method (PHWE). This method is probably the most commonly used method for extraction of hemicelluloses. The term "PHWE" refers to the processes of extraction that use superheated or subcritical water. Theoretically, subcritical water is defined as liquid water in equilibrium with saturated vapour. The pressure of the vapour phase is in the range of 1 bar to 217 bar, and the temperature ranges from 100 °C to 374 °C. As a system in thermodynamic equilibrium, the pressure is exerted by the vapour in equilibrium with the liquid phase of the system, at the given temperature. PHWE processes can be carried out both in flow-through or static batch modes.[23,24] The chemical properties of the xylans that are extracted in the liquid phase from the hardwood chips, obviously depend on the chemical composition of the native polymer. In addition to the characteristics of the raw material, the experimental conditions of the PHWE method and the fractionation method used for the recovery of the polymer from the extracted liquor also play an important role. Besides the extraction mode, the initial pH of the aqueous media, and temperature of the extraction, there are several parameters that have a remarkable influence on the final composition of the high molar mass xylan extracted with the PHWE method. These parameters are summarized in Table 4.1, and they will be discussed briefly herein.

4.3.1.1 Wood chips

A deep knowledge of the interactions between the natural polymers present in hardwood, and the wood's ultrastructure, is of the utmost importance when considering the isolation of xylan from this source. Hardwood is a complex anisotropic material composed of cellulose, lignin, hemicelluloses,

Table 4.1 Identification of several parameters affecting the final composition of xylan isolated with PHWE.

Category	Parameter
• Wood chips	- Botanical origin
	- Storage conditions
	- Particle size
	- Moisture content
• Impregnation stage	- Impregnation conditions
• PHWE, experimental conditions	- pH
	- Time
	- Temperature and pressure
	- Liquor to wood ratio, LWR
	- Mixing (batch mode)
• Method of separation from liquor	- Addition of an "anti-solvent" (precipitation)
	- Type of membrane (ultrafiltration)

and minor components. The cell walls of hardwood are described as consisting of a network of cellulose microfibrils embedded in an amorphous arrangement of pectin, hemicelluloses, lignin, and minor components. Generally, hardwoods contain 15–30% of acetylated 4-O-methyl glucuronoxylan, and 2–5% of glucomannan. The formulae of the hemicelluloses are represented in Scheme 4.6.[25]

The cell walls are organized in different layers with distinct levels of crystallinity and porosity.[26] The main polymers encountered in these layers are interconnected with each other through different chemical bonds (*e.g.*, hydrogen, esters, glycosidic, and ether bonds).[27] The linkages between hemicelluloses and lignin were suggested a long time ago by Erdman in 1866 (cited in Merewether, 1957[28]). The association between lignin and hemicelluloses yields so-called lignin-carbohydrate complexes (LCCs),[28–30] and the existence of these linkages represents one of the challenges in getting pure hemicelluloses from wood using PHWE methods. The xylan content among the different layers of the cell wall might be significantly different for the same type of wood. For example, in the case of young fibres of birch (*Betula verrucosa* Ehrh.), it was found that the glucuronoxylan content is higher in the secondary wall in comparison to the primary wall.[31] As Figure 4.1 shows, the xylan content reaches its maximum weight percentage in the middle layer of the secondary wall (S2). Apparently, the rigid structure of the chipped birch wood needs to be open to extract as high an amount of xylan as possible from it. The storage conditions, the moisture content of the wood, and the particle size might play a significant role in liquor penetration.[32,33] Thus, these parameters are also important for the extraction of hemicelluloses.

4.3.1.2 Impregnation stage

The impregnation of the wood chips prior to PHWE might be critical for the success of the extraction process. For instance, removal of entrapped air

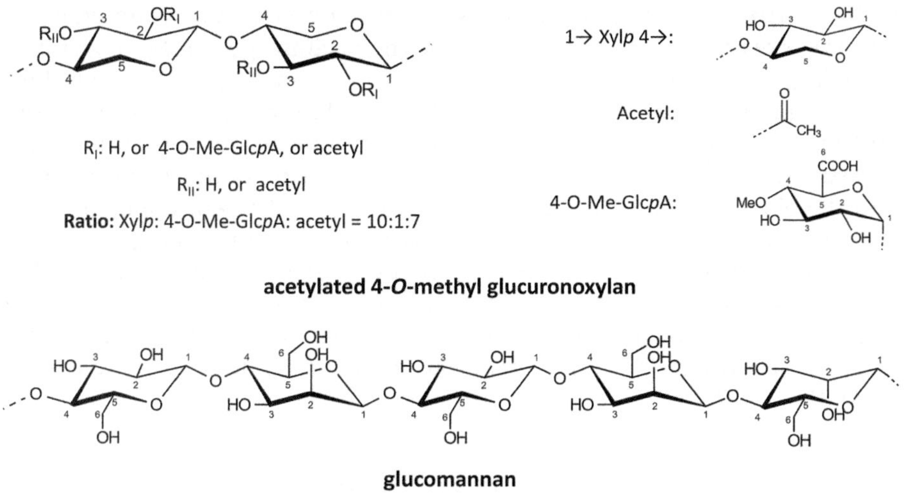

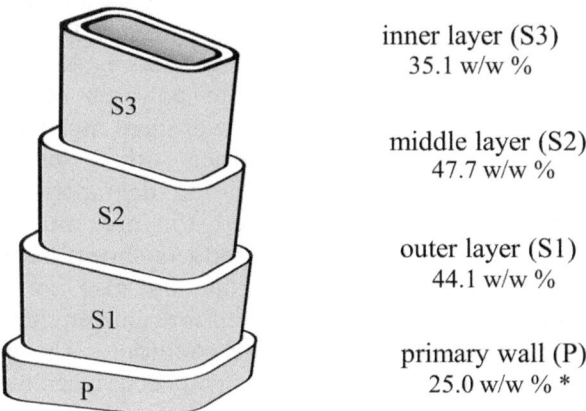

R$_I$: H, or 4-*O*-Me-Glc*p*A, or acetyl

R$_{II}$: H, or acetyl

Ratio: Xyl*p*: 4-*O*-Me-Glc*p*A: acetyl = 10:1:7

1→ Xyl*p* 4→:

Acetyl:

4-*O*-Me-Glc*p*A:

acetylated 4-*O*-methyl glucuronoxylan

glucomannan

Scheme 4.6 Primary structure of acetylated 4-*O*-methyl glucuronoxylan and gluco-mannan (adapted from Sjöström *et al.*[25]). Xyl*p*-, xylopyranose; Glc*p*-, glucopyranose; Man*p*-, mannopyranose; Me-, methyl.

inner layer (S3)
35.1 w/w %

middle layer (S2)
47.7 w/w %

outer layer (S1)
44.1 w/w %

primary wall (P)
25.0 w/w % *

Figure 4.1 A schematic illustration of the fibre cell wall layers of young fibres of birch (*Betula verrucosa* Ehrh.), with the primary wall (P) and the three layers of the secondary wall (S1–S3). The numbers on the right express the xylan content as a weight percentage of the corresponding layer/wall.[31] *As a weight percentage of the primary layer and middle lamella.

facilitates penetration of the extraction liquor, therefore, increasing the process yield.[34]

Removal of air from wood chips can be facilitated using vacuum. However, an efficient removal of air from wood chips with normal moisture content is difficult to achieve using this method. This is most probably because the capillaries from the wood cell walls are sealed by extractives. Besides that, the high surface tension that occurs at the liquid–air interfaces in the narrow capillaries of a partially saturated chip may counteract the negative pressure gradient generated by the vacuum pump. Moreover, vacuum impregnation is difficult to implement on a large scale. For these reasons, pre-steaming is often preferred as a tool for removing the entrapped air from wood chips. Pre-steaming consists of a flow of steam at atmospheric pressure (or higher pressures). The thermal expansion of the woody material and the elevated temperatures facilitate the removal of entrapped air from the wood chips. Independent of the method used for impregnation, the removal of oxygen from the extraction medium is highly recommended. Apparently, the impregnation helps not only for a better penetration of the liquor into the chips but also to prevent the undesired oxidative degradation of polysaccharides.[29,35] Other "penetration aid" techniques were discussed by Malkov *et al.*[36]

4.3.1.3   Experimental Conditions of PHWEs

Some of the properties of the extracted xylan, such as degree of polymerization (DP) and degree of acetylation (DS_{Ac}), can be tuned by selecting adequate extraction conditions. In that sense, the initial pH of the aqueous media and temperature play a critical role as these parameters have a direct control over the chemical reactions with carbohydrates. For example, the use of extreme pH values and high temperatures (above 180 °C) might lead to strong hydrolysis and degradation of the polyoses of interest.[29,37] In consequence of this, xylans with a very low average molar mass or DP are dissolved in the extraction liquor together with a lot of undesirable degradation products. Typical polysaccharide degradation products are formic acid, methanol, furfural, hydroxymethylfurfural, mucic acid, levulinic acid, humins, *etc.*[29,38] The degradation products shown in Scheme 4.7 are originated not only from the polysaccharides but also from lignin, which makes the isolation of xylan from the liquor more challenging when extreme conditions are used. Under such extreme conditions, the more extensive degradation of the wood chip components will occur with longer extraction

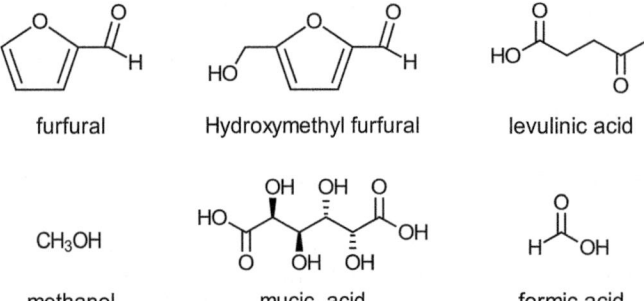

furfural	Hydroxymethyl furfural	levulinic acid

methanol	mucic acid	formic acid

Scheme 4.7   Degradation products originated at elevated temperatures (adapted from Fengel *et al.*[29]).

times. Therefore, extreme conditions with long extraction times are usually not recommended when the purpose of the extraction is to recover xylans with high DP value. On the other hand, low temperatures (below 150 °C) lead to very low extraction yields.[23,24] Therefore, the extraction conditions selected will always be a balance between all the factors mentioned here.

It is worth reminding that, besides the fact that the temperature of the system controls the kinetics of the chemical reactions of the wood components, the properties of water are significantly affected by the changes in temperature and pressure. At high temperatures and increased pressure conditions, liquid water is a highly efficient solvent for the extraction of xylan from wood chips. Under these conditions, water has a lower dielectric constant, decreased density and viscosity, and less ionization capacity compared to water at room temperature. The lower dielectric constant of subcritical water allows this solvent to dissolve less polar compounds.[39] A less viscous solvent will certainly penetrate more easily into the porous material. The lower ionization capacity will affect the strength of acids in the solution. Several properties of water at two different temperatures are compared in Table 4.2.

Common operating temperatures in PHWE can range from 150 °C to 180 °C, independently of the initial pH of the solution. In neutral or acidic conditions, two different xylan fractions can be recovered in the extraction liquor. The highly acetylated xylan fraction is dissolved at the early stage.[33] The hydrolysis of acetyl groups from the acetylated hardwood xylans, leads to the formation of acetic acid solution. The released acid drops the pH of the aqueous media, and catalyzes the hydrolysis of polysaccharides.[29] Several strategies have been intensively studied during the last decades to avoid the strong degradation of hemicelluloses caused by a high concentration of acetic acid. For example, short extraction times or the addition of salts and buffers at the beginning of the process were tested.[44] Recently, external control of the pH of the liquor during the extraction process was attempted by Krogell *et al.*[45-47] The main purpose of this work was to find the experimental conditions that allow the highest extraction yield of high molar mass hemicelluloses from softwood. The positive results from the work have shown the benefits of having precise control of the solution pH during the extraction of softwood hemicelluloses.

Apparently, the liquor-to-wood ratio (LWR) and intensity of mixing do not play a critical role for the chemical composition of the recovered material.

Table 4.2 Dielectric constant, viscosity, and ionization capacity of pure liquid water under different conditions.

	Liquid water at 20 °C	Liquid water at 160 °C	Ref.
Static dielectric constant	80.27	41.95	39
Vapour pressure (kPa)	2.34	618.23	40
Density $(Kg\ m^{-3})$	998.3	907.0	41
Din. Viscosity (µPa.s)	1002.0	187.8	42
Ionization capacity, pK_w	14.5	11.5	43

However, these factors cannot be ignored. The intensity of mixing facilitates mass transfer and might increase the yield, and LWR determines the concentration and volume of the recovered liquor.

4.3.1.4   *Methods of Separations from Liquor*

The addition of an anti-solvent such as ethanol is a suitable method for the fractionation of the extraction liquor. Ethanol disrupts the hydrogen bonds between hemicelluloses turning them insoluble, whereas lignin fragments and undesirable compounds remain dissolved in the ethanol–water solution. The volume of liquor has a direct impact on the amount of ethanol needed for the precipitation of hemicelluloses. According to Brillouet *et al.*, the ratio of ethanol to water that is needed to precipitate heteroxylans from solution should be at least 2 : 1.[48] When a higher ethanol-to-water ratio is used, polysaccharides with a lower DP can be precipitated.

Another strategy for the fractionation of the extraction liquor might be the use of membranes. In this case, it is preferable to use diluted liquors in order to prevent the membrane clogging-up. A set of membranes with different polarity and molecular cut-off were successfully used on a large scale for the separation of hemicelluloses from solution.[49] In combination with membranes, polymeric adsorbents have been tested for reducing membrane fouling.[49]

4.3.2   Extraction of Xylan from Chemical Hardwood Pulp

Bleached hardwood pulp fibres can be utilised as a source of xylans. The main advantage is that the raw material, which has been subjected to (Kraft) pulping with subsequent bleaching steps, is very pure regarding the amount of lignin and extractives. Therefore, the xylans extracted from bleached hardwood pulp can be expected to contain very little contaminant, which is seldom possible when using hardwood chips as the raw material. The advantage of alkali extraction of hardwood elemental chlorine free bleached pulp (ECF) is the upgrade of paper pulp to dissolving pulp, since the latter usually has a higher market price than the former. Extracted dissolving pulps can be used for the preparation of cellulose derivatives or regenerated fibres.

Eucalyptus globulus has been subjected to cold caustic extraction (CCE) with the aim of removing a pure fraction of xylans while preserving the strength properties of the remaining pulp.[50] CCE was conducted at 30 °C for 25 minutes at a consistency of 10% and a NaOH charge of 0.29 g g^{-1} dry pulp. Xylans were dissolved in NaOH solution and precipitated by decreasing the pH of the filtrate and washing with ethanol. The extraction procedure removed an average of 5.0% of the pulp material, which was analysed by size exclusion chromatography (SEC), high-performance ion chromatography (HPIC) and Fourier transform infrared spectroscopy (FT-IR). HPIC measurements revealed that the precipitate consisted mainly of xylose (>99%) whereas FT-IR verified that there was no uronic acid or lignin present in the extracted xylans. The material was found to be polymeric with an average molar mass (Mw) of 17 500 g mol^{-1}, which corresponds to a DP of *ca.* 98, with a narrow size distribution (polydispersity *ca.* 1).[50] The remaining *Eucalyptus globulus* pulp after extraction of xylans was analysed for certain pulp properties to verify the effect of extraction. It was found that *ca.* 30% more refining was needed for the extracted pulp to achieve the same drainability properties. In comparison with non-extracted eucaliptus pulp, the extracted eucalyptus pulp has the same or higher tensile strength at the same refining degree, and requires less energy for drying.[50]

4.3.3 Xylan Derivatives

Xylan isolated from plant materials can be used for the preparation of xylan derivatives. These derivatives can be synthesized either in homogeneous or heterogeneous conditions.[51] The novel functionality (*e.g.*, ester and ether functional groups) introduced to the polymeric chain imparts novel properties to the designed polymer. Thus, the properties of the xylan derivatives (XDs) can be tuned by the type of moieties introduced and by their degree of substitution (DS). However, the structure of the xylan derivatives is also strongly dependent on the characteristics of the xylan used to prepare these derivatives. As an example, the antiviral activity of xylan sulfates (XS) depends both on the molar mass of the original xylan, and the DS of the sulfate groups. XS of high molar mass and a $DS_{Sulfate}$ above 1.2 showed antiviral activity, whereas XS of low molar mass and $DS_{Sulfate}$ as high as 1.7 did not show this property.[51] Although a lot of effort is being made to unveil the exact mechanism responsible for these differences in the chemical properties, the explanations remain unclear yet.

Xylan derivatives have potential applications in many different fields, such as health care pharmaceuticals, hygiene and water purification products, food industry, and adhesives, among others.[1,8,9,51–53] Examples of xylan derivatives prepared and their potential applications are shown in Table 4.3.

4.4 Surface Engineering of Fibres Using Xylan and Xylan Derivatives

The application of biomass-derived polysaccharides as surface modifying agents is an attractive approach for engineering fibre-based materials. Additionally, it provides new market opportunities for underutilized polysaccharides. Direct self-assembly of biopolymers and their derivatives on the fibre surfaces is a tool aimed to keep the fibre wall structure intact and endow the fibres and fibre networks with new properties.[54,55] Traditionally, xylan has been added to fibres during pulp processing to facilitate beating and to improve the mechanical properties of the fibre network.[56,57] Beating is a process involving the mechanical treatment of pulp fibres in an aqueous media with the aim of improving their papermaking properties. By means of

Table 4.3 Examples of functional groups introduced to the polymeric chain of xylans.

Name	Structure of the functional group	Potential application
Hydroxypropyltrimethylammonium xylan (HPMAX)		Paper additive flocculation aid, antimicrobial agent[51]
Carboxymethyl xylan		Pulp and paper,[51] anti-tumor drug[1]
Xylan furoate		Film formation[52]
Xylan sulfate		Biologically active component in drugs[1]
Xylan ibuprofen		Biologically active component in drugs[51]
FTIC[a] xylan		Mapping of wood fibres[62]

[a]FTIC: fluorescein isothiocyanate.

polysaccharide chemistry, different functional groups can be introduced to xylan to facilitate the interaction with the fibres and to add new functionalities.[58] A lot of research was carried out to get a deeper understanding of the adsorption phenomenon, the effect on the fibres' properties and to enhance the performance of this biopolymer as a functional polymer.[59–63]

The properties of the substrate and sorbate are of great importance for fibre modification *via* adsorption of biopolymers. Although the properties of pulp fibres are not the focus of this chapter, it is necessary to mention a few important characteristics. Pulp fibres are a porous heterogeneous substrate.[29] As mentioned earlier, they have a complex chemical composition, mainly consisting of three polymers (cellulose, hemicelluloses, and lignin) and low molecular weight extractives.[29] The content of these compounds varies across the fibre wall and, therefore, the composition of fibres in the bulk differs from that of the surface.[29] Another important feature of the fibres is the negative electrostatic charge. The cellulose fibres carry negative charge in slightly acidic and neutral conditions, which originates mostly from the uronic acid groups of hemicelluloses.[64] All these properties can vary considerably depending on the source of the fibres and the manufacturing process, and they have to be taken into consideration when designing the modification of the fibres by xylan or xylan derivatives. On the other hand, the properties of xylan, such as chemical composition and purity, also play a significant role in the modification process. The effect of these properties and the reaction conditions on xylan adsorption are discussed in more detail below.

4.4.1 Mechanism of Xylan Adsorption to Cellulose Fibres

In aqueous solutions, xylan tends to self-associate, and it is believed that most of the xylan adsorbs in the form of aggregates.[61] Such aggregates are formed due to interactions between or within the polysaccharide chains through intra- and intermolecular hydrogen bonds, as well as through so-called hydrophobic interactions of the hydrophobic substituents (*i.e.* lignin).[61,65,66] Aggregation favors the xylan transfer from solution to the fibre surfaces and, therefore, its adsorption.[61,65,66] There is no commonly accepted theory about the types of forces retaining xylan on the cellulose surfaces. However, different research groups have proposed that such forces are hydrogen bonds or a van der Waals attraction.[60,67,68] The formation of hydrogen bonds between the xylan and cellulose fibres can be supported by the natural arrangement of xylan in the fibre cell wall, where xylan it is tightly bonded with cellulose *via* hydrogen bonds formed between the available hydroxyl groups of these polysaccharides.[29] Figure 4.2 illustrates xylan self-association and the mechanism of its adsorption onto the cellulose fibres.

Adsorption of xylan onto the cellulose material reaches equilibrium slowly.[59,67–69] This can be attributed to the hindered diffusion of xylan molecules caused by a steric effect (conformation of molecules) and electrostatic repulsion between the fibre surfaces and the polymeric molecules.[69] Several studies showed that the adsorption of xylan takes place across the entire fibre cell wall and is not limited to the surfaces.[62,70] However, the highest concentration of xylan was observed on the outer fibre layers where xylan formed nano- and micrometre-sized particles.[62,64,71] A good visualization of the distribution of adsorbed xylan across the fibre cell wall was demonstrated by Köhnke *et al.*[62] It is depicted in Figure 4.3. The images were obtained by confocal laser scanning microscopy studies on bleached Kraft softwood fibres modified by birch glucuronoxylan previously labeled with fluorescent molecules.

4.4.2 Effect of Xylan Structure and Purity on its Adsorption

The xylan structure and xylan purity depends on the biomass source as well as on extraction conditions, method of recovery and purification methods

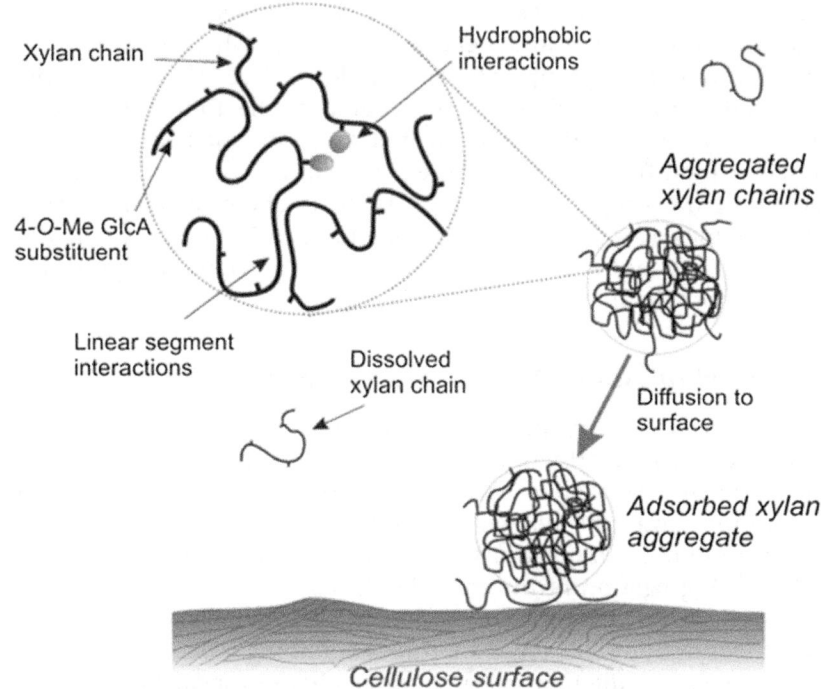

Xylan chain

Hydrophobic
interactions

*Aggregated
xylan chains*

4-O-Me GlcA
substituent

Linear segment
interactions

Dissolved
xylan chain

Diffusion to
surface

*Adsorbed xylan
aggregate*

Cellulose surface

Figure 4.2 Mechanism of xylan adsorption onto the cellulose surface of fibers.[61]
Reprinted with permission from ref. 61. Copyright (2003) American
Chemical Society.

employed. Side groups of xylan have a significant effect on the solubility of
the polymer and its interactions with other materials. Arabinosyl and acetyl
side-groups attached to the backbone hinder xylan aggregation. As reported
by Kabel *et al.*[72] who studied adsorption of various xylans onto bacterial
cellulose at a pH of 5 and at 40 °C, the removal of acetyl groups and
arabinose units favored agglomeration of xylan due to improved intra- and
inter molecular interactions of the xylan chains. This in turn increased the
adsorption of xylan onto the cellulose material.

The presence of uronic acid groups makes xylan molecules negatively
charged. These groups cause electrostatic repulsion of the xylan chains
preventing their association in aqueous solution. Furthermore, they impair
interactions between xylan and cellulose fibres due to the same reason.[61]
Besides that, uronic acids exhibit strong hydrophilic character, and xylan
molecules that contain many uronic acid groups are surrounded by a larger
amount of water molecules.[59] This restricts the close contact of the polysaccharide with the cellulose surfaces impeding, as a result, the adsorption
of xylan onto cellulose surfaces. Hansson and Hartler[59] compared
adsorption of pine and birch xylan onto cotton fibres. The pine xylan showed
worse adsorption in comparison with birch xylan. The difference was explained by a higher content of uronic acid groups, and also by the presence

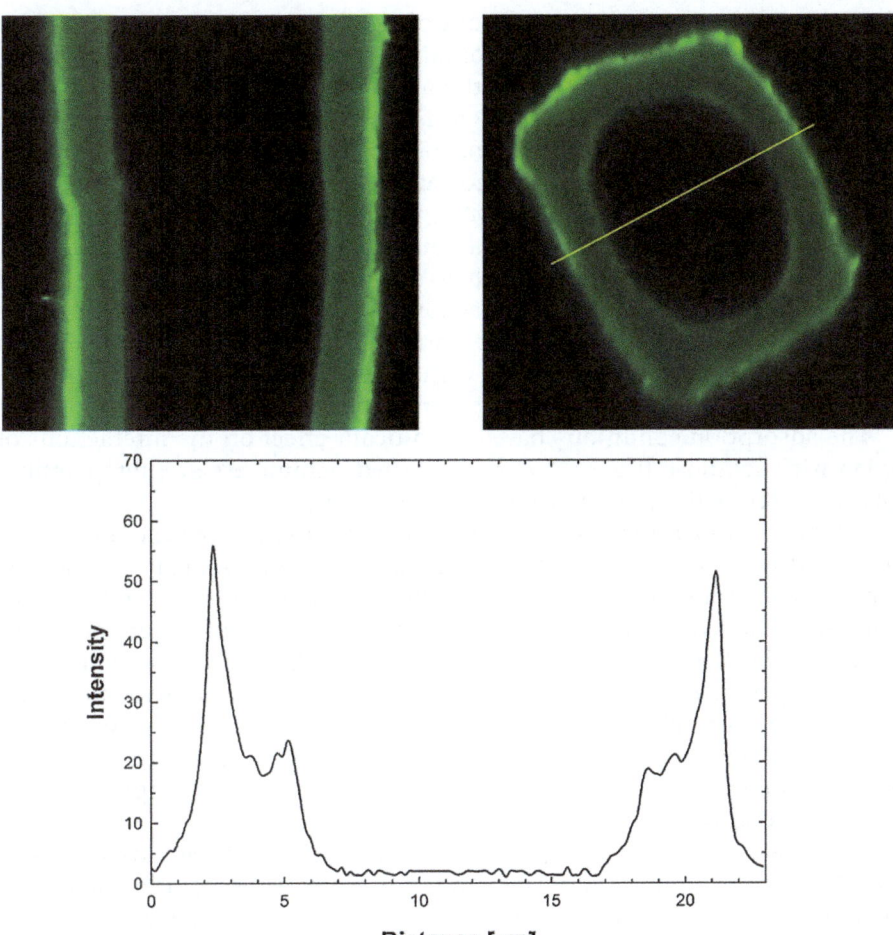

Figure 4.3 Distribution of birch glucuronoxylan labeled with a fluorescent molecule across the bleached kraft softwood fiber. Adsorption conditions: 80 °C and 0.1 M NaCl.[62]
Reprinted from ref. 62. T. Köhnke, K. Lund, H. Brelid and G. Westman, Kraft pulp hornification: A closer look at the preventive effect gained by glucuronoxylan adsorption, *Carbohydr. Polym.*, **81**, 226–233, Copyright (2010), with permission from Elsevier.

of arabinosyl substituents in the pine xylan. The cleavage of uronic acid groups from birch xylan increased two-fold the adsorption.[59] The authors of another study[73] showed that hexenuronic acid groups formed during a cooking reaction also decreased the affinity of xylan to cellulose fibres.

Lignin might be present as an impurity and has an effect on xylan solubility, as well as on the sorption characteristics.[66,73] Linder *et al.*[74] performed ozonation treatment to alter the structure of lignin linked to xylan. The opening of the aromatic ring introduced hydrophilic groups into the lignin, and this improved xylan solubility. As a result, xylan macromolecules formed fewer aggregates, and a lower amount of xylan nanoparticles was found on the fibres.[74] Xylan fractions that contain high amounts of lignin associate into much bigger aggregates in neutral aqueous solutions compared to xylan fractions with a lower content of lignin.[66] Westbye *et al.*[65] prepared several xylan fractions with different contents of uronic acid groups and lignin, and studied their agglomeration behavior at pH 7. The authors also conducted adsorption experiments onto softwood fibres at pH 10 and 100 °C. Among these fractions, xylans with a lower amount of uronic acid groups and higher lignin content tended to agglomerate to a greater extent. Such xylans showed a five-fold increase in the amount adsorbed onto the substrate than xylans that remained soluble at pH 7. It is worth mentioning that lignin itself also agglomerates in aqueous solutions with different pH depending on the pK_a value of lignin. An increase in temperature increases lignin aggregation.[66]

The adsorption conditions have a significant effect on the interactions of xylan with cellulose fibres. The most studied parameters are: temperature, pH, ionic strength, xylan concentration, and time.

A higher temperature during the modification enhances the amount of adsorbed xylan on the fibres.[61,73] The increase in temperature intensifies the vibration of water molecules that results in a decreased water layer surrounding the xylan macromolecules. A thinner layer of water molecules improves the close contact of xylan with cellulose fibres.[59] A combination of high temperature and alkaline medium, at a certain point leads to the removal of uronic acid groups and therefore to agglomeration of xylan molecules.[61,73] However, there is a critical pH range of 13 and higher at which xylan solubility increases due to ionization of the hydroxyl groups negatively affecting adsorption.[68,75]

An increase of ionic strength decreases repulsion between xylan chains, and between xylan and fibres, facilitating polymer agglomeration as well as its interaction with the fibres.[62] Generally, at higher xylan concentration, the amount of xylan covering the surface increases.[70,76] Xylans of different origin show different correlations between the concentration in the solution and the adsorbed amount, as shown by Hansson and Hartler.[59] Birch xylan exhibited a linear correlation, and the adsorption increased significantly as the concentration of polysaccharide increased. In contrast, adsorption of pine xylan showed a minor dependence on the concentration of xylan in the solution. The adsorption time has a more pronounced effect at higher xylan concentration and higher temperatures.[70,73]

4.4.3 Fibre Engineering with Xylan Sorption in Bleaching

Recently, Grigoray *et al.*[57] performed xylan adsorption onto Kraft pine pulps during an oxygen delignification stage. This stage is performed after pulping to remove a part of residual lignin from pulp fibres. The aforementioned study was carried out at an oxygen pressure of 6 bar and at a temperature of 95 °C in an alkaline medium (NaOH dosage 2 wt%, pH 11). Two types of birch xylans were applied: xylan extracted from wood by PHWE (WXyl) and xylan extracted from bleached birch kraft pulp by CCE (PXyl). The chemical structure of WXyl was less altered during the extraction process compared to PXyl. WXyl contained acetyl groups, a high amount of uronic acid units, and a significant amount of bound lignin (up to 5 wt%). In contrast, PXyl did not contain lignin and acetyl groups, and it had a lower amount of uronic acid groups. At the applied dosages of 5 wt%, the addition of PXyl did not interfere with the efficiency of delignification. However, WXyl had a negative impact on the pulp bleachability due to the lignin impurities. Bleachability shows how easily pulp can be bleached and in practice, it is reflected by the amount of bleaching reagent that is needed to remove or destroy colored compounds (chromophores). The kappa number, an indirect indicator of lignin content, increased by 6 units and brightness value decreased by 6% ISO for the WXyl treated pulps compared to the reference sample after the treatment. The addition of PXyl at the dosage of 5 wt% had a more positive effect on the mechanical properties of the fibre network than the addition of WXyl. Tensile index, tensile energy absorption, and stretch of the sheets made of the bleached PXyl-modified fibres were higher on average, by around 23%, 69%, and 40%, respectively, than for the corresponding reference sample.

The results of the study showed that xylan can be applied as a dry strengthening agent to improve the mechanical properties of a final fibre product. The modification step of pulp fibres with xylan can be integrated in the oxygen delignification stage of the existing fibre lines of the pulp and paper industries. The purity of the xylan is important to consider in order to obtain better mechanical properties of the fibrous material as well as to avoid a negative effect on pulp bleachability.

4.4.4 Xylan Derivatives for Pulp Fibre Modification

Ionic groups introduced in xylan chain imparts high solubility to the XDs.[77] Up to date, water soluble XDs with ionic functional groups have been used mainly as dry strengthening agents for wood pulp fibres.[63,73,78] However, new xylan derivatives with unique properties have been prepared recently, and the possibility of using those polymers as fibre-modifying agents might open the doors for new types of functional fibres. The cationic charge facilitates the adsorption of the XDs on the negatively charged cellulosic fibres. In the case of anionic XDs, the electrostatic repulsion with the fibres can be superseded by using blends of cationic and anionic polymers (*i.e.*, polyelectrolyte complexes).[79] Basic research with polyelectrolyte complexes and bleached Kraft pulp (BKPP) fibres was conducted by Vega *et al.*[79] In this work, two different types of polyelectrolyte complexes (PECs) were prepared by mixing aqueous solutions of a cationic cellulose derivative (CN^+) with anionic xylan derivatives in aqueous solutions. The complex [CN^+CMX^-] solution was prepared by mixing carboxymethylxylan (CMX^-) solution with CN^+ solution, whereas [CN^+XS^-] solution was prepared by mixing the corresponding solutions of xylan sulfate (XS^-) and CN^+. The prepared PEC solutions were mixed with the fibres at room temperature. Figure 4.4 shows

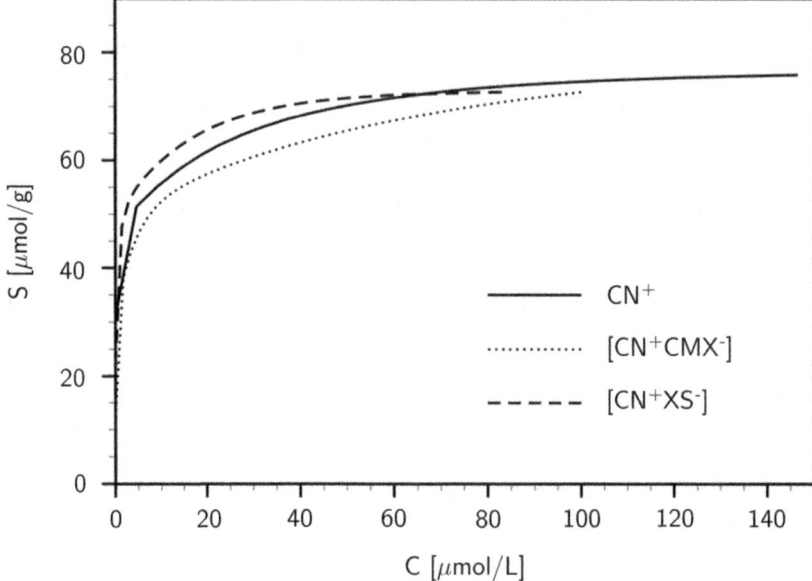

Figure 4.4 Sorption isotherms obtained from polyelectrolyte titrations of $[CN^+\text{-}CMX^-]$ (system composed of carboxymethylxylan (CMX^-) and cellulose (3-carboxypropyl)trimethylammonium chloride ester (CN^+)), $[CN^+XS^-]$ (system composed of xylan sulfate (XS^-), and CN^+). S, amount of adsorbed charge $[\mu mol\ g^{-1}]$; C, equilibrium concentration in the solution $[\mu mol\ L^{-1}]$.[79]
Reprinted with permission from ref. 79. B. Vega, H. Wondraczek, C. S. Pinto Zarth, E. Heikkilä, P. Fardim and T. Heinze, *Langmuir*, 2013, **29**(44), 13388–13395. Copyright (2013) American Chemical Society.

the sorption isotherms of CN^+, $[CN^+CMX^-]$, and $[CN^+XS^-]$ on BKPP. According to the results, the electrostatic interactions between the cationic net charge of the PECs, and the anionic charge of the fibres, has a strong influence on the adsorption process. Moreover, the PECs remain adsorbed on the fibres after adsorption. The possibility of using PECs might be useful in two different cases. In the first case, the polyanion is used as an "enhancer"; *i.e.*, the polyanion is used to increase the amount of the target polycation adsorbed on the fibres. In the second case, the polycation is used as a "carrier"; *i.e.*, the polycation is added to help with the adsorption of the target polyanion on the anionic fibre surfaces.

In addition to the ionic groups, xylan can be decorated with other functional groups. Such multifunctional xylan derivatives (MXDs) can be prepared using the same approach as that published for multifunctional cellulose derivatives.[54,80] The possibility to apply MXDs, which are decorated both with cationic moieties and target functions, represents an important simplification of the modification process of cellulose fibres. The cationic moiety improves the affinity of the MXDs to the fibres, whereas the target function endows the fibres with a new property (*e.g.* fluorescence). As the modification is performed in aqueous solutions under mild conditions, it ensures the preservation of the ultrastructure of the original wood pulp fibres.

4.4.5 Properties of Xylan-modified Fibres

Addition of xylan improves interactions of the fibres with water. As an example, a contact angle with water of chemo-thermomechanical single fibres with a high lignin content and a fibre network (sheets) made of these fibres was significantly reduced with addition of xylan.[71] This trend can be explained by the prominent hydrophilic character of xylan caused by a high amount of available hydroxyl groups. When cellulose fibres modified with xylan are placed in water, xylan increases the amount of bound water and this facilitates fibre swelling and water penetration.[57,73,81] As a result, fibres covered by xylan can be easily beaten (fibrillated). Therefore, the modification of pulp fibres by xylan will reduce the energy consumption of a highly energy intensive stage of pulp processing.[73] Uronic acid groups attached to the xylan backbone also have a positive effect on the water interaction and fibre fibrillation. The modification of pulp fibres with xylan bearing uronic acid groups will increase a negative charge. The fibrils of such fibres with an excess of anionic groups will separate easily due to their repulsion because of the same charge.[81,82]

Fibres undergo hornification during drying, which means that they lose their ability to swell.[70] This is caused by the collapse of the fibre wall, closure of the fibre pores and crystallization of fibrils.[62,70] This phenomenon leads to stiffening of the fibre surfaces, which reduces the contact of the fibres during sheet formation and deteriorates the properties of the final fibre network. Hornification is especially an actual problem in the case of recycling fibres and fibres dried before shipment to paper mills. Köhnke *et al.*[62] showed that the negative effect of drying can be offset by xylan adsorption. Particularly, the addition of xylan before drying increases swelling of the fibres and preserves most of the original surface area. Xylan also acts as an interfibrillar spacer and prevents crystallization of fibrils.[70]

The strength of fibrous materials is mostly determined by the strength of the individual fibres and by bonding between adjacent fibres (number, distribution and strength of bonds).[83] In the pulp and paper industry, socalled dry strengthening agents are added before web formation to improve the bonding between the fibres. Xylan can also be potentially applied for this purpose. Flexible chains of the polymer with available hydroxyl groups create more bonding areas between the fibres. Consequently, the network of xylan modified fibres exhibits superior mechanical properties such as internal bonding, tensile index, tensile energy absorption, stretch, and bursting strength.[56,57,62,70,76,81,84,85] The location of adsorbed xylan is also significant for the performance of the fibre network. It is expected that xylan located on the fibre surfaces makes a more substantial contribution to the internal bonding ability of fibres. Schönberg[81] found that internal bonding directly correlated with the amount of xylan located on the fibre surfaces. It is worth mentioning that the amount of xylan in original fibres and cellulose to xylan ratio are also important when fibrous materials are designed.[73]

Novel functionalities can be introduced to the fibres using MXDs, namely fluorescence, bio-catalytic activity, antimicrobial activity, selective barrier properties and stimuli-responsive properties.[80] The possibility to prepare functional fibres from pulp fibres represents an excellent alternative to some petroleum products, and provides opportunities for innovations in the area of the traditional pulp and paper industry.

4.5 Conclusions

Xylans are heteropolysaccharides that are one of the ubiquitous components found in plant cell walls. They are the main hemicellulose type in hardwoods and many annual crops. Interest towards the isolation of a pure fraction of high molar mass xylans, which can be utilised for the preparation of novel bio-based materials, is steadily increasing. However, there is still a lack of technological solutions for the extraction of such polymers on the large scale. This can be explained by the fact that xylan is tightly associated with other polymers of the plant cell wall (*i.e.*, cellulose and lignin) and also by the hindered accessibility of the cell wall. According to the published data, pressurised hot-water extraction (PHWE) and cold caustic extraction (CCE) are the most suitable methods for the extraction of high molar mass xylan from plant materials and pulps, respectively.

The xylan obtained from bleached pulp by CCE is highly deacetylated and does not show bound lignin, in contrast to the xylan extracted from wood by PHWE. Pure fractions of isolated xylan with a high molar mass can be applied to fibres as dry strengthening agents. Moreover, different functionalities can be introduced to the polymeric chain of xylan by derivatisation. These xylan derivatives can be used for different applications such as paper additive flocculation aids, antimicrobial agents, anti-tumor drugs, among others. In addition, the adsorption of xylan derivatives can endow the cellulose materials with new functionalities, namely fluorescence and bioactivity, among others.

Although a few applications have been addressed in this chapter, there are plenty of opportunities to be discovered for xylan and its derivatives.

References

1. A. Ebringerová and T. Heinze, *Macromol. Rapid Commun.*, 2000, **21**(9), 542–556.
2. A. Ebringerová, Z. Hromádková and T. Heinze, in *Polysaccharides I. Structure, Characterization and Use*, ed. T. Heinze, Springer-Verlag, Berlin Heidelberg, 1st edn, 2005, ch. 1, pp. 1–67.
3. R.-C. Sun, *BioResources*, 2008, **4**(2), 452–455.

4. R. Deutschmann and R. F. Dekker, *Biotechnol. Adv.*, 2012, **30**(6), 1627–1640.

5. F. Peng, P. Peng, F. Xu and R.-C. Sun, *Biotechnol. Adv.*, 2012, **30**(4), 879–903.

6. P. Peng and D. She, *Carbohydr. Polym.*, 2014, **112**, 701.

7. J. Hartman, *Hemicellulose as Barrier Material*, Tk.L Thesis, Royal Institute of Technology, 2006.

8. A. E. Da Silva, H. Rodrigues, M. C. Gomes, E. E. Oliveira, T. Nagashima and E. Sócrates, in *Hemicellulose for Pharmaceutical use, Products and Applications of Biopolymers*, ed. J. Verbeek, InTech, Croatia, 2012, pp. 61–84.

9. F. B. Sedlmeyer, *Food Hydrocolloids*, 2011, **25**, 1891–1898.

10. F. Carvalheiro, L. C. Duarte and F. M. Gírio, *J. Sci. Ind. Res.*, 2008, **67**, 849–864.

11. V. B. Agbor, N. Cicek, R. Sparling, A. Berlin and D. B. Levin, *Biotechnol. Adv.*, 2011, **29**, 675–685.

12. E. C. Bensah and M. Mensah, *Int. J. Chem. Eng.*, 2013, **2013**, 1–20.

13. J. Rissanen, H. Grénman, S. Willför, D. Murzin and T. Salmi, *ChemSusChem*, 2014, **7**(10), 2947–2953.

14. H. Grénman, K. Eränen, J. Krogell, S. Willför, T. Salmi and D. Murzin, *Ind. Eng. Chem. Res.*, 2011, **50**(7), 3818–3828.

15. P. Reyes, R. Teixeira Mendonça, J. Rodríguez, P. Fardim and B. Vega, *J. Chil. Chem. Soc.*, 2013, **58**(1), 1614–1618.

16. P. Reyes, R. Texeira Mendonça, M. Aguayo, J. Rodriguez, B. Vega and P. Fardim, *Rev. Árvore*, 2013, **37**(1), 175–180.

17. F. Gírio, F. Fonseca, F. Carvalheiro, L. Duarte, S. Marques and R. Bogel-Lukasik, *Bioresour. Technol.*, 2010, **101**, 4775–4800.

18. R.-C. Sun, X. F. Sun and J. Tomkinson, in *Hemicelluloses: Science and Technology*, ed. P. Gatenholm and M. Tenkanen, ACS Symposium series, American Chemical Society, Washington DC, 1st edn, 2003, ch. 1, vol. 864, pp. 2–22.

19. J. Rowley, S. R. Decker, W. Michener and S. Black, *3 Biotech*, 2013, **3**, 433–438.

20. P. Immerzeel, P. Falck, M. Galbe, P. Adlercreutz, E. Nordberg Karlsson and H. Stålbrand, *LWT – Food Sci. Technol*, 2014, **56**, 321–327.

21. X.-N. Nie, J. Liu, D. She, R.-C. Sun and F. Xu, *BioResources*, 2013, **8**(3), 3817–3832.

22. W. Lan, C.-F. Liu and R.-C. Sun, *J. Agric. Food Chem.*, 2011, **59**, 8691–8701.

23. P. Kilpeläinen, PhD, Thesis, Åbo Akademi University, 2015.

24. P. Kilpeläinen, K. Leppänen, P. Spetz, V. Kitunen, H. Ilvesniemi, A. Pranovich and S. Willför, Pressurised hot water extraction of acetylated xylan from birch sawdust, *Nord. Pulp Pap. Res. J.*, 2012, **27**, 682–688.

25. E. Sjöström, in *Fundamentals and Applications*, ed. E. Sjöström, Wood chemistry, Academic Press, New York, 2nd edn, 1981, ch. 65.

26. L. J. Gibson, *J. R. Soc., Interface*, 2012, **9**, 2749–2766.

27. M. Balakshin, E. Capanema, H. Gracz, H. M. Chang and H. Jameel, *Planta*, 2011, **233**, 1097–1110.
28. J. W. T. Merewether, *Holzforschung*, 1957, **11**(3), 65–80.
29. D. Fengel and G. Wegener, Wood, *Chemistry, Ultrastructure, Reactions*, W. de Gruyter Inc., Berlin and New York, 1984.
30. T. Koshijima and T. Watanabe, in *Association Between Lignin and Carbohydrates in Wood and Other Plant Tissues*, Springer, Berlin Heidelberg, 2003.
31. H. Meier, *J. Polym. Sci.*, 1961, **51**, 11–18.
32. R. Nelson and C. Schuerch, *J. Polym. Sci.*, 1956, **22**(102), 435–448.
33. R. Jara, PhD thesis, University of Mayne, 2010.
34. K. J. Brown, Forest Products Laboratory, U.S. Department of Agriculture in Cooperation with the University of Wisconsin, 1957, Report no. 2101, pp. 1–25.
35. S. Von Schoultz, Method for extracting biomass, *Pat.* WO2014009604 A1, *CA pat.*, 2015.
36. S. Malkov, P. Tikka and J. Gullichsen, *Pap. Puu*, 2001, **83**(8), 1–14.
37. M. Borrega, K. Nieminen and H. Sixta, *Bioresour. Technol.*, 2011, **102**, 10724–10732.
38. H. Sixta, A. Potthast and A. W. Krotschek, in *Handbook of Pulp*, ed. H. Sixta, Wiley-VCH Verlag GmbH & Co, KGaA, Weinheim, 2006, vol. 1, pp. 109–510.
39. M. Uematsu and E. U. Franck, *J. Phys. Chem. Ref. Data*, 1980, **9**, 1291–1306.
40. NPL. Vapor Pressure of Water at Temperatures Between 0 and 360, http://www.kayelaby.npl.co.uk/chemistry/3_4/3_4_2.html., accessed: May 2015.
41. The international association for the properties of water and steam IAPWS-IF7, http://www.peacesoftware.de/einigewerte/calc_dampf.php5, accessed: May 2015.
42. Viscosity tables, http://www.viscopedia.com/viscosity-tables/substances/water, accessed: May 2015.
43. W. L. Marshall and E. U. Franck, *J. Phys. Chem. Ref. Data*, 1981, **10**(2), 295–304.
44. X. Chen, M. Lawoko and A. van Heiningen, *Bioresour. Technol.*, 2010, **101**(20), 7812–7819.
45. J. Krogell, PhD thesis, Åbo Akademi University, Åbo, Finland, 2015, ISBN: 978-952-12-3202-2 (online).
46. J. Krogell, K. Eränen, K. Granholm, A. Pranovich and S. Willför, *Ind. Cop Prod.*, 2014, **61**, 9–15.
47. J. Krogell, K. Eränen, A. Pranovich and S. Willför, *Ind. Cop Prod.*, 2015, **67**, 114–120.
48. J. M. Brillouet, J. P. Joseleau, J. P. Utile and D. Leliévre, *J. Agric. Food Chem.*, 1982, **30**, 488–495.
49. E. Koivula, M. Kallioinen, T. Sainio, E. Antón, S. Luque and M. Mänttäri, *Bioresour. Technol.*, 2013, **143**, 275–281.
50. C. Sieveking, M.Sc. Thesis, Åbo Akademi University, Finland, 2014.

51. K. Petzold-Welcke, K. Schwikal, S. Daus and T. Heinze, *Carbohydr. Polym.*, 2014, **100**, 80–88.
52. S. Hesse, T. Liebert and T. Heinze, *Macromol. Symp.*, 2006, **232**, 57–67.
53. I. Simkovic, O. Gedeon, I. Uhliariková, R. Mendichi and S. Kirsfnerová, *Carbohydr. Polym.*, 2011, **83**, 769–775.
54. O. Grigoray, H. Wondraczek, E. Heikkilä, P. Fardim and T. Heinze, *Carbohydr. Polym.*, 2014, **111**, 280–287.
55. O. Grigoray, H. Wondraczek, S. Daus, K. Kühnöl, S. K. Latifi, P. Saketi, P. Fardim, P. Kallio and T. Heinze, *Macromol. Mater. Eng.*, 2015, **300**, 277–282.
56. M. C. S. Muguet, C. Pedrazzi and J. L. Colodette, *Holzforschung*, 2011, **65**, 605–612.
57. O. Grigoray, J. Järnström, E. Heikkilä, P. Fardim and T. Heinze, *Carbohydr. Polym.*, 2014, **112**, 308–315.
58. B. Vega, K. Petzold-Welcke, P. Fardim and T. Heinze, *Carbohydr. Polym.*, 2012, **89**, 768–776.
59. Å. Hansson and N. Hartler, *Sven. Papperstidn.*, 1969, **72**, 521–548.
60. F. Mora, K. Ruel, J. Comtat and J. P. Joseleau, *Holzforschung*, 1986, **40**, 85–91.
61. Å. Linder, R. Bergman, A. Bodin and P. Gatenholm, *Langmuir*, 2003, **19**, 5072–5077.
62. T. Köhnke, K. Lund, H. Brelid and G. Westman, *Carbohydr. Polym.*, 2010, **81**, 226–233.
63. K. Schwikal, T. Heinze, B. Saake, J. Puls, A. Kaya and A. Esker, *Cellulose*, 2011, **18**, 727–737.
64. E. Sjöström, *Nord. Pulp Pap. Res. J.*, 1989, **4**, 90–93.
65. P. Westbye, C. Svanberg and P. Gatenholm, *Holzforschung*, 2006, **60**, 143–148.
66. P. Westbye, T. Köhnke, W. Glasser and P. Gatenholm, *Cellulose*, 2007, **14**, 603–613.
67. A. Paananen, M. Österberg, M. Rutland, T. Tammelin, T. Saarinen, K. Tappura and P. Stenius, *Hemicelluloses: Science and Technology*, ed. P. Gatenholm and M. Tenkanen, ACS Symposium Series, American Chemical Society, Washington DC, 1st edn, 2004, ch. 18, vol. 864, pp. 269–290.
68. M. Mitikka, R. Teeäär, M. Tenkanen, J. Laine and T. Vuorinen, in *8th Int. Symp. Wood Pulping Chem.*, 1995, 3, pp. 231–236.
69. G. Ström, P. Barla and P. Stenius, *Sven. Papperstidn.*, 1982, **85**, R 100–R 106.
70. T. Köhnke and P. Gatenholm, *Nord. Pulp Pap. Res. J.*, 2007, **22**, 508–515.
71. Å. Henriksson and P. Gatenholm, *Holzforschung*, 2001, **55**, 494–502.
72. M. A. Kabel, H. van den Borne, J.-P. Vincken, A. G. J. Voragen and H. A. Schols, *Carbohydr. Polym.*, 2007, **69**, 94–105.
73. T. C. F. Silva, J. L. Colodette, L. A. Lucia, R. C. de Oliveira, F. N. Oliveira and L. H. M. Silva, *Ind. Eng. Chem. Res.*, 2011, **50**, 1138–1145.

74. Å. Linder, J. P. Roubroeks and P. Gatenholm, *Holzforschung*, 2003, **57**, 496–502.
75. S. A. Rydholm, *Pulping Processes*, Interscience Publishers, New York, 1965, ch. 9, pp. 439–692.
76. W. Han, C. Zhao, T. Elder, K. Chen, R. Yang, D. Kim, Y. Pu, J. Hsieh and A. J. Ragauskas, *Carbohydr. Polym.*, 2012, **88**, 719–725.
77. K. Schwikal, T. Heinze, A. Ebringerová and K. Petzold, *Macromol. Symp.*, 2006, **232**, 49–56.
78. T. Heinze, S. Hornig, N. Michaelis and K. Schwikal, in *Model Cellulosic Surfaces*, ed. M. Roman, ACS Symposium Series 1019, American Chemical Society, Washington DC, 2009.
79. B. Vega, H. Wondraczek, C. Salomão Pinto Zarth, E. Heikkilä, P. Fardim and T. Heinze, *Langmuir*, 2013, **29**(44), 13388–13395.
80. B. Vega, H. Wondraczek, L. Bretschneider, T. Näreoja, P. Fardim and T. Heinze, *Carbohydr. Polym.*, 2015, **132**, 261–273.
81. T. Schönberg, A. Oksanen, H. Suurnäkki, Kettunen and J. Buchert, *Holzforschung*, 2001, **55**, 639–644.
82. J. Laine and P. Stenius, *Pap. Puu*, 1997, **79**, 257–266.
83. J. Marton, in *Paper Chemistry*, ed. J. C. Roberts, Blackie Academic & Professional, Glasgow, 2nd edn, 1996, ch. 6, pp. 83–97.
84. O. Dahlman and J. Sjöberg, *Nord. Pulp Pap. Res. J.*, 2003, **18**, 310–315.
85. F. Ramirez, J. Puls, V. Zuniga and B. Saake, *Holzforschung*, 2008, **62**, 329–337.

CHAPTER 5

Recent Advances in the Synthesis of Sugar-based Surfactants

JOSÉ KOVENSKY*[a,b] AND ERIC GRAND[a,b]

[a] Laboratoire de Glycochimie, des Antimicrobiens et des Agroressources FRE CNRS 3517 CNRS, Université de Picardie Jules Verne, 33 rue Saint Leu, 80039 Amiens cedex, France; [b] Institut de Chimie de Picardie FR CNRS 3085, Université de Picardie Jules Verne, 33 Rue Saint-Leu, 80039 Amiens cedex, France
*Email: jose.kovensky@u-picardie.fr

5.1 Enzymatic Synthesis of Sugar-based Surfactants

The enzymatic synthesis of sugar-based surfactants, mainly sugar esters, has been developed in the last decade as an alternative (or sometimes complementary) to chemical synthesis, in an environmentally friendly approach. In comparison, enzymatic processes are more selective, do not require high temperatures, and fewer (or no) byproducts are formed.

5.1.1 Lipases

Lipases are the most widely used enzymes for the synthesis of sugar-based surfactants, as a result of their high stability and activity in a broad diversity of solvents. The reaction of unprotected sugars with lipases leads to sugar esters, widely used non-ionic and non-toxic biosurfactants, where the primary hydroxyls are selectively substituted by fatty acids.

RSC Green Chemistry No. 44
Biomass Sugars for Non-Fuel Applications
Edited by Dmitry Murzin and Olga Simakova
© The Royal Society of Chemistry 2016
Published by the Royal Society of Chemistry, www.rsc.org

Karmee has reviewed the uses of lipases for the synthesis of surfactants from glycerol, L-ascorbic acid, and sugars, with long or medium chain fatty acids.[1] Examples of the synthesis of sugar esters in organic solvents such as *tert*-butanol, acetone or ionic liquids were presented. Annuar and co-workers[2] compared different conditions for the enzymatic synthesis of sugar esters in non-aqueous media. They analyzed the influence of the organic solvent on the kinetic parameters of the enzyme, as a consequence of changes in their three-dimensional conformation relative to that in aqueous solutions. A higher $\log P$ of the solvent appears to correlate well with a higher lipase activity. Ionic liquids are superior to conventional non-aqueous solvents for lipase-mediated synthesis, provided that the stability of the enzyme is not affected. Water activity can also affect lipase performance. Rodrigues and co-workers[3] discussed the applications of the lipase synthesis of sugar esters in the food industry. They highlighted the advantages of enzyme immobilization and protein engineering for improving lipase stability. Cauglia and Canepa studied the synthesis of glucose myristate by Novozyme 435 (the lipase B from *Candida antarctica*, immobilized on macroporous acrylic resin) as a model reaction for general considerations on sugar ester production.[4] They analyzed the influence on the synthesis of the pre-reaction treatment of the enzyme and reaction mixture, water adsorbents, reaction solvents, products and reagents, in terms of control of the water activity often negatively affecting the enzyme efficiency. Molecular sieves have shown the best efficiency in water removal and seemed essential to achieve good ester yields. The same conclusions about the benefits of molecular sieves have been reported by Basri and co-workers[5] for the synthesis of xylitol stearate, palmitate and caprate, in hexane using Novozyme 435 at 60 °C (88–96% conversion). The main products were diesters. Rahman and co-workers[6] reported the microwave assisted synthesis of glucose oleate in a biphasic system comprising *tert*-butanol and the ionic liquid 1-butyl-3-methylimidazolium tetrafluoroborate [BMIM][BF$_4$], catalyzed by Novozyme 435, showing the best activity at 60 °C (30 min, 90% conversion). Oligofructose (degree of polymerization (DP) 2–8) fatty (caprylic, lauric, palmitic and stearic) acid esters have been synthesized by van Kempen *et al.* using Novozyme 435 in 1:9 DMSO–*tert*-butanol[7,8] (see Scheme 5.1). Three different reaction procedures were considered: (1) esterification in the presence of molecular sieves, (2) transesterification in the absence of molecular sieves, and (3) transesterification in the presence of molecular sieves. No major differences in the yields were found between the three different reaction procedures. The main compounds were mono- and diesters.

Fructose, sucrose and lactose oleic esters were synthesized by Gonçalves and co-workers using *Candida antarctica* type B lipase immobilized on two different supports, namely acrylic resin and chitosan.[9] The enzyme immobilized on chitosan showed the highest yield of lactose ester (84.1%). Additionally, the production of fructose ester was found to be higher for the enzyme immobilized on the acrylic resin support (74.3%) as compared with the one immobilized on chitosan (70.1%). The same trend was observed for

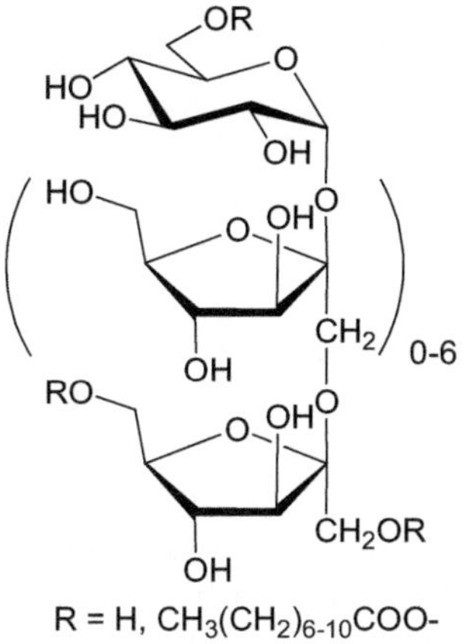

R = H, CH$_3$(CH$_2$)$_{6-10}$COO-

Scheme 5.1 Oligofructose fatty acid esters.[7,8]

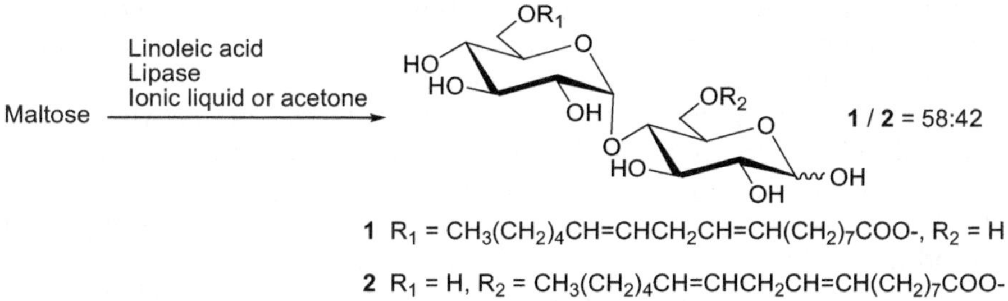

1 / 2 = 58:42

1 R$_1$ = CH$_3$(CH$_2$)$_4$CH=CHCH$_2$CH=CH(CH$_2$)$_7$COO-, R$_2$ = H

2 R$_1$ = H, R$_2$ = CH$_3$(CH$_2$)$_4$CH=CHCH$_2$CH=CH(CH$_2$)$_7$COO-

Scheme 5.2 Lipase catalyzed acylation of maltose.[11]

the sucrose ester, although with lower yields. The use of the enzyme immobilized on chitosan is interesting due to the possibility of enzyme reuse. Monolauroyl maltose, palatinose, trehalose and sucrose were synthesized by Kobayashi and co-workers using *Candida antarctica* lipase to catalyze the condensation in a 8:2 mixture of *tert*-butanol and pyridine;[10] their surfactant properties were measured. Maltose was acylated by Fischer and co-workers with linoleic acid on the primary *O*-6,6′ hydroxyl functions;[11] see Scheme 5.2. The ionic liquid 1-ethyl-3-methylimidazolium methanesulfonate [emim][MeSO$_3$] and the potentially renewable acetone enabled the best conversions. The lipases from *Pseudomonas cepacia* and immobilized *Candida antarctica* allowed the highest yields in a screening with 10 different lipases. Molecular sieves improved maltose transformation further up to 82%. Analysis indicated the formation of mono-6 or 6′-*O*-linoleyl-α-D-maltose as a mixture of two regioisomers in a 1.4:1 ratio.

Methyl α-D-glucopyranoside was used by Annuar and co-workers for the enzymatic (CAL B) synthesis of esters of fatty acids (C12, C14 and C16) aiming at physico–chemical studies of liquid crystal properties.[12] The ketal protected 1,2:5,6-di-*O*-isopropylidene-α-D-glucofuranose was used as the substrate for the lipase-catalyzed synthesis of secondary alcohol esters of

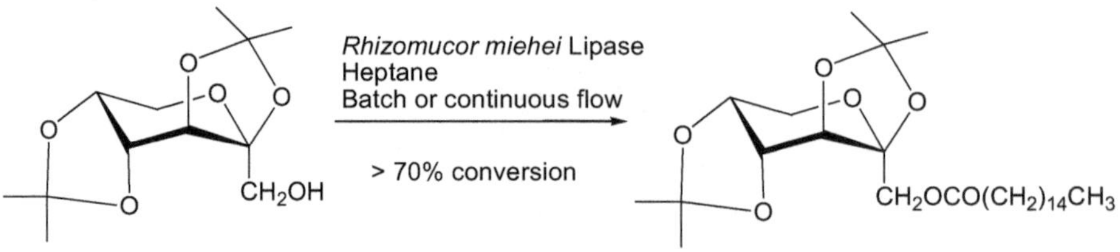

Scheme 5.3   Esterification of 2,3:4,5-di-*O*-isopropylidene-β-D-fructopyranose.[14]

glucose in continuous flow by de Souza and co-workers using lipozyme RM IM (commercial lipase preparation of *Rhizomucor miehei*, immobilized in an anion-exchange resin) in organic solvents (heptane, *tert*-butylmethylether).[13] Very high conversions (>94%) were obtained at short resident times (0.8–5.4 min). The same group reported the esterification of 2,3:4,5-di-*O*-iso-propylidene-β-D-fructopyranose derivative (see Scheme 5.3) where important variables for batch conditions were identified and then translated to the continuous flow regime.[14] The results presented indicate that *R. miehei* IM is a better immobilized lipase for a continuous flow environment, leading to high conversions at short residence times.

Sugar esters were prepared by Hayes and co-workers[15] in high yields using immobilized *Rhizomucor miehei* lipase in solvent-free systems at 65 °C. A two-step process was developed to produce a solvent-free supersaturated solution of 1.5–2.0 wt% sugar in oleic acid and fructose oleate. Using this approach, a product mixture containing 88% fructose oleate was formed, of which 92% was monoester, within 6 days. This equates to a productivity of 0.2 $mmol_{ester}$ h^{-1} g^{-1}_{lipase}, which is similar to values reported for synthesis in the presence of a solvent. The same group reported optimization of the fructose oleate synthesis using a process that involves programming of water removal.[16] 92.6 wt% fructose oleate was produced within 132 h, yielding a productivity of 0.297 $mmol_{ester}$ h^{-1} g^{-1}_{lipase}. They also studied the effect of the acyl donor (oleic, caprylic, lauric and myristic acids) and acceptor (fructose, sucrose, glucose and xylose) on the synthesis in a bioreactor filled with immobilized *Rhizomucor miehei* lipase.[17] The initial rate and final conversion of the acyl donor were found to be linearly dependent on the initial saccharide concentration with the relationship applying universally to all donors and acceptors studied. Fructose, due to the relatively small size of its crystals, yielded the highest concentration in the suspension-based me-dium among the esters, and hence the highest reaction rate and ester yield. The activity of the immobilized enzyme did not decrease appreciably after four successive runs for solvent-free fructose-oleate esterification, or equivalently, a 22 day reaction period. The synthesis of ascorbyl oleate and ascorbyl palmitate esters with the immobilized *Thermomyces lanuginosus* lipase lipozyme TL IM and triglycerides as a source of fatty acids was in-vestigated by Plou and co-workers.[18] Lipozyme TL IM gave rise to a lower yield of 6-*O*-ascorbyl oleate than Novozyme 435 when using triolein (64 *vs.* 84%) and olive oil (27 *vs.* 33%) as acyl donors. Both 6-*O*-ascorbyl oleate and

6-*O*-ascorbyl palmitate displayed excellent surfactant and antioxidant properties. The enzymatic synthesis of xylose esters has recently been reported by Bidjou-Haiour and Klai,[19] together with the determination of the critical micelle concentration (CMC) of the products. Annuar and co-workers[20,21] studied the kinetics of the lipase-catalyzed synthesis of 6-*O*-glucosyldecanoate. The highest conversion (65%) was achieved in 1:1 DMSO–*tert*-butanol at 20–60 °C. Kinetic studies showed that the apparent maximum reaction rate is not affected by the polarity of the solvent, but it increases with increasing temperature up to about 313 K.

5.1.2 Glycosidases

Rhamnolipids have been prepared by Deleu and co-workers by two biocatalyzed steps.[22] First, a primary alcohol function was introduced onto rhamnose by glycosylation of 1,3-propanediol catalyzed by naringinase (a rhamnosidase). Then immobilized lipase B from *Candida antarctica* catalyzed the esterification of the primary hydroxyl group with mono- and dicarboxylic fatty acids of increasing chain length (from C8 to C14, see Scheme 5.4). The new rhamnolipid obtained with tetradecanoic acid showed very good surface properties: its CAC (critical aggregation concentration) and γ_{CAC} (surface tension at CAC) are particularly low (1.70 µM and 27.6 mN m^{-1}, respectively) and it can form insoluble monolayers.

Aspergillus awamori K4 β-xylosidase has been used by Kurakake and co-workers to synthesize a sugar fatty acid ester.[23] Hexamethylene glycol was glycosylated with xylose. The resulting hydroxyhexyl xyloside was esterified with linoleic acid by a lipase. The binding with the hydrophobic and less hindered hexamethylene chain improved the esterification reaction. Mladenoska explored the use of microemulsions as bioreactors for the *trans*-glycosylation reaction catalyzed by three microbial β-galactosidases: fungal *Aspergillus oryzae*, yeast *Kluyveromyces marxianus* and bacterial *Escherichia coli* β-galactosidase.[24] Thus, the sugar moiety of the donor *p*-nitrophenyl-β-D-galactoside or *p*-nitrophenyl-β-D-glucoside is transferred to hexanol to afford hexylglycosides.

Sophoroselipids (SL) are bolaform biosurfactants that are abundantly produced by microorganisms from renewable resources. These derivatives were chemoenzymatically synthesized by Kitamoto and co-workers[25] from "acid form" diacetylated SL (SLdiAc), which are preferentially produced by *Candida floricola* TM 1502. After microbial production of SLdiAc from oleic acid/glucose, SLdiAc was converted into acetylated glucoselipid. Among twelve species of glucosidases, pectinase, pectolyase (see Scheme 5.5) and polygalacturonase were found to cleave the β-1,2-glycosidic linked disaccharide; pectolyase in particular was effective at 40 °C and pH 4.0. The deacetylated counterparts were obtained by alkaline hydrolysis.

Alkyl sophorosides have been produced using *Candida bombicola* cultures with glucose as the main carbon source and dodecan-2-ol as the co-substrate. The main product *sec*-dodecylsophoroside was used by Lang

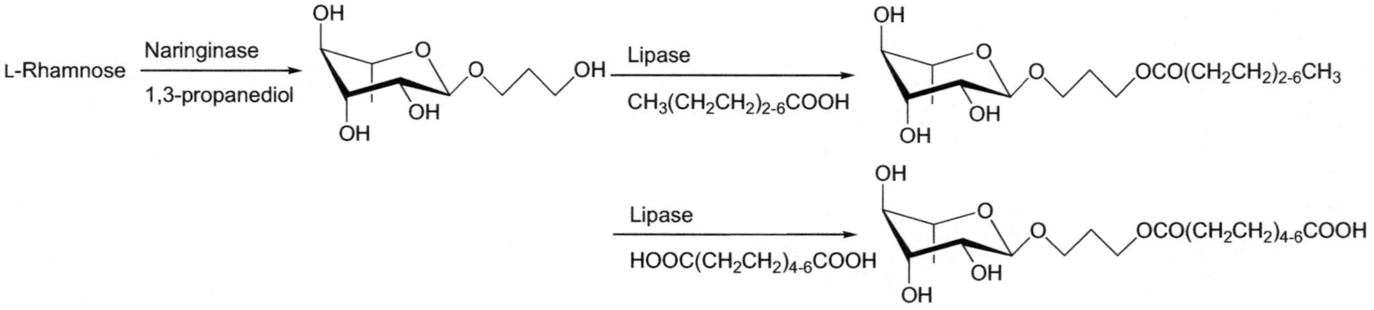

Scheme 5.4 Two-enzymatic-step synthesis of rhamnolipids.[22]

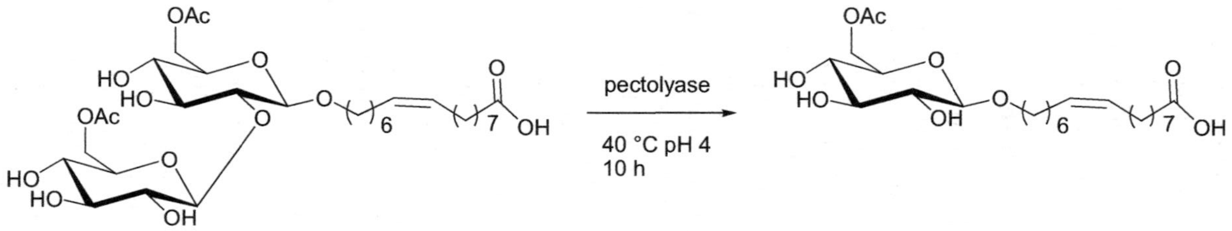

Scheme 5.5 Sophoroselipids as starting materials for the synthesis of bolaform.[25]

and co-workers as a substrate for further enzymatic modifications.[26] β-Glucuronidase cleaved the acetyl glucose unit leading to *sec*-dodecylglucoside; see Scheme 5.6. In a subsequent lipase-catalyzed acylation with sebacic acid in toluene, this compound was functionalized regioselectively at the primary C-6 position of glucose. In addition, *sec*-dodecylsophoroside was esterified by lipase-catalyzed reactions in toluene with 3-hydroxydecanoic acid and 17-hydroxystearic acid, resulting in mono- and diacylations of primary hydroxyl positions of the sophorose unit (C-6 and C-6′).

5.1.3 Cyclodextrin Glycosyltransferase

The unique ability of cyclodextrin glycosyltransferase (GGTase) to form and utilize the cyclic maltooligosaccharide cyclodextrin (CD) makes this enzyme an attractive catalyst for the synthesis of alkyl glycosides. Börner and co-workers[27] studied the sugar headgroup elongation of alkyl glucosides (acceptor) *via* two transglycosylation reactions from either a linear (maltohexose) or a cyclic (CD) glycosyl donor. The hydrophobic complexation between CD and alkyl glucoside, characterized by isothermic titration calorimetry, promotes the sugar transfer. On the other hand, Adlercreutz and co-workers[28] used *Bacillus macerans* CGTase to convert commercially available dodecyl β-D-maltoside to dodecyl β-D-maltooctaoside in a single step with α-cyclodextrin as a glycosyl donor (Scheme 5.7). Yields up to 80% were obtained. The *Thermoanaerobacter* enzyme catalyzed disproportionation reactions leading to a broader product range.

Plou and co-workers[29] reported the synthesis of a series of α-glucosyl derivatives of resveratrol (3,5,4′-trihydroxystilbene) by a transglycosylation reaction catalyzed by CGTase using starch as a glucosyl donor, in order to improve the bioavailability of the antioxidant. Several reaction parameters (temperature, solvent composition, enzyme concentration and starch–resveratrol ratio) were optimized. The yield of α-glucosylated products reached 50% in 24 h. Three families of products were obtained: glucosylated at 3-OH, at 4′-OH and at both 3-OH and 4′-OH. The bonds between glucoses were basically $\alpha(1 \rightarrow 4)$.

5.2 Chemical Synthesis of Sugar-based Surfactants

5.2.1 Chemical Synthesis of Alkyl Glycoses, Glycosides and Polyglycosides

Alkyl glycosides and alkyl polyglycosides (APG) are among the simplest class of sugar-based surfactants available. One alkyl chain is linked to a mono-, di- or oligosaccharide. The source and nature of the carbohydrate, the length of the alkyl chain and its position on the sugar backbone offer a huge variety of surfactant structures and properties. In the literature, the most used process is the well-known glycosylation or transglycosylation reaction under Fisher's

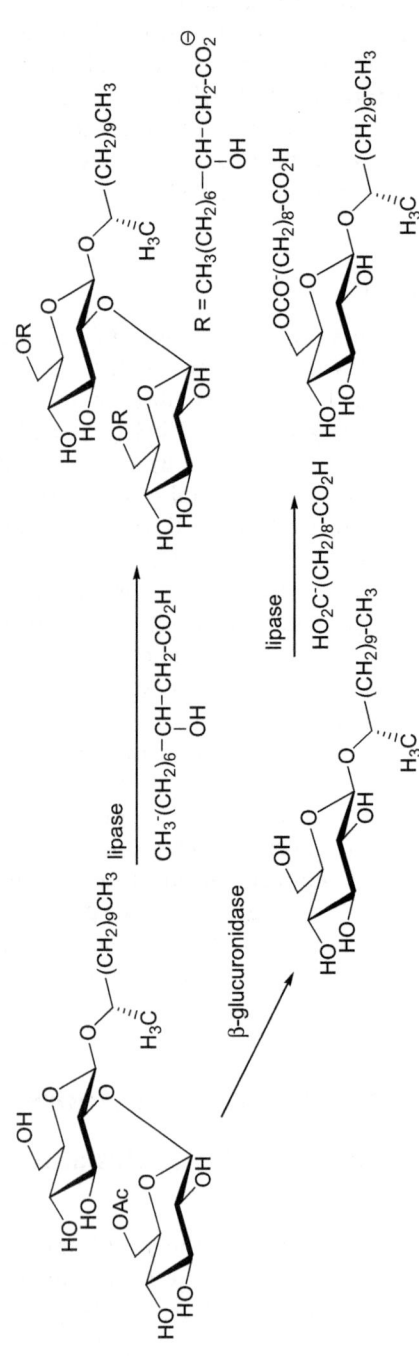

Scheme 5.6 Enzymatic synthesis from sophoroselipids.[26]

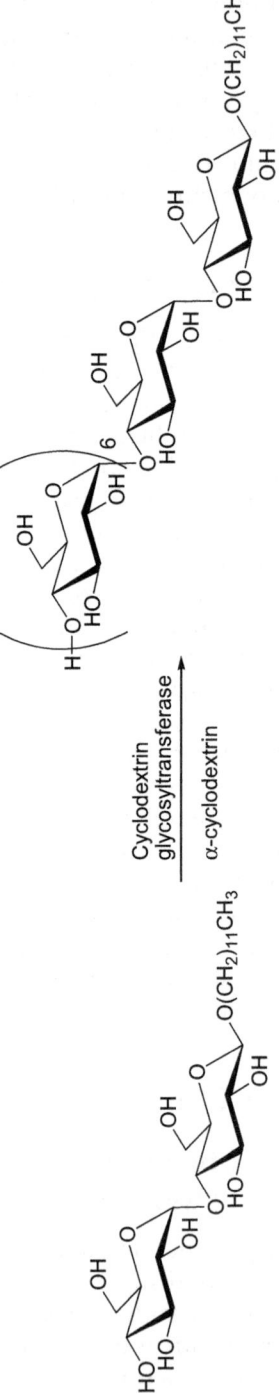

Scheme 5.7 CGTase synthesis of dodecyl β-D-maltooctaoside.[28]

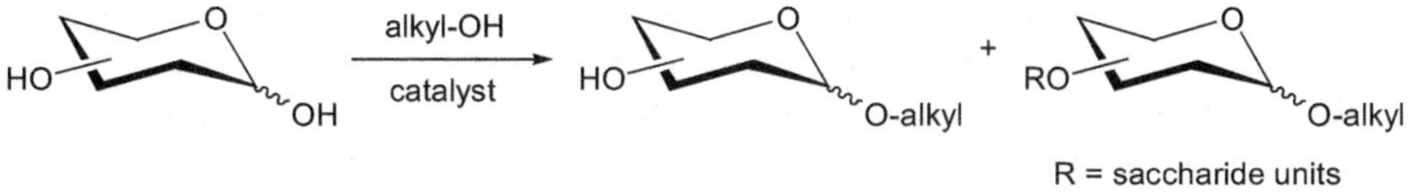

R = saccharide units

Scheme 5.8 Fischer glycosylation.

conditions and related ones; see Scheme 5.8. Most of the time, the products obtained are mixtures of mono-, di- and oligosaccharides.

APG can be classically obtained in one or two steps depending on the sugar solubility. Thus Aiad and co-workers obtained APG from glucose and different alcohols from octanol to tetradecanol by a two-step procedure.[30] First, butanol is glucosylated (*p*-toluenesulfonic acid (PTSA), 105 °C) and the more soluble intermediate is transformed in a transglucosylation reaction with the fatty alcohol (PTSA, 120 °C reduced pressure, 35–45% yield). Villandier and Corma synthesized the methyl glucoside as an intermediate ($H_3PW_{12}O_{40}$, 195 °C, 30 bar, 65% yield). After filtration, transglucosylation was performed with octanol or decanol to give a mixture of pyranoside and furanoside (Amberlyst 15Dry or H-Beta zeolite, 105 °C, reduced pressure, 39% yield for octyl glucofurano/pyranoside).[31] A modification of this process has been mentioned in a review of Lattes and co-workers with a one-step *in situ* glucosylation (butanol) and transglucosylation (dodecanol) (PTSA).[32] APG were also obtained by Li and co-workers in one step by glucosylation of dodecanol (PTSA, 120 °C, reduced pressure).[33] Similar conditions were used by Estrine and co-workers for the glucosylation of decanol (H_2SO_4, 95 °C, reduced pressure and additional furandicarboxylic acid, 59% yield in monoglucoside).[34] In the presence of furandicarboxylic acid, a less colored APG was obtained containing a higher content of monoglucoside. A comparison between oil bath heating and microwave irradiation was made by Martínez-Palou and co-workers[35] for the glucosylation of dodecanol (H_2SO_4, 70 °C) and Hricovíniová and Hricovíni[36] for the glycosylation of various primary (C4 to C18), secondary and tertiary alcohols by rhamnose (PTSA). Martínez-Palou and co-workers observed higher conversion and DP with conventional heating at 70 °C, probably due to longer reaction times and because the temperature was reached smoothly. Hricovíniová and Hricovíni obtained better yields in mono-rhamnoside (45% yield for decyl rhamnoside) under microwave irradiation than with classical heating (35% yield) even with a longer reaction time. The lack of a primary hydroxyl group on rhamnose tends to limit higher DP. Richel and co-workers reviewed the use of microwaves for the glycosylation reaction.[37] The glycosylation has been done in ionic liquids. Augé and Sizun reported the reaction between various monosaccharides and octanol or other alcohols.[38] The reaction was performed in [BMIM][OTf] (ionic liquid) or neat. Octyl glucoside was obtained with better yields in [BMIM][OTf] (ScOTf, 80 °C, 24 h, 74% yield) than neat (44% yield). Villandier and Corma detailed a synthesis of alkyl glucosides and alkyl xylosides from cellulose using an ionic liquid.[39] Cellulose was first hydrolysed ([BMIM][Cl], Amberlyst 15Dry, water, 100 °C). Alcohol (butanol to octanol) was then added to obtain the desired glucoside (90 °C, reduced pressure, 40 mol% of octyl glucoside). Goursaud and Benvegnu used a heterogeneous medium to obtain octyl β-D-fructopyranoside (THF, FeCl$_3$, 30% yield after acetylation).[40]

A three-step synthesis (acetylation, glycosylation, deacetylation) can afford monoglycosides without oligomers; see Scheme 5.9.

Activation of the anomeric position is classically realized by a Lewis acid. Hashim and co-workers glycosylated secondary alcohols (octan-2-ol to undecan-2-ol, BF$_3$OEt$_2$, DCM) with peracetylated galactose.[41] Modification of the reaction time allowed beta and alpha anomers to be obtained. Lehmler and co-workers used similar conditions (BF$_3$OEt$_2$, DCM) to glycosylate different primary alcohols (C14 to C19) with peracetylated glucose, galactose or maltose (45–78% yield).[42] Both anomers were obtained, the β-anomer under kinetic conditions (0 °C, 2 h then rt) and the α-anomer under thermodynamic conditions (0 °C, 20 min then 30 °C, 10 h). Aburto and co-workers used another catalyst (ZnCl$_2$, toluene, reflux, 24 h) to glycosylate alcohols (C8 to C18) with glucose or cellobiose.[43] Grand and co-workers studied the evolution of yields and the anomer ratio of the glycosylation of alcohols (C6 to C16) using fusion conditions with classical heating and under microwave irradiation (ZnCl$_2$, heating).[44] Krausz and co-workers have used similar conditions to glycosylate ω-undecenol with peracetylated glucose, galactose or lactose (58–85% yield).[45] Anomeric carbon can be activated differently as done by Liu and co-workers who carried out the glucosylation of fatty alcohols (C10 to C18) using a trichloroacetimidate.[46] A three-step synthesis was used by Obreza and co-workers[47] to substitute the 6-OH of galactose (R = C8 to C16); see Scheme 5.10.

The same reactions were used by Chaveriat and co-workers (R = octyl, dodecyl).[48] Moreover, they added a more or less flexible spacer arm (butyl, butynyl and phenyl) between the alkyl chain and the galactose unit to modify their properties. The introduction of an aliphatic spacer arm increased the amphiphilic properties of the compounds and the CMC values were 40–500 times lower than their analogs without a spacer arm.

5.2.2   Chemical Synthesis of Alkyl *N*-Glycosides

As in APG, the sugar unit is usually linked to the anomeric carbon *via* an oxygen atom (a classical glycosidic bond). One of the possible structural variations is to replace the oxygen atom by a nitrogen atom to obtain amino or amido derivatives; (Scheme 5.11).

Othman and co-workers realized a reaction between different heterocyclic amines and unprotected sugars (xylose, glucose and arabinose) to give glycosylamine-based surfactants.[49] Coma and co-workers synthesized different *N*-alkylglycosylamines by reaction of an alkylamine (alkyl = C6 to C18) with glucose or galactose (48–96% yield).[50] The authors studied their antifungal activities. Among these two series prepared from D-glucose and D-galactose,

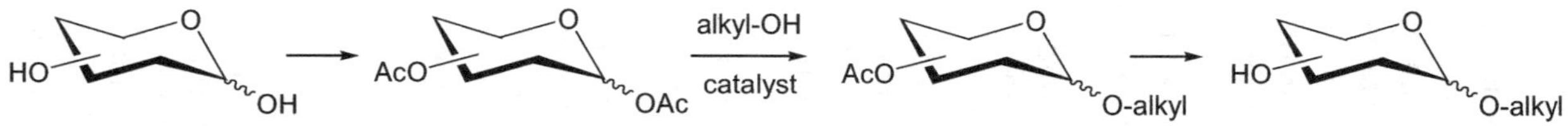

Scheme 5.9 Glycosylation with acetyl activation.

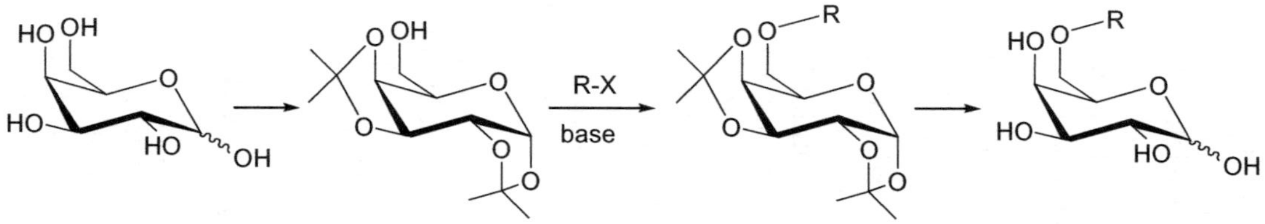

Scheme 5.10 Synthesis of 6-*O*-alkyl-D-galactose.[47,48]

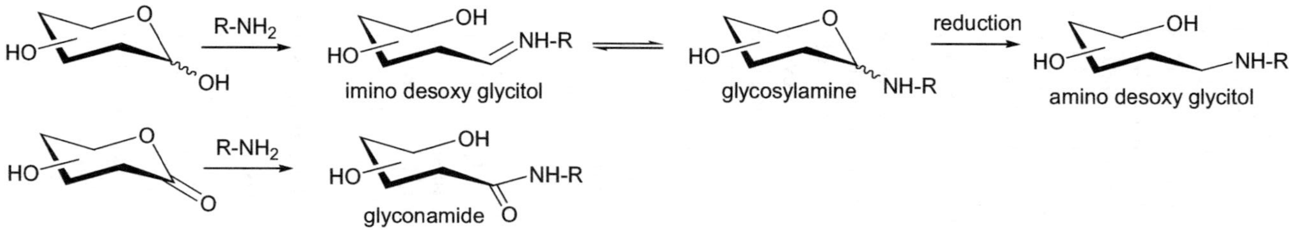

Scheme 5.11 *N*-Glycoside surfactants.

four of them, namely the ones prepared by using dodecylamine and hex-adecylamine as *N*-alkylating agents, fully inhibited the mycelium growth of *A. niger*. The galactosidic derivative C12Gal had a lower inhibiting concentration than the glucosidic derivative C12Glu, but both of these C12 compounds showed only a biostatic (and no biocidal) effect.

Abdel-Raouf and co-workers carried out the reaction between glucose and various alkylamines (alkyl = C8 to C12), but they reduced *in situ* the imino intermediate with $ZnCl_2$ to obtain the corresponding 1-(*N*-alkylamino)-1-desoxyglucitol.[51] Then, the authors ethoxylated the resulting surfactants; see Scheme 5.12 ($n = 9$, 13 and 22).

Dax and co-workers made an amphiphilic galactoglucomannan derivative by reductive amination of the reducing end of the polysaccharide using $NaBH_3CN$;[52] see Scheme 5.13.

The click reaction of glycosylazide and alkyne led to glycosyl alkyltriazole surfactants as reported by Mohammed and co-workers[53,54] (Scheme 5.14a) and Krausz and co-workers[45] (Scheme 5.14b).

Wang and co-workers obtained a trisiloxane-tailed lactobionamide surfactant by reaction of lactobionic acid and a substituted amine;[55] see Scheme 5.15a. They also obtained amphiphilic lactobionamide-grafted polysiloxanes in a similar way;[56] see Scheme 5.15b ($R = \beta$-Gal, $k = 0$). The polymers obtained have a higher surface activity in water solution than those of conventional carbon–carbon chain glycopolymers. Zeng and co-workers have synthesized similar silicone gluconamide ($R = H$, $k = 1$) and lactobionic acid ($R = \beta$-Gal, $k = 1$) surfactant derivatives;[57] see Scheme 5.15b.

Similar surfactants and cationic surfactants have been synthesized by Han and co-workers from D-glucono-1,5-lactone (alkyl = C8 to C12);[58] see Scheme 5.16.

Han and co-workers also reviewed carbohydrate-modified silicone surfactants.[59] The amidation reaction was used by other groups to obtain glyconamide-based surfactants: Blanzat and co-workers[60] (Scheme 5.17a), Yoshimura, Torigoe and co-workers[61,62] (Scheme 5.17b), Zeng and co-workers[63] (Scheme 6.17c), Zhang and Qiao,[64] Du and co-workers[65] (Scheme 5.17d).

5.2.3   Chemical Synthesis of Alkyl *C*-Glycosides

In the case of *C*-glycosides, the sugar moiety is linked to the hydrophobic chain without a glycosidic bond. Derivatives of 5-hydroxymethylfurfural, isosorbide and related structures prepared from sugar are not included in this paragraph.[66] A one-step synthesis of *C*-glycoside surfactants has been described by Benvegnu and co-workers using a Horner–Wadsworth–Emmons reaction;[67] see Scheme 5.18. The reaction was conducted in water for water soluble phosphonates or neat (alkyl = C1 to C17 and (*Z*)-heptadec-8-enyl) with lactose, galactose, xylose or glucose to give *C*-glycosides with various ratios of *C*-pyranoside/*C*-furanoside structures and alpha/beta configurations (25–85% yield). The crude product isomerized in a basic water medium to highly thermodynamically favored beta *C*-pyranoside structures.[68]

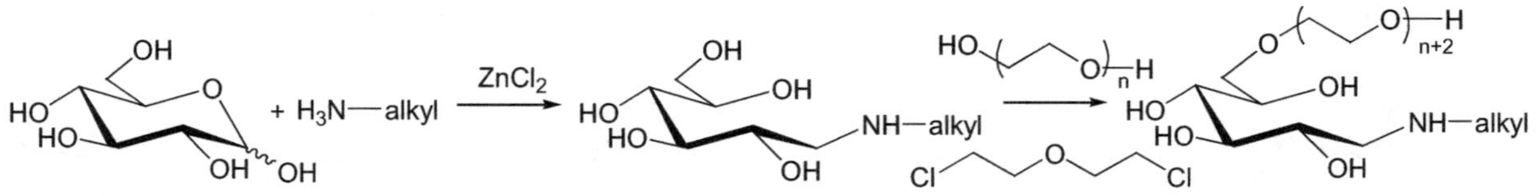

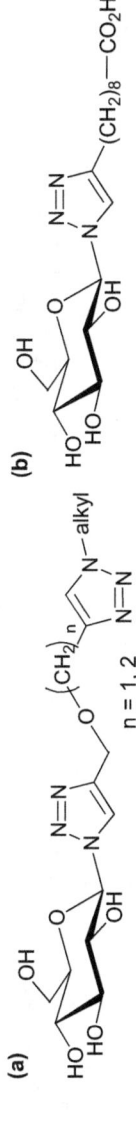

Scheme 5.12 Ethoxylated alkyl *N*-glycosides.[51]

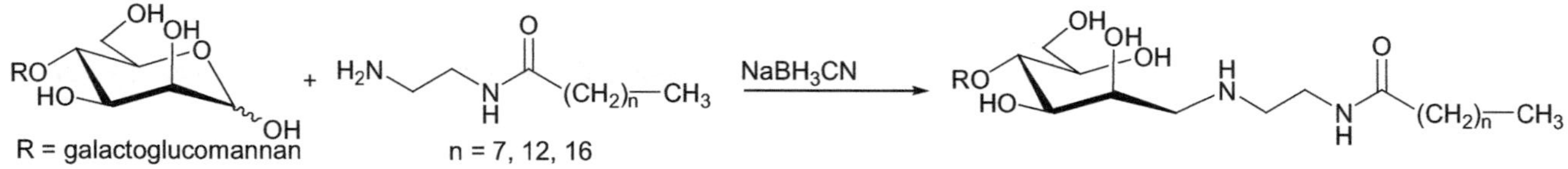

Scheme 5.13 Synthesis of an amphiphilic galactoglucomannan derivative.[52]

Scheme 5.14 Click reaction.[45,53,54]

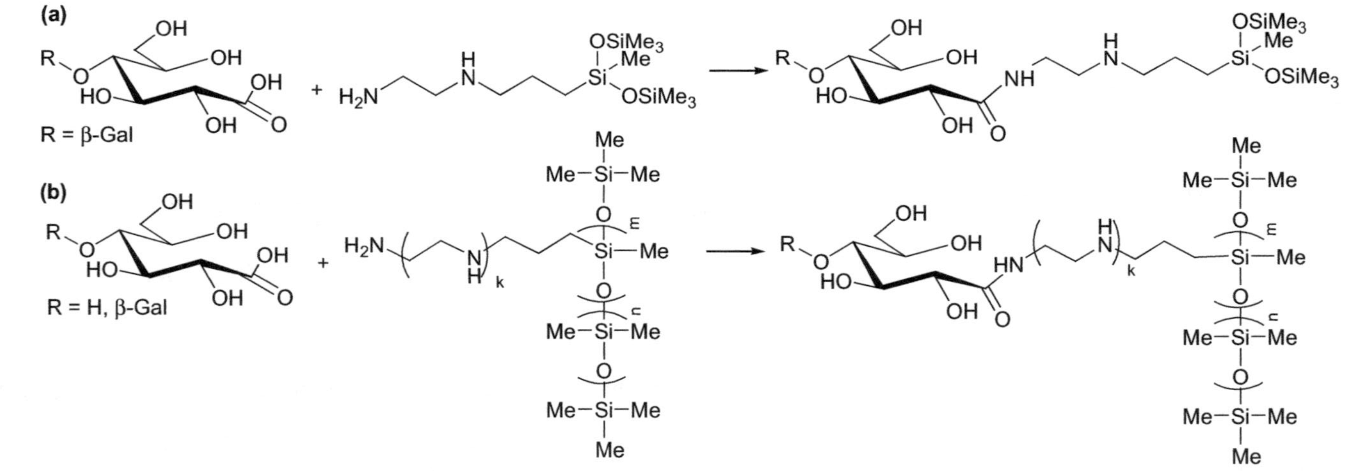

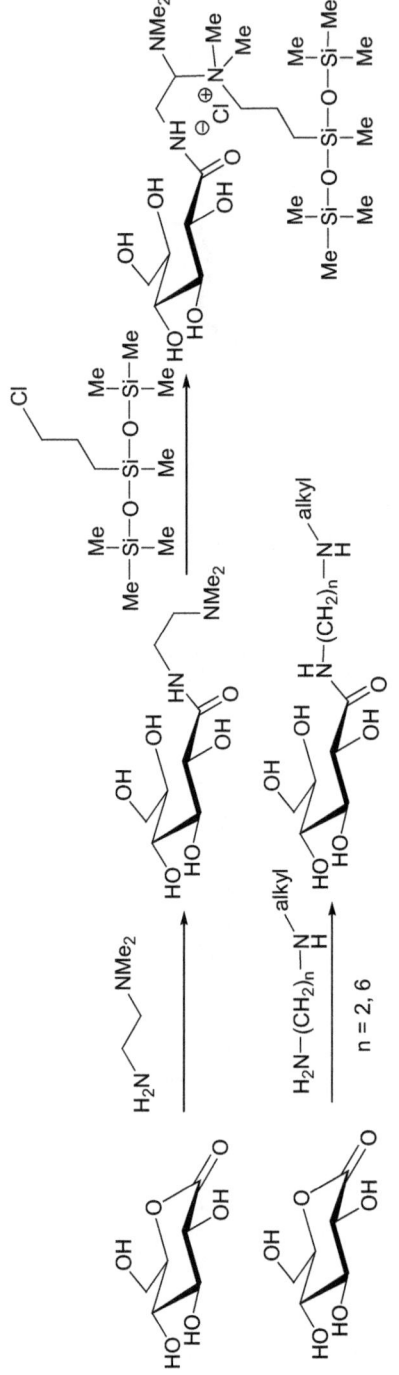

Scheme 5.15 Glyconamide-modified siloxane and polysiloxanes.[55-57]

Scheme 5.16 Gluconamide-modified silicone surfactants.[58]

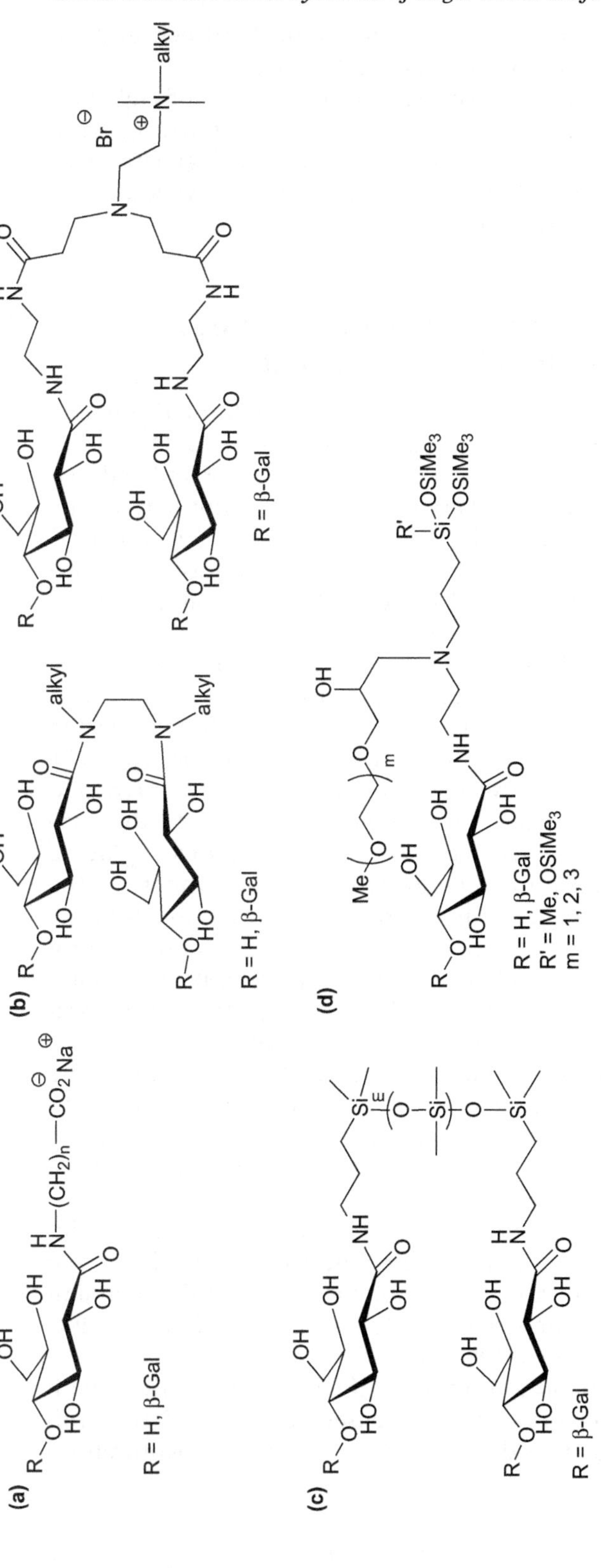

Scheme 5.17 Gluconamide-based surfactants.[60–65]

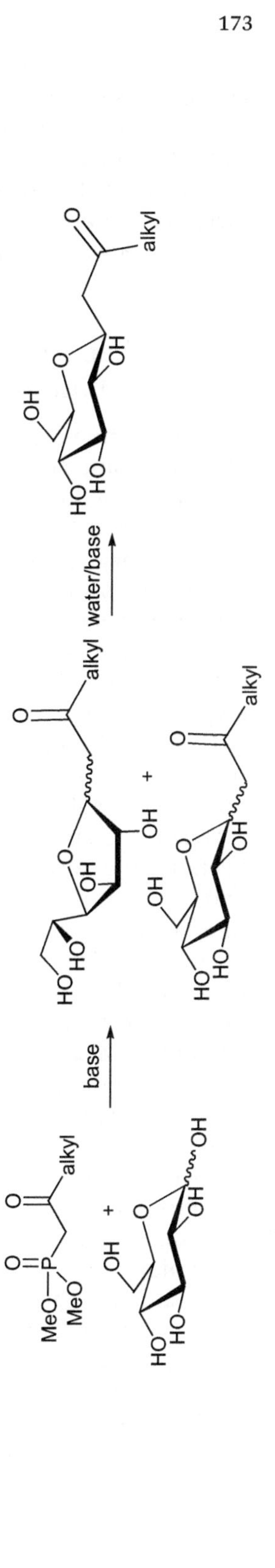

Scheme 5.18 Synthesis of *C*-glycoside surfactants from glucose.[68]

The same ketone was used in an aldol condensation by Beach and co-workers;[69] see Scheme 5.19 (alkyl = C5 to C11, 38–60% yield).

Corma and co-workers studied the synthesis of monoesters of D-glucitol with the objective of reducing by-products (di, triesters, anhydro);[70] see Scheme 5.20. The reaction of 1,2:3,4,5,6-tri-O-isopropylidene-D-glucitol with oleic acid catalyzed by beta zeolite gave the monoester (46% yield), a few diester (1% yield) and anhydro compounds.

5.2.4   Chemical Synthesis of Non-classical Structures: Gemini, Bolaform, Dicephalic, Catanionic

Among the non-classical surfactant structures, gemini, bolaform, dicephalic and catanionic surfactants can be differentiated (see Scheme 5.21).

Gemini are duplicated surfactants linked together by their polar heads *via* a spacer arm. Mohammed and co-workers synthesized gemini surfactants based on bis-(alkyltriazolyl)mannitol (alkyl = C7 to C12);[71] see Scheme 5.22. The gemini structure was obtained by a double click cycloaddition on 3,4-di-O-propargyl-D-mannitol (80–89% yield). The aim of using mannitol is to improve water solubility and to reduce toxicity.

Negm and Mohamed also used a sugar backbone (D-glucose or D-fructose) as a spacer arm between two ammonium-alkanoate structures (alkanoate = C12 to C18 and oleate);[72] see Scheme 5.23. Glucose or fructose was first diacylated by bromoacetyl groups, which were then converted to the final surfactants by nucleophilic substitution with the corresponding amines. The authors tried to correlate the surfactant structure and the interfacial properties to the anti-bacterial and antifungal activities. Octadecyl and oleate derivatives of glucose and fructose showed the higher free energy change of adsorption at the interfaces; this could be correlated to their adsorption on the cell membranes. These derivatives showed higher bactericidal activities than that of the control (cetyl trimethyl ammonium bromide). The synthesized sugar-based amphiphiles showed equal or greater antifungal efficacy on *Aspergillus niger* and *Aspergillus flavus* strains compared to the control grisofulvine.

The spacer chain of gemini surfactants can contain ether groups as for the structures synthesized by Liu and co-workers;[46] see Scheme 5.24 (alkyl = C10 to C18). Only the beta anomer was obtained in the glycosylation conditions used by the authors by activation of a glucosyl trichloroacetimidate (89% yield for C12).

Yoshimura and co-workers[61] synthesized gemini lactobionamide or gluconamide based surfactants to study their interfacial behaviour, see Scheme 5.25. The gemini surfactants were obtained by a double condensation reaction between lactobionic acid or gluconic acid with 1,2-di(alkylamino)ethane (alkyl = C8 to C14; 9–20% yield).

Lequart and co-workers proposed to improve the water solubility of benzimidazolones by substitution with a glucose based surfactant.[73] Benzimidazolone structures were substituted twice with a 3-O-alkyl-D-glucofuranose

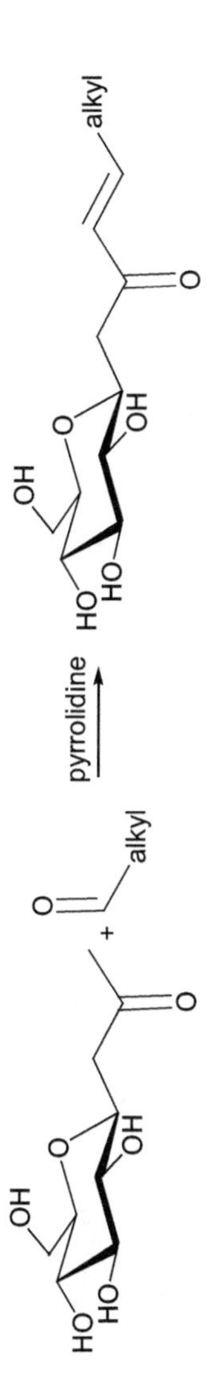

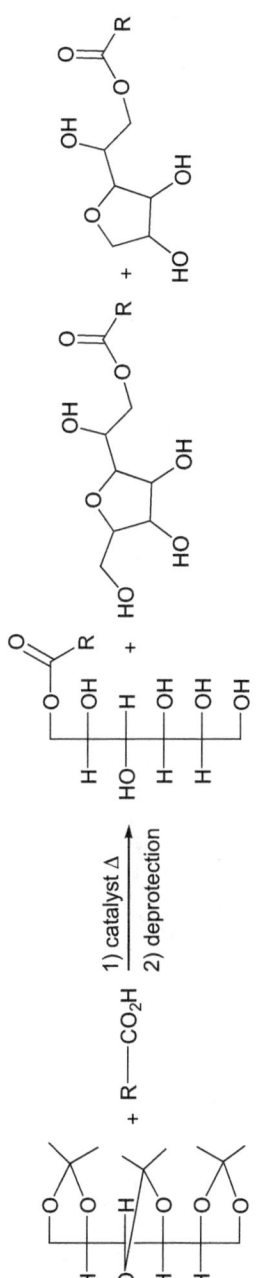

Scheme 5.19 *C*-Glucoside obtained by aldol condensation.[69]

Scheme 5.20 Synthesis of monoesters of sorbitol.[70]

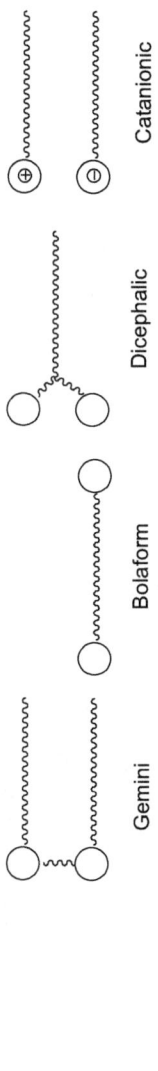

Scheme 5.21 Non-classical surfactant structures.

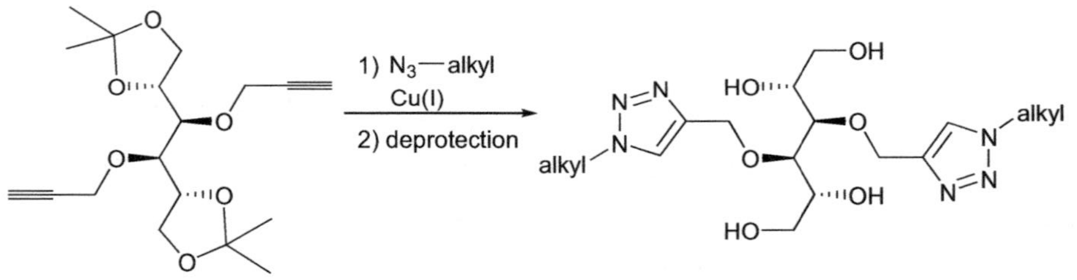

Scheme 5.22   Gemini surfactants obtained by click chemistry.[71]

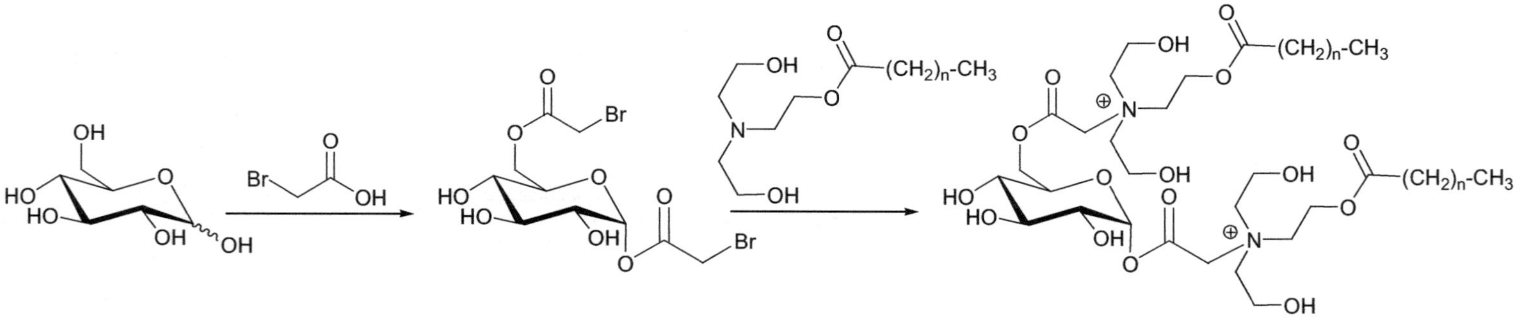

Scheme 5.23   Synthesis of cationic gemini surfactants.[72]

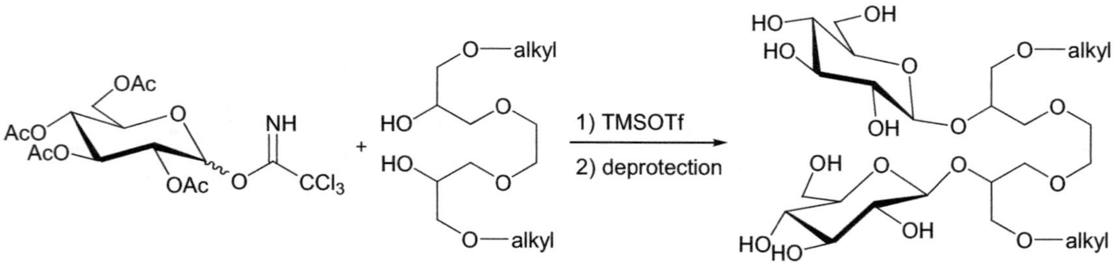

Scheme 5.24   Preparation of gemini surfactants through glycosylation.[46]

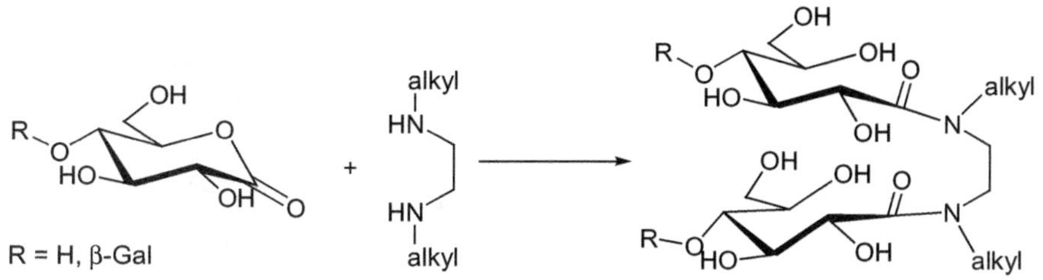

Scheme 5.25 Synthesis of amide gemini surfactants.[61]

derivative (alkyl = C4 to C12, 70–93% yield) before a deprotection step (70–95% yield); see Scheme 5.26.

Dicephalic structures, which contain two polar heads linked together at one end of a hydrophobic chain, were synthesized by Rodríguez-Abreu and co-workers[74] to modify the physico–chemical properties of benzimidazolones. Epichlorohydrin was first substituted by two protected benzimidazolone units; see Scheme 5.27. Reaction of the intermediate with an alkyl bromide (alkyl = C10 to C16) allowed the introduction of the hydrophobic chain. After deprotection of the benzimidazolone moiety, a reaction with an anhydro glucose derivative (92–95% yield) and deprotection (78–83% yield), gave the final dicephalic surfactants.

Pucci and co-workers synthesized various fluoroalkyl surfactants.[75] Some of them have a dicephalic structure. The fluoroalkyl chain was added to the sugar structure by Michael addition of a thiol to an acrylamide (56% yield); see Scheme 5.28a. The authors also used a radical addition on a but-3-enamide derivative (47% yield); see Scheme 5.28b. The authors made other structural variations such as polar heads comprising three glucose residues, disaccharides, and open-chain sugar units.

Yoshimura and co-workers[76] prepared dimers (Scheme 5.29), or trimers of a methacrylate glucoside by radical polymerisation and added the hydrophobic chain by an amidation reaction (alkyl = C11, C17).

Zhang and Qiao[64] and Du and co-workers[65,77] mixed a sugar head (lactobionamide, gluconamide) and oligo(ethylene-oxide) as the polar heads of a dicephalic surfactant; see Scheme 5.30.

Torigoe and co-workers synthesized dicephalic and dendrimer structures from a poly(amidoamine) dendron.[62] Lactobionic acid reacted with the amino groups located at each end of the dendron; see Scheme 5.31.

Heidelberg and co-workers used a double Staudinger reaction to introduce fatty acyl chains to a bis-amino glucoside or lactoside;[78] see Scheme 5.32 (alkyl = C7 to C11). In the case of the lactoside derivative, the reaction failed to give the dicephalic structure.

Debuigne and co-workers worked on the synthesis of various structures with surfactant properties obtained by photochemical thiol-ene or thiol-yne reactions.[79] The reaction between tetradec-1-yne and a thiol derivative of mannose gave, after UV irradiation, the corresponding dicephalic surfactant (78% yield); see Scheme 5.33.

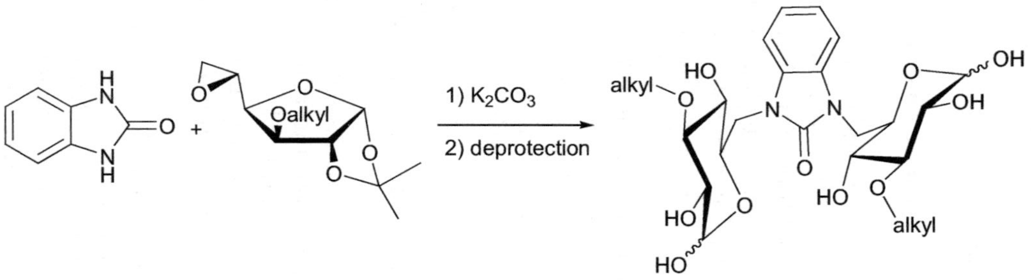

Scheme 5.26   Benzimidazolone gemini structures.[73]

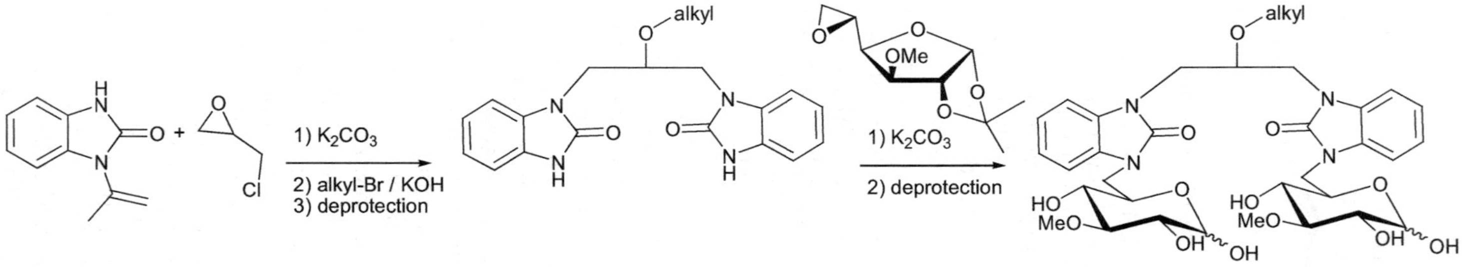

Scheme 5.27   Synthesis of dicephalic benzimidazolone surfactants.[74]

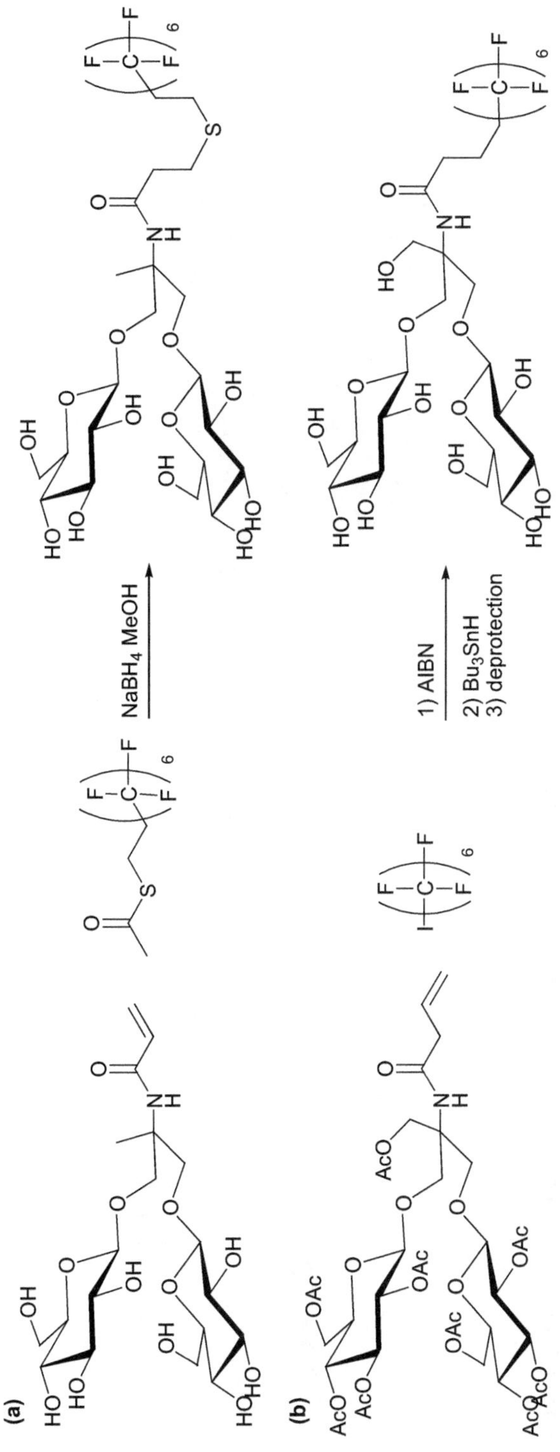

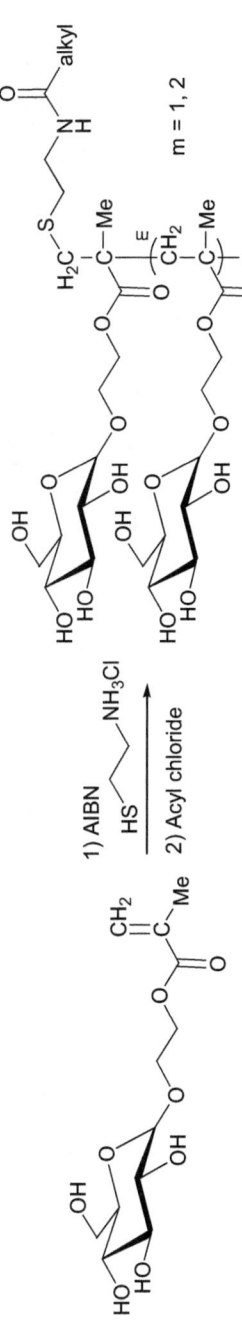

Scheme 5.28 Synthesis of fluoroalkyl dicephalic surfactants.[75]

Scheme 5.29 Synthesis of dicephalic surfactants by radical polymerisation.[76]

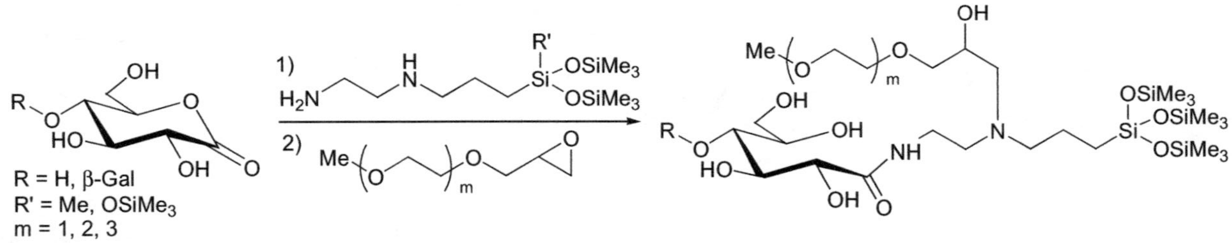

Scheme 5.30 Dicephalic surfactants obtained by epoxide opening.

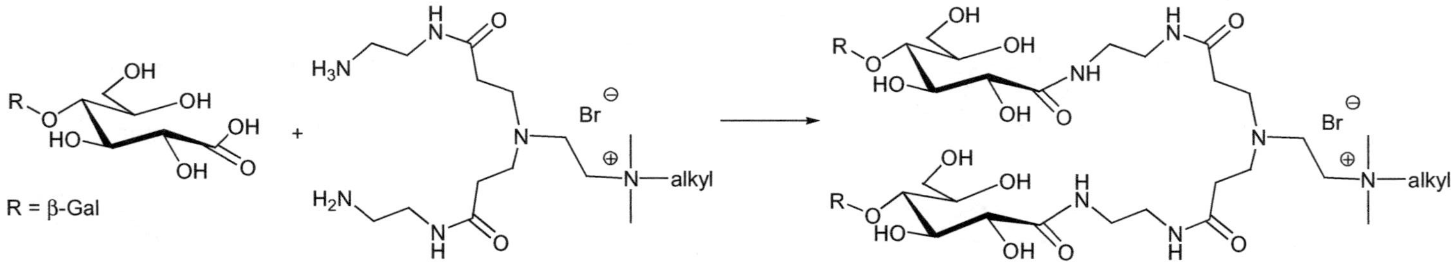

R = H, β-Gal
R' = Me, OSiMe₃
m = 1, 2, 3

R = β-Gal

Scheme 5.31 Poly(amidoamine) structures. [62]

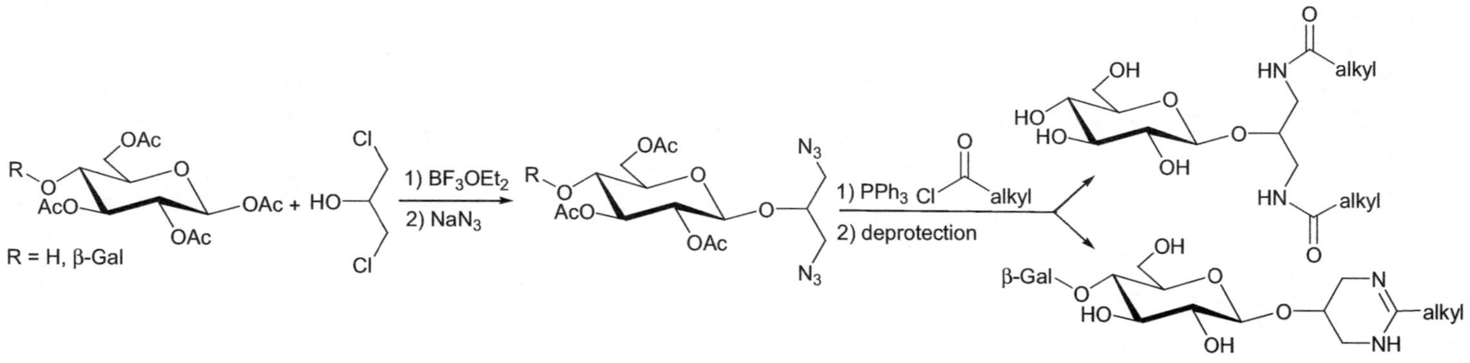

Scheme 5.32 Dicephalic surfactants by a double Staudinger reaction.[78]

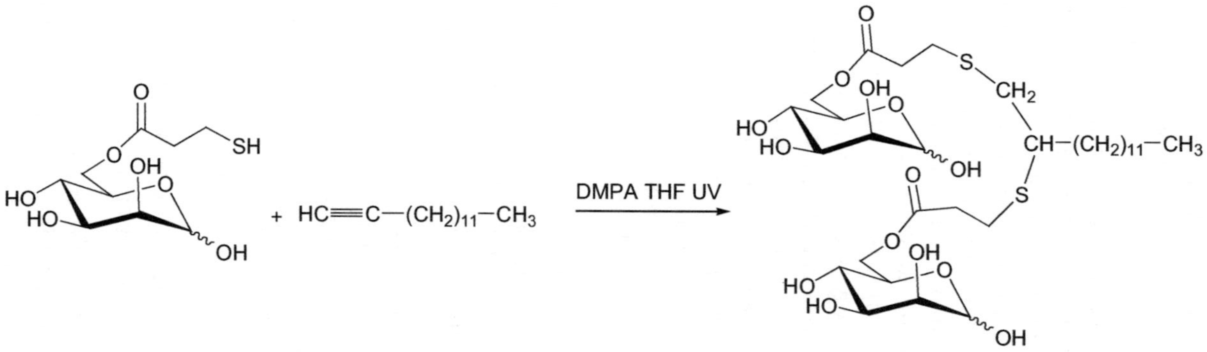

Scheme 5.33 Dicephalic structures by thiol-yne addition.[79]

Using tetradeca-1,13-diene, the authors obtained a structure with the two sugar units located at each end of the aliphatic chain (91% yield); see Scheme 5.34. Such structures are called bolaforms or bolaamphiphiles.

Bola surfactants with a polysiloxane spacer chain were synthesized by condensation of a diaminopolysiloxane with lactobionic acid by Zeng and co-workers[63] or condensation of glucose with an epoxide by Racles and Cozan;[80] see respectively Scheme 5.35a and b. Racles and Cozan also synthesized a polymeric silicone glucose-based surfactant. Other structures of silicone-containing gemini sugar-based surfactants can be found in reviews by Han and co-workers.[59,81]

Bouquillon and co-workers synthesized bolaamphiphiles using an olefin cross-metathesis reaction with xylose to obtain a dixylosylalkene or a dixylosylalkane after hydrogenation;[82] see Scheme 5.36a ($n = 3$, 7, 8). Olefin cross-metathesis gave the dimeric structures (69–90% yield, mixture of E/Z isomers). The authors tried to synthesize the dicarboxylate derivative in the same way (see Scheme 5.36b),[82b] but hydrogenolysis of the benzyl protecting groups was difficult and incomplete.

Krausz and co-workers synthesized other bolaform structures with two sugar units using the same reaction.[45] Olefin cross-metathesis with ω-undecenyl gluco, -galacto or -lactoside (alpha or beta anomers) gave the bolaform structure (61–85% yield); see Scheme 5.37a. The authors also realized the synthesis of bolaamphiphiles using a click cycloaddition; see Scheme 5.37b (53% yield). The authors employed the same reaction to synthesize a bolaamphiphile with one sugar unit and one carboxylate as the polar heads; see Scheme 5.37c (90% yield) and Scheme 5.37d (43% yield).

This reaction has already been employed by the authors to build "star-like" surfactants.[83] Blanzat and co-workers synthesized similar sugar-based bolaamphiphiles.[60] Different glyconamides were obtained by reaction of lactobionic or gluconic acids with various ω-amino acids (50–90% yield); see Scheme 5.38a ($n = 4$, 6, 10). The authors used them to make catanionic aggregates; see Scheme 5.38b.

The same group also prepared other catanionic surfactants from a phosphinic acid derivative and *N*-hexadecylamino-1-deoxylactitol;[84] see Scheme 5.38c.

5.2.5 Chemical Synthesis of Functionalized Sugar-based Surfactants

To change the interfacial properties, physico–chemical properties or biological activities, the sugar moiety is usually modified as well as the hydrophobic chain. The primary alcohol found in most of the saccharides can be oxidized into an aldehyde or a carboxylic acid and transformed into their derivatives. Richel and co-workers studied the glycosylation of glucuronic acid and galacturonic acid under microwave irradiation using sulfuric acid loaded onto silica and other catalysts;[85] see Scheme 5.39. Reaction with

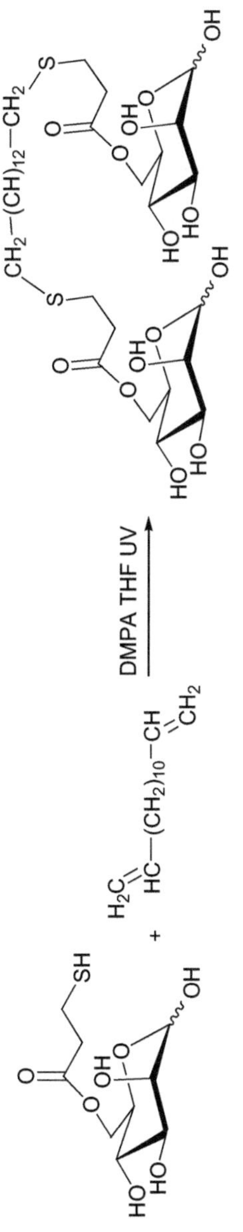

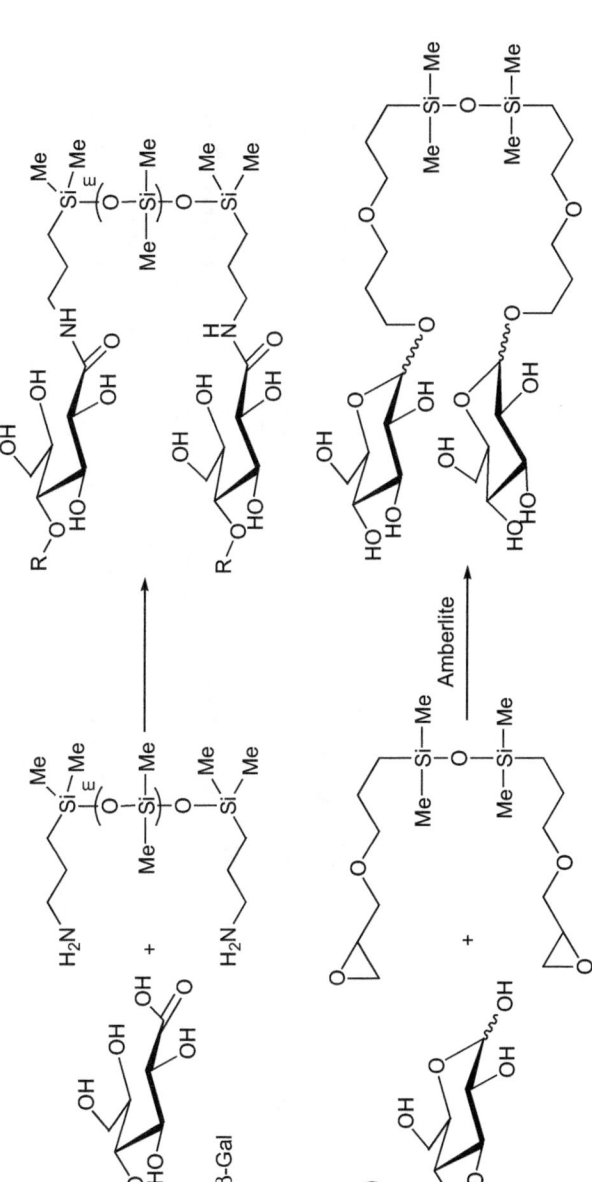

Scheme 5.34 Sulfur-containing bolaform surfactants.[79]

Scheme 5.35 Synthesis of siloxane bolaform surfactants.[63,80]

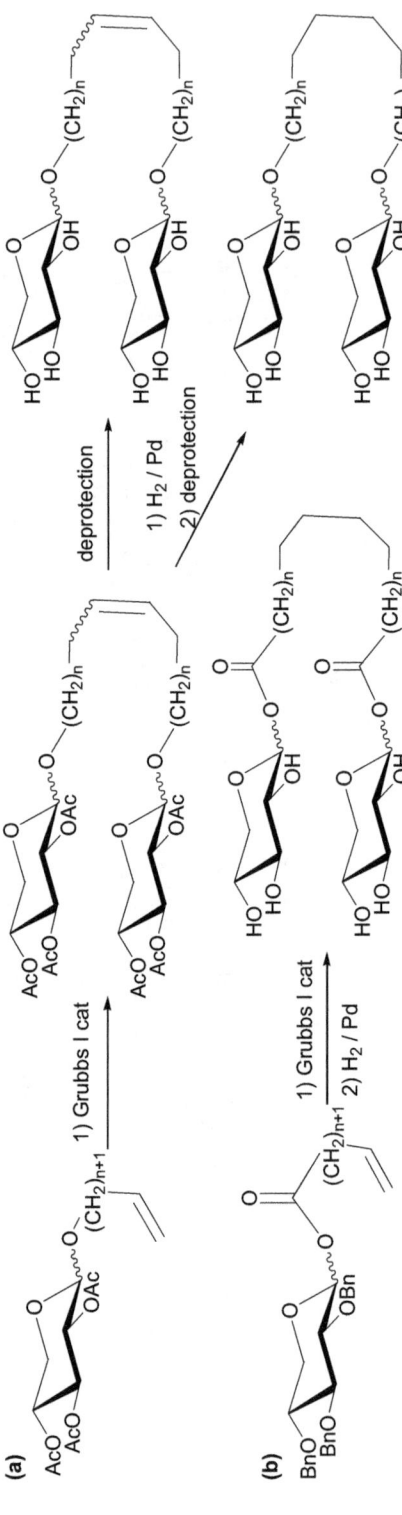

Scheme 5.36 Synthesis of bolaform structures by metathesis.[82]

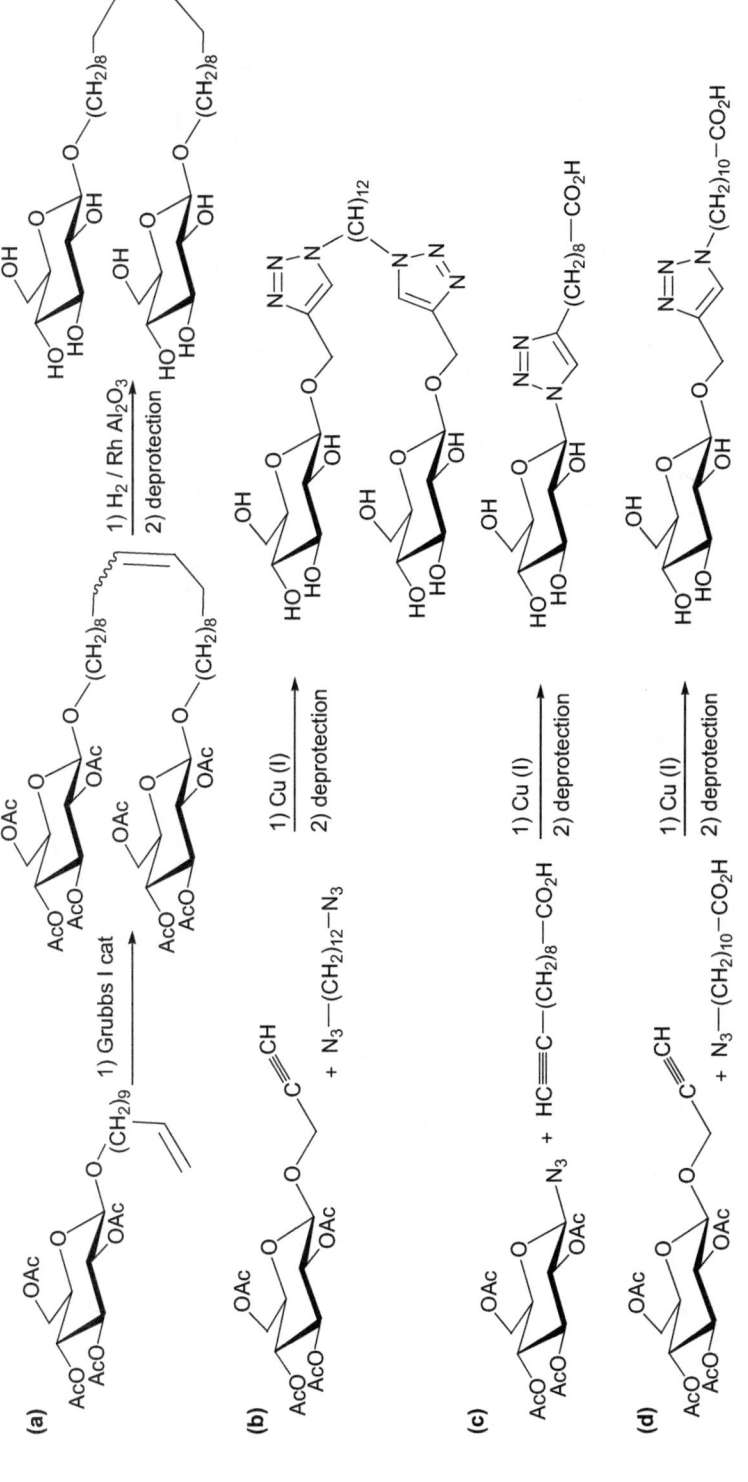

Scheme 5.37 β-D-Glucoside-based bolaamphiphiles.[45]

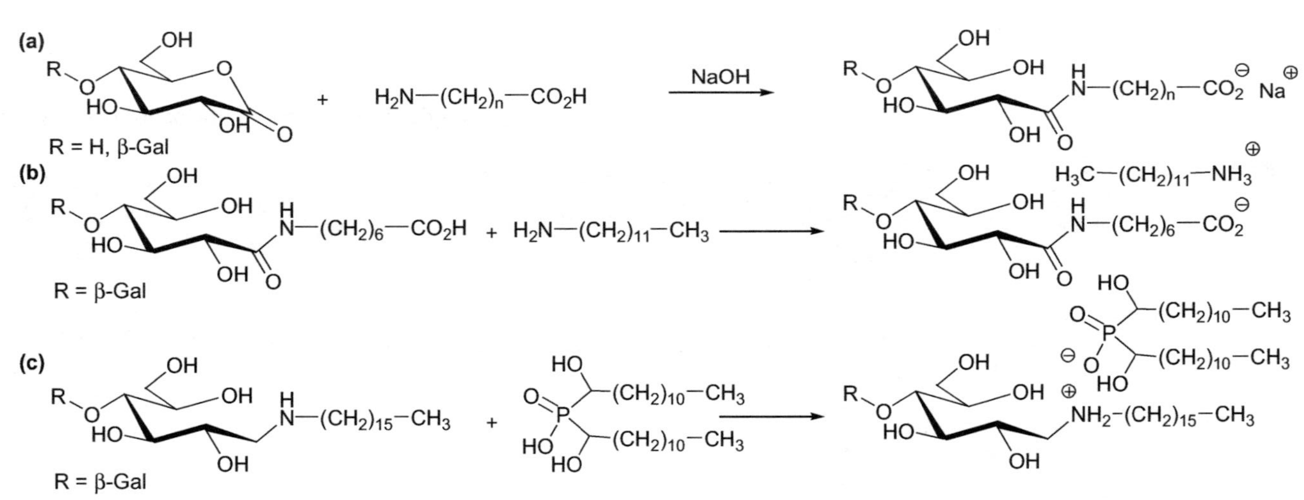

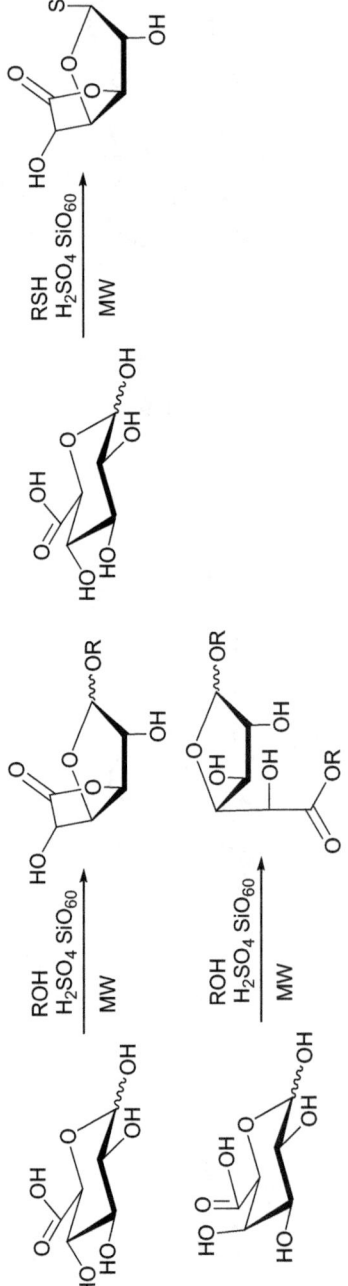

Scheme 5.38 Catanionic surfactants.[60,84]

Scheme 5.39 Uronate- and lactone-based surfactants.[85]

various alcohols (C4 to C12 and other alcohols) and glucuronic acid led to the corresponding alkyl glucofuranosidurono-6,3-lactones (62–98% yield, beta major). The same reaction with octanethiol or tetradecanethiol led to the corresponding lactone (97% and 68% yields respectively). Reaction with alcohols (C4 to C12) and galacturonic acid led to the corresponding major alkyl (alkyl galactofuranosid)uronate (49–86% yield, beta major).

Laurent and co-workers synthesized glucuronamide and galacturonamide derivatives using different activation modes of the carboxylic acid (80–93% yield with oxalyl chloride);[86] see Scheme 5.40 (alkyl = C4 to C18).

Kovensky and co-workers synthesized glucuronic-based surfactants from glucose for metal chelation and flotation. Selective oxidation of octyl D-glucoside led to octyl D-glucosiduronic acid (alpha 85%, beta 74% yield). Activation of the uronic acid and reaction with hydroxylamine or various amino acids allowed the synthesis of the corresponding hydroxamic acid (alpha 48%, beta 78% yield)[87] and glycoamino acids (75–90% yield)[88] to study their chelation ability (see Scheme 5.41).

Goursaud and Benvegnu[40] have added a glycine betain moiety on a fructopyranoside through an amide linkage (66% yield); see Scheme 5.42.

Mravljak and Pečar synthesized a glucosamine-based surfactant containing a spin probe for EPR spectroscopy investigations of the extracellular matrix;[89] see Scheme 5.43.

Obreza and co-workers[47] synthesized oxime derivatives from 6-*O*-alkyl-D-galactose and *N*-acyl-D-glucosamine to evaluate their antioxidant activities; see Scheme 5.44 (alkyl and acyl = C8 to C16). A proportional relationship of antioxidant properties and concentration was noted for the glycolipid mimetics, except for derivatives of galactose, which expressed no antioxidant activity. Amphiphilic derivatives of glucose express significant differences in antioxidant properties, depending on whether they consist of two or only one oxime functionality. The compounds with two oxime groups showed higher antioxidant activity compared to compounds with only one oxime group. Among compounds with two oxime groups, the highest scavenging activity was observed for the C14 derivative.

Palandoken and co-workers synthesized oximes in one step from sugar (glucose, maltose, glyceraldehyde) and *O*-alkylhydroxylamine;[90] see Scheme 5.45 (alkyl = C10, C12 and adamantyl).

Heidelberg and co-workers made surfactants derived from 6-azido-6-desoxy-D-glucose by a Staudinger reaction and acylation;[91] see Scheme 5.46 (alkyl = C7 to C15, 48% yield). Reaction with branched acyl chloride is also described. The authors studied their ability to be used as water in oil emulsifiers.

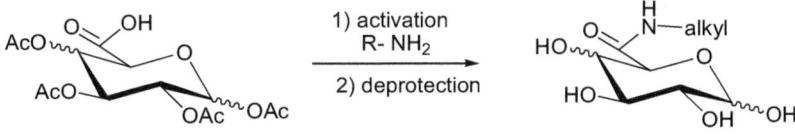

Scheme 5.40 Uronamide surfactants.[86]

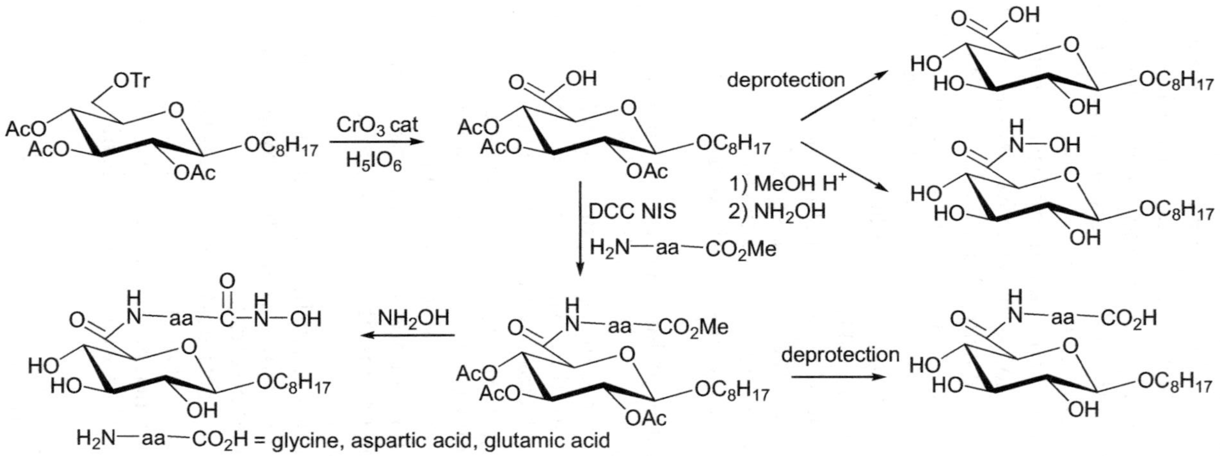

Scheme 5.41 Synthesis of uronic acid and amino acid containing surfactants.[87,88]

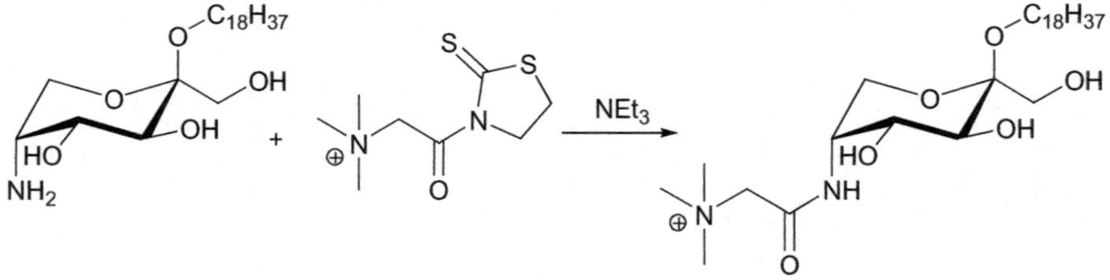

H₂N—aa—CO₂H = glycine, aspartic acid, glutamic acid

Scheme 5.42 Fructopyranoside glycine betain surfactant.[40]

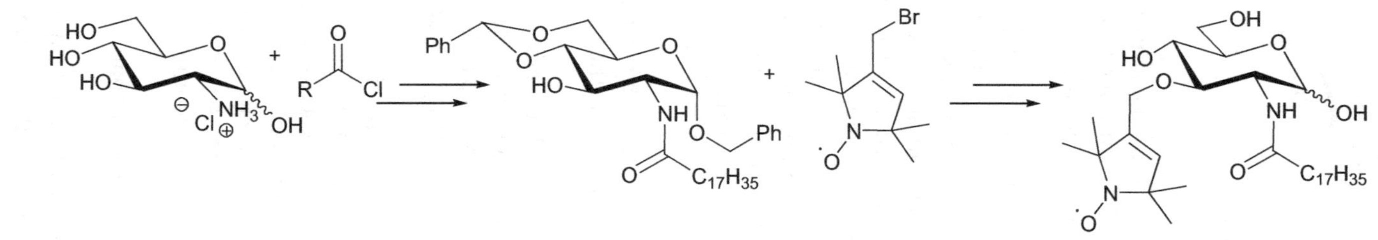

Scheme 5.43 Synthesis of a spin probe.[89]

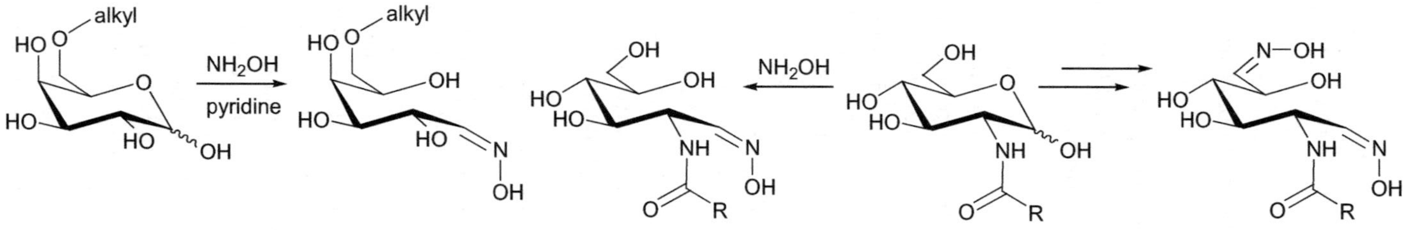

Scheme 5.44 Oxime surfactants.[47]

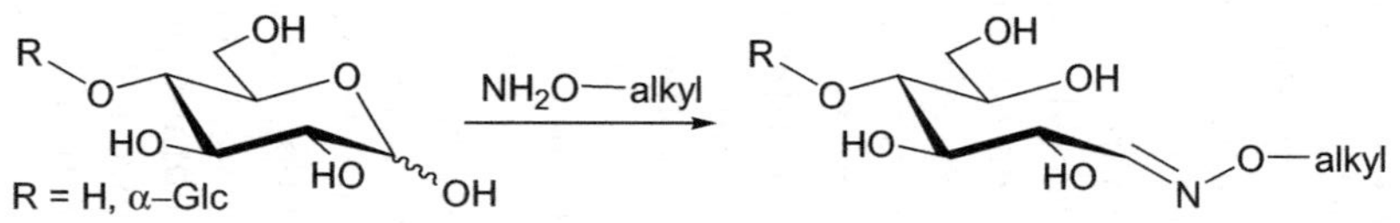

Scheme 5.45 One-step synthesis of oxime surfactants.[90]

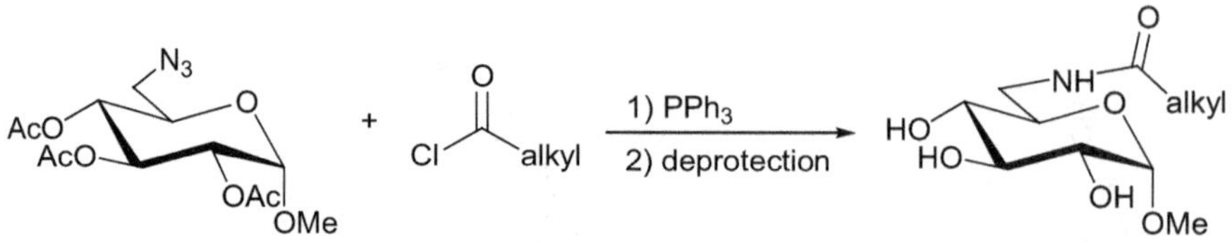

Scheme 5.46 Surfactant obtained from 6-azido-6-desoxy-D-glucose.[91]

Abdel-Raouf and co-workers modified the polar head, adding to a glucose unit a polyethylene glycol chain (400–4000 g mol^{-1}).[92] The authors introduced this chain after acylation with a fatty acid, see Scheme 5.47 (R = C15, C17), or adipic acid.

Fluoroalkyl-based surfactants behave differently from the classical alkyl-based surfactants as the chain is stiffer and interacts more weakly with the hydrophobic part of membrane proteins, therefore these surfactants with decreased detergent properties can be used to extract membrane proteins without denaturing them. They are also interesting for drug delivery as they self-assemble in more stable supramolecular structures. Lehmler and co-workers[42] synthesized fluoroalkyl glucoside from peracetylated glucose, galactose or maltose (BF$_3$OEt$_2$) to obtain alpha or beta anomers depending on the experimental conditions; see Scheme 5.48. Activation of glucose by the bromide derivative led to the orthoester instead of the glucoside.

Debuigne and co-workers introduced the fluorinated chain by enzymatic acylation of mannose;[93] see Scheme 5.49.

Lehmler and co-workers[94] introduced a fluoroalkyl chain by click cycloaddition on a propargyl β-D-xylopyranoside (43% yield for the two steps); see Scheme 5.50.

Pucci and co-workers[75] synthesized and studied fluoroalkyl glucose-based surfactants with one, two or three glucose units and a fluor atom or an ethyl or propyl group at the end of the fluoroalkyl chain; see Scheme 5.51. Addition of an alkyl group, *i.e.* ethyl or propyl, increases the affinity toward membrane proteins.

Silicone-based surfactants have special behavior as silicone is both hydrophobic and oleophobic. Han and co-workers reviewed such surfactants with a sugar as a polar head.[59] The same group had already described the synthesis of glycoside-based trisiloxane surfactants;[95] see Scheme 5.52a. The glycosylation led to the major mono glycoside together with a small amount of APG. Hydrosilylation of the resulting alkene led to a siloxane-based surfactant. Racles and co-workers used cyclosiloxane-based surfactants obtained by similar reactions;[96] see Scheme 5.52b.

Debuigne and co-workers synthesized a siloxane derivative of mannose (79% yield);[93] see Scheme 5.53.

Shmendel and co-workers and Maslov and co-workers synthesized different mannose[97] (see Scheme 5.54) and galactose[98] neoglycolipids for DNA delivery.

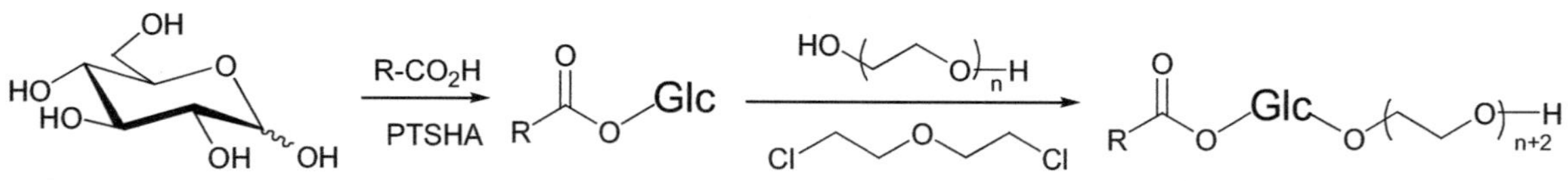

Scheme 5.47 Polyethylene glycol modified sugars.[92]

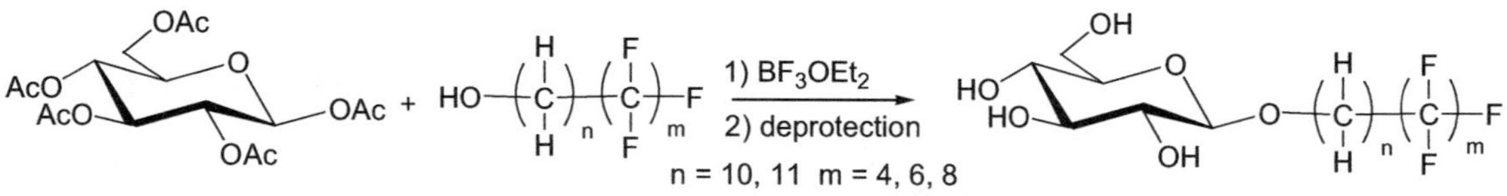

Scheme 5.48 Fluoroalkyl glycosides.[42]

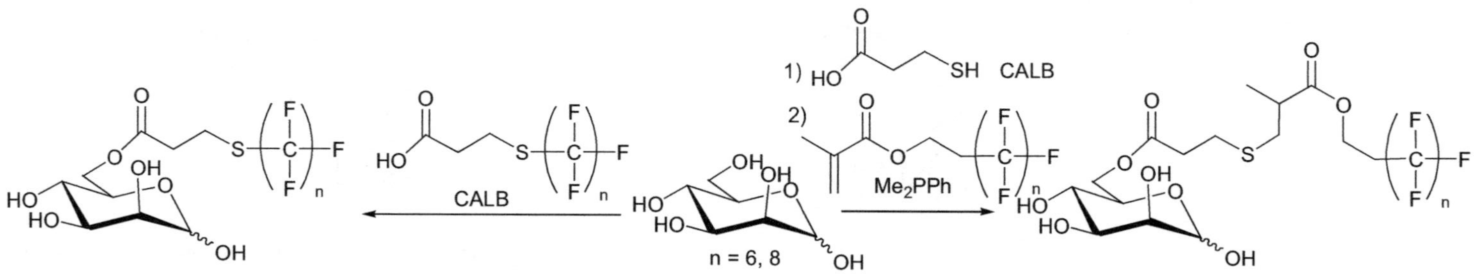

Scheme 5.49 Fluorinated sugar-based surfactants.[93]

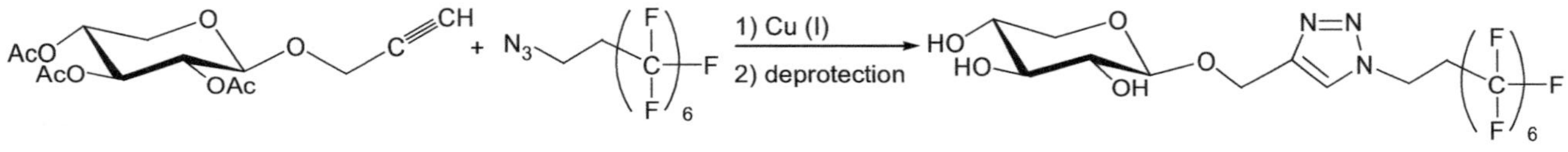

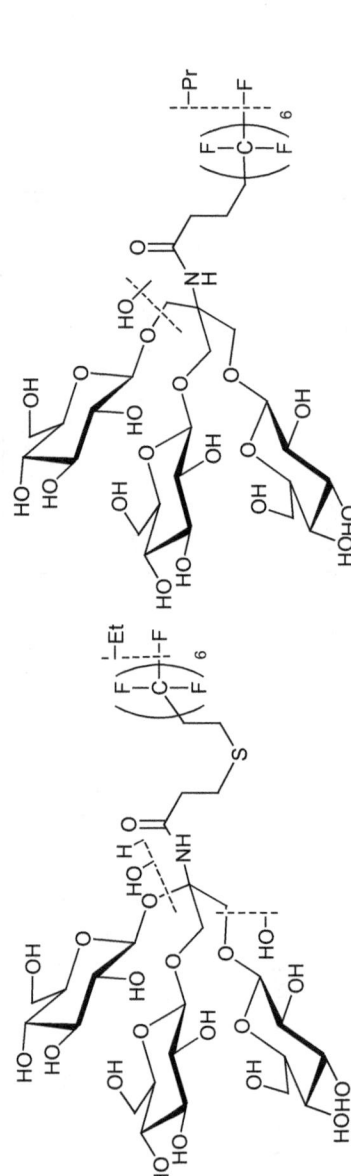

Scheme 5.50 Fluoroalkylated click surfactants.[94]

Scheme 5.51 Fluoroalkyl oligosaccharides.[75]

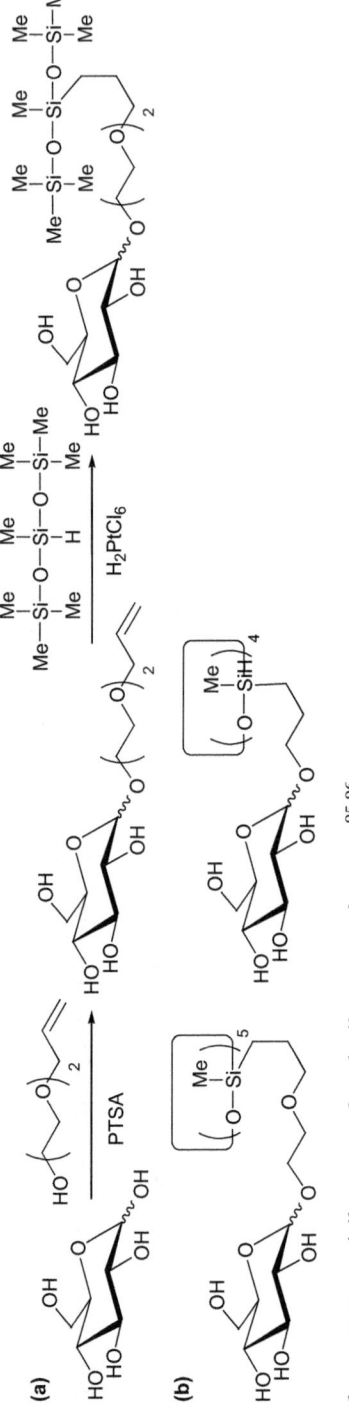

Scheme 5.52 Trisiloxane and cyclosiloxane surfactants.[95,96]

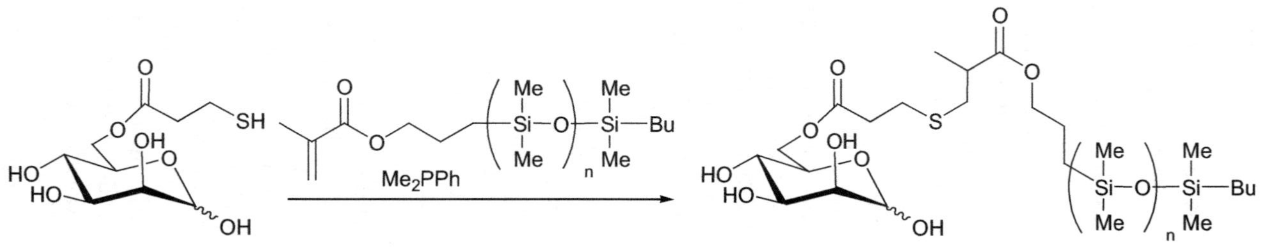

Scheme 5.53 Siloxane mannose structures.[93]

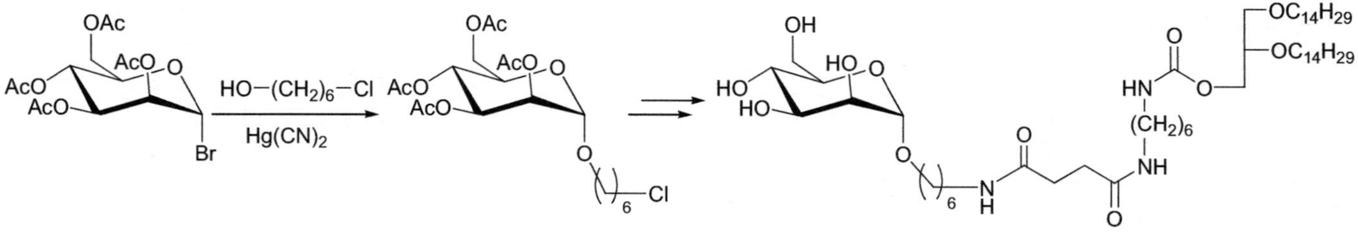

Scheme 5.54 Mannose neoglycolipids.[97]

Morales and co-workers[99] synthesized modified glucosyl and glucuronosyl alkyl gallates to obtain new antioxidants; see Scheme 5.55 (alkyl = C4 to C18). Glycosylation was realized in conditions that gave selectively the beta anomer with the glucosyl trichloroacetimidate (67–93% yield) or glucuronosyl trichloroacetimidate (42–63% yield).

Kamitakahara and co-workers synthesized numerous trisaccharide surfactants. One saccharide unit was hydrophobic as it is persubstituted by alkyl groups (R = Me, Et);[100] see Scheme 5.56.

To modify the interfacial properties and biological activities of sugar-based surfactants, a heterocycle can be added as a spacer arm between the hydrophilic head and the hydrophobic chain. Most of the examples found in the literature covered contain an azole heterocycle and more often a triazole. Othman and co-workers[49] synthesized a range of sugar-based surfactants containing a heterocycle *i.e.* 1,2,4-oxadiazole, 1,3,4-thiadiazole and 1,2,4-triazole. The hydrophobic heterocycle derivative was glycosylated in the final step (45% yield) to give glycosylamine-based surfactants; see Scheme 5.57. The authors evaluated their antibacterial activities. The thiadazole-glucose and triazole-arabinose derivatives affected all the tested Gram-positive and Gram-negative bacteria. The oxadiazole-xylose derivative showed a selective effect on Gram-positive and Gram-negative microbes. Gram-positive *B. cereus* and *S. aureus* were affected by all glycosyl compounds. Gram-negative *P. aeruginosa* was resistant to oxadiazole-xylose and triazole-arabinose, while all glycosyl derivatives showed antibacterial activity against *P. fluorescens* and *E. coli.*

Bis(1,2,3-triazole) derivatives were prepared by Mohammed and co-workers[53,54] by reaction of a glucosylazide with an alcyne-substituted alkyl 1,2,4-triazole by a click cycloaddition (73–84% yield); see Scheme 5.58a (alkyl = C7 to C12). Mohammed described the same sequence using 6-azido-6-desoxy-D-galactose (71–75% yield for the click cycloaddition);[54] see Scheme 5.58b (alkyl = C8, C9). The same group also synthesized a bis(1,2,3-triazol) gemini surfactant;[71] see Scheme 5.22.

Starting from a propargyl xyloside, Lehmler and co-workers[94] synthesized 1-*N*-alkyl-1,2,3-triazolyl substituted D-xylopyranosides by click cycloaddition (50–73% yield); see Scheme 5.59 (alkyl = C6 to C16).

The same sequence of reactions was used by Heidelberg and co-workers with alkyl azide (alkyl = C8 to C18 branched and (*E*)-octadec-9-enyl) and pure peracetylated alpha (or beta) propargyl D-glucopyranoside (73 to 79% yield for the click cycloaddition step);[101] see Scheme 5.60. The reaction with unprotected propargyl D-glucopyranoside (anomeric mixture) gave the same compounds (anomeric mixture of a technical grade) in one step.

Krausz and co-workers used the same click reaction to build surfactant structures with two polar heads or "star-like";[83] see Scheme 5.37. Lequart and co-workers[73] and Rodriguez-Abreu and co-workers[74] synthesized benzimidazolone-based surfactants; see respectively Schemes 5.26 and 5.27. Halila and co-workers[102] used a triazole as a link between a hydrophilic polysaccharide and a hydrophobic acetylated polysaccharide.

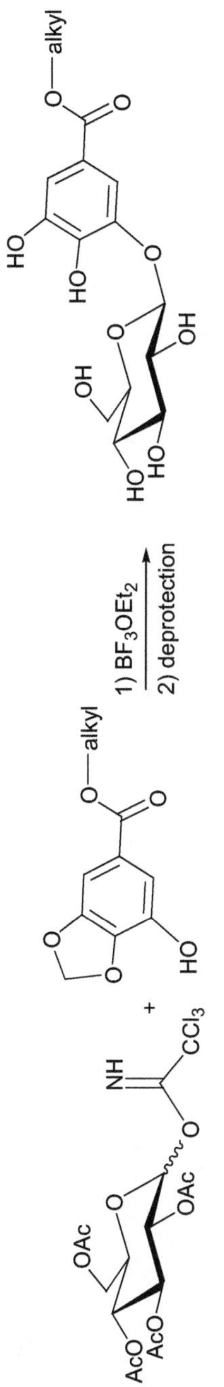

Scheme 5.55 Glucosyl gallate derivative.[99]

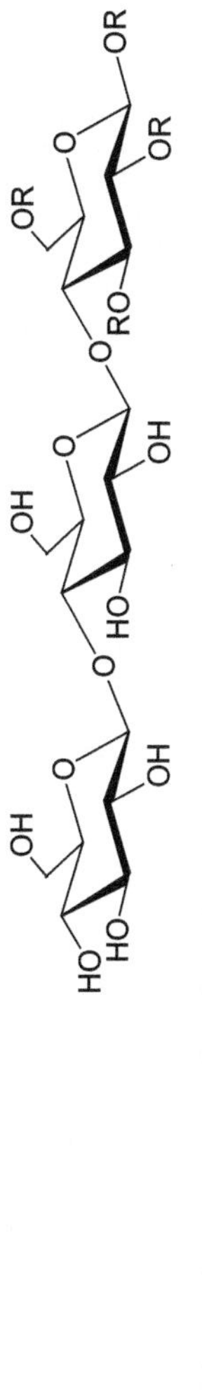

Scheme 5.56 Trisaccharide surfactants.[100]

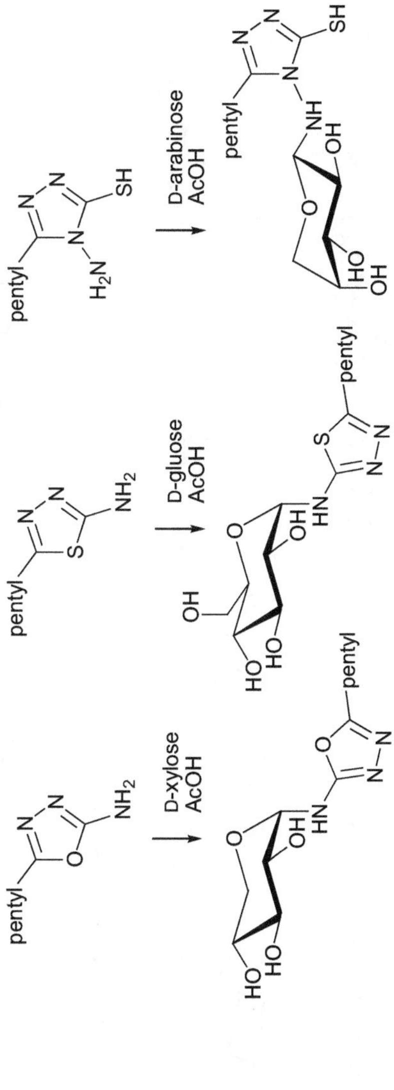

Scheme 5.57 Oxadiazole-, thiadiazole- and triazole-containing surfactants.[49]

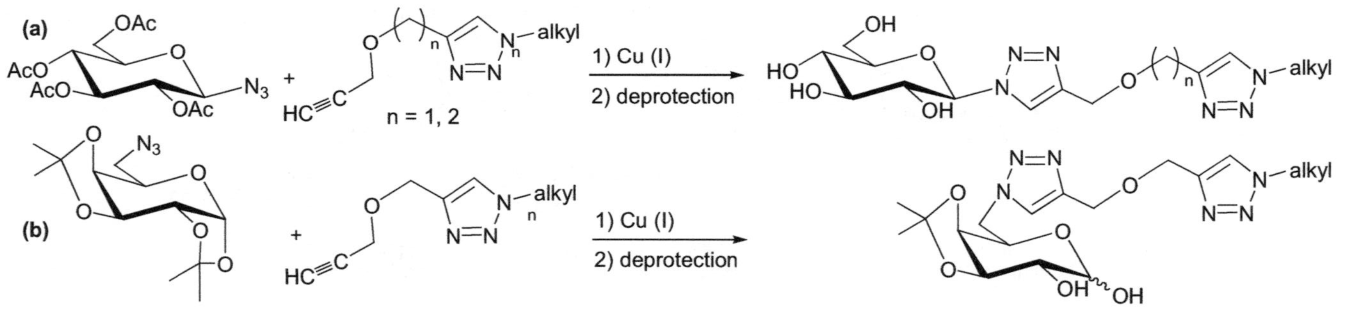

Scheme 5.58 Double-clicked surfactants.[53,54]

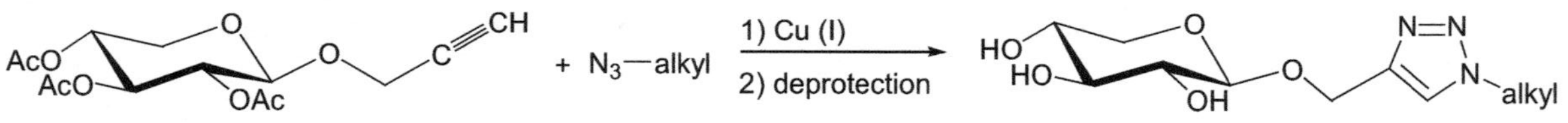

Scheme 5.59 Xylose-clicked surfactants.[94]

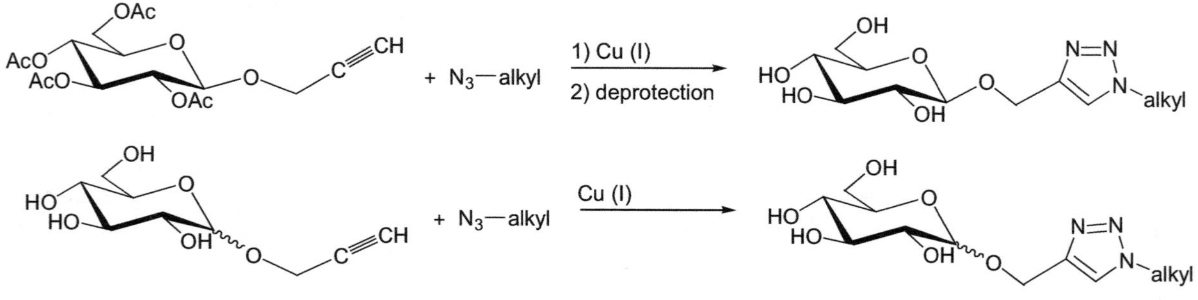

Scheme 5.60 Click reaction on protected and unprotected sugars.[101]

5.2.6 Polysaccharides

O-Acetyl galactoglucomannan (GGM) consists of a linear backbone of randomly distributed (1 → 4)-linked β-D-mannopyranosyl (acetylated at C2 or C3) and (1 → 4)-linked β-D-glucopyranosyl units, with α-D-galactopyranosyl units as single side units. To this polysaccharide, saturated fatty acids $CH_3(CH_2)_nCOOH$ ($n = 7$, 12 and 16) were coupled by two different routes by Dax and co-workers.[52] Fatty acids activated with 1,1′-carbonyldiimidazole (CDI) were grafted to GGM on their hydroxyl groups. Alternatively, amino-activated fatty acids synthesized using ethylenediamine were reacted with the reducing end of GGM; see Scheme 5.13.

A novel type of oligosaccharide block copolymer was synthesized by Halila and co-workers[102] by coupling a free xyloglucan oligosaccharide (as the hydrophilic part) to a peracetylated xyloglucan oligosaccharide (as the hydrophobic part) by click chemistry (Scheme 5.61).

5.3 Surfactants from Microorganisms

An extracellular mannosylerithritol lipid (MEL) is secreted by the yeast *Candida (Pseudozyma) antarctica* (ATCC 32657). Value added MEL was biologically synthesized by using two different types of honey as a natural and newer water-soluble carbon source. The MEL produced, characterized by Pratap and co-workers, showed surfactant properties.[103] A group of sugar fatty acid non-ionic surfactants, ethyl di-rhamnolipids, was synthesized by Ju and co-workers.[104] Di-rhamnolipids were isolated from a *Pseudomonas aeruginosa* fermentation broth containing a rhamnolipid mixture with about 64% di-rhamnolipids, and then purified by silica gel chromatography. Di-rhamnolipids were successfully ethylated at 0 °C for 24 h. Preliminary emulsification tests indicated that ethyl di-rhamnolipids (Scheme 5.62) are a potential class of useful sugar fatty acid non-ionic surfactants.

In addition, ethyl rhamnolipids have been used by the same group as a renewable source to produce biopolyurethanes.[105]

5.4 Conclusion and Perspectives

Sugar-based surfactants can be divided into two types. The less expensive ones can be synthesized in one or two steps, while those requiring more complicated syntheses should present improved specific properties. Sugar-based surfactants have recently been prepared either by enzymatic or chemical synthesis. Lipases are still the most often employed; the use of immobilized enzymes and ionic liquid solvents are increasingly chosen. Glycosidases and glycosyltransferases have also been reported, but their scope is still narrow. Enzymatic synthesis is useful for the preparation of simple structures in one or two steps, but it can lead to mixtures when multiple primary hydroxyl groups are available. Chemical synthesis is generally preferred for the synthesis of simple alkyl glycosides and APG. Most of

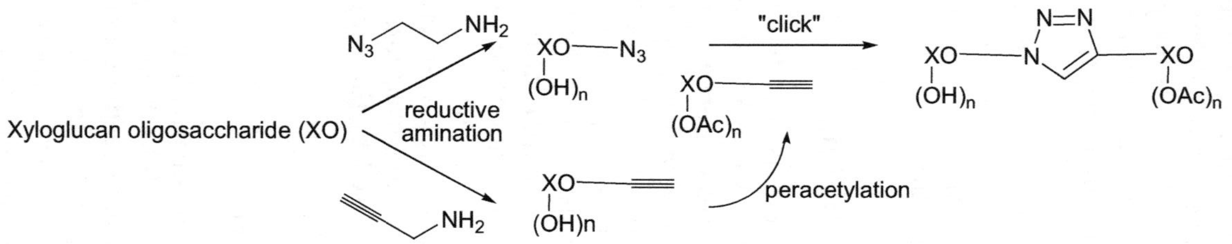

Scheme 5.61   Synthesis of amphiphilic block xyloglucans.[102]

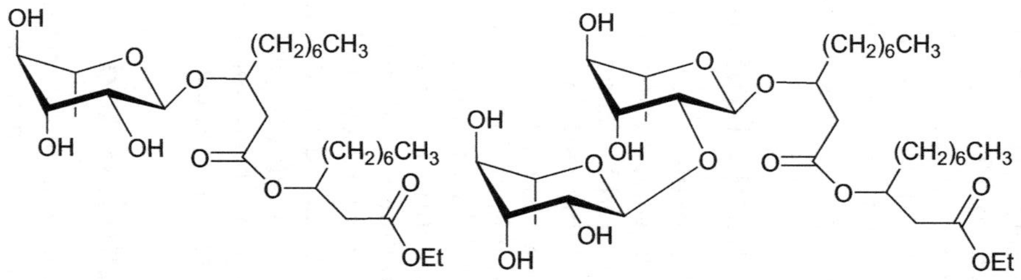

Scheme 5.62   Ethyl rhamnolipids.[105]

the recent papers involve more efficient catalysts or improved conditions such as the use of ionic liquid, which seems to be promising. *N*- and *C*-Glycoside analogues have also been prepared in one or two steps. Still, most of the modified structures that have been published recently have been synthesized in several steps such as for amino acid, oxime, fluor or silicone-containing derivatives. Perhaps the most promising molecules arise from non-classical structures such as gemini, bolaform or dicephalic surfactants, due to their generally lower CMC, among other properties. Ester, amide, triazols or thioether linkages have been used in their preparation, however, shorter and high yielding syntheses still need to be developed in the future.

References

1. S. K. Karmee, *Biofuels, Bioprod. Bioref.*, 2008, **2**, 144.
2. A. M. Gumel, M. S. M. Annuar, T. Heidelberg and Y. Chisti, *Process Biochem.*, 2011, **46**, 2079.
3. N. S. Neta, J. A. Teixeira and L. R. Rodrigues, *Crit. Rev. Food Sci. Nutr.*, 2015, **55**, 595.
4. F. Cauglia and P. Canepa, *Bioresour. Technol.*, 2008, **99**, 4065.
5. A. Adnani, M. Basri, N. Chaibakhsh, A. B. Salleh and M. B. A. Rahman, *Asian J. Chem.*, 2011, **23**, 388.
6. M. B. A. Rahman, M. Arumugam, N. S. K. Khairuddin, E. Abdulmalek, M. Basri and A. B. Salleh, *Asian J. Chem.*, 2012, **24**, 5058.
7. S. E. H. J. van Kempen, C. G. Boeriu, H. A. Schols, P. De Waard, E. Van der Linden and L. M. C. Sagis, *Food Chem.*, 2013, **138**, 1884.
8. S. E. H. J. van Kempen, H. A. Schols, E. van der Linden and L. M. C. Sagis, *Food Funct.*, 2014, **5**, 111.
9. N. A. S. Neta, J. C. Santos, S. O. Sancho, S. Rodrigues, L. R. B. Gonçalves, L. R. Rodrigues and J. A. Teixeira, *Food Hydrocolloid*, 2012, **27**, 324.
10. T. Takahashi, T. Kobayashi and S. Adachi, *Food Sci. Technol. Res.*, 2012, **18**, 167.
11. F. Fischer, M. Happe, J. Emery, A. Fornage and R. Schütz, *J. Mol. Catal. B: Enzym.*, 2013, **90**, 98.
12. M. F. K. Ariffin, M. S. M. Annuar and T. Heidelberg, *J. Surfactants Deterg.*, 2014, **17**, 683.
13. H. S. Ruela, F. K. Sutili, I. C. R. Leal, N. M. F. Carvalho, L. S. M. Miranda and R. O. M. A. de Souza, *Eur. J. Lipid Sci. Technol.*, 2013, **115**, 464.
14. F. K. Sutili, H. S. Ruela, S. G. F. Leite, L. S. M. Miranda, I. C. R. Leal and R. O. M. A. de Souza, *J. Mol. Catal. B: Enzym.*, 2013, **85–86**, 37.
15. R. Ye, S.-H. Pyo and D. G. Hayes, *J. Am. Oil Chem. Soc.*, 2010, **87**, 281.
16. R. Ye and D. G. Hayes, *J. Am. Oil Chem. Soc.*, 2011, **88**, 1351.
17. R. Ye and D. G. Hayes, *J. Am. Oil Chem. Soc.*, 2012, **89**, 455.
18. D. Reyes-Duarte, N. Lopez-Cortes, P. Torres, F. Comelles, J. L. Parra, S. Peña, A. V. Ugidos, A. Ballesteros and F. J. Plou, *J. Am. Oil Chem. Soc.*, 2011, **88**, 57.
19. C. Bidjou-Haiour and N. Klai, *Asian J. Chem.*, 2013, **25**, 4347.

20. J. Han, M. S. M. Annuar, M. F. K. Ariffin, A. M. Gumel, S. Ibrahim, T. Heidelberg, B. Bakar, A. B. M. S. Hossain and Y. Sharifuddin, *Biotechnol. Biotechnol. Equip*, 2011, **25**, 2642.

21. A. M. Gumel, M. S. M. Annuar, T. Heidelberg and Y. Chisti, *Bioresour. Technol.*, 2011, **102**, 8727.

22. K. Nott, G. Richard, P. Laurent, C. Jerome, C. Blecker, J.-P. Wathelet, M. Paquot and M. Deleu, *Process Biochem.*, 2013, **48**, 133.

23. M. Kurakake, Y. Ito and T. Komaki, *Biotechnol. Lett.*, 2011, **33**, 2453.

24. I. Mladenoska, *Food Technol. Biotechnol.*, 2012, **50**, 420.

25. T. Imura, Y. Masuda, H. Minamikawa, T. Fukuoka, M. Konishi, T. Morita, H. Sakai, M. Abe and D. Kitamoto, *J. Oleo Sci.*, 2010, **59**, 495.

26. V. K. Recke, M. Gerlitzki, R. Hausmann, C. Syldatk, V. Wray, H. Tokuda, N. Suzuki and S. Lang, *Eur. J. Lipid Sci. Technol.*, 2013, **115**, 452.

27. T. Börner, K. Roger and P. Adlercreutz, *ACS Catal.*, 2014, **4**, 2623.

28. D. Svensson, S. Ulvenlund and P. Adlercreutz, *Biotechnol. Bioeng.*, 2009, **104**, 854.

29. P. Torres, A. Poveda, J. Jimenez-Barbero, J. L. Parra, F. Comelles, A. O. Ballesteros and F. J. Plou, *Adv. Synth. Catal.*, 2011, **353**, 1077.

30. M. M. A. El-Sukkary, N. A. Syed, I. Aiad and W. I. M. El-Azab, *J. Surfactants Deterg.*, 2008, **11**, 129.

31. N. Villandier and A. Corma, *ChemSusChem*, 2011, **4**, 508.

32. I. Rico-Lattes, E. Perez, S. Franceschi-Messant and A. Lattes, *C. R. Chim.*, 2011, **14**, 700.

33. J. Li, Y. Liu, G. Zheng, Y. Sun, Y. Hao and T. Fu, *Advanced Materials Research*, ed. Z. Liu, F. Peng and X. Liu, Trans Tech Publications, 2012, **550–553**, 1, 75.

34. D. S. van Es, S. Marinkovic, X. Oduber and B. Estrine, *J. Surfactants Deterg.*, 2013, **16**, 147.

35. R. Ceron-Camacho, J. A. Aburto, L. E. Montiel and R. Martínez-Palou, *C. R. Chim.*, 2013, **16**, 427.

36. Z. Hricovíniová and M. Hricovíni, Tetrahedron, *Asymmetry*, 2014, **25**, 1008.

37. (a) A. Richel, P. Laurent, B. Wathelet, J.-P. Wathelet and M. Paquot, *Catal. Today*, 2011, **167**, 141; (b) A. Richel, P. Laurent, B. Wathelet, J.-P. Wathelet and M. Paquot, *C. R. Chim.*, 2011, **14**, 224.

38. J. Augé and G. Sizun, *Green Chem.*, 2009, **11**, 1179.

39. N. Villandier and A. Corma, *Chem. Commun.*, 2010, **46**, 4408.

40. F. Goursaud and T. Benvegnu, *Carbohydr. Res.*, 2009, **344**, 136.

41. N. Z. B. M. Rodzi, T. Heidelberg, R. Hashim, A. Sugimura and H. Minamikawa, *Phys. Procedia*, 2011, **14**, 91.

42. (a) X. Li, J. Turanek, P. Knoetigova, H. Kudlackova, J. Masek, D. B. Pennington, S. E. Rankin, B. L. Knutson and H.-J. Lehmler, *New J. Chem.*, 2008, **32**, 2169; (b) X. Li, J. Turanek, P. Knoetigova, H. Kudlackova, J. Masek, D. S. Parkin, S. E. Rankin, B. L. Knutson and H.-J. Lehmler, *Colloids Surf. B*, 2009, **73**, 65.

43. R. Cerón-Camacho, R. Martínez-Palou, B. Chávez-Gómez, F. Cuéllar, C. Bernal-Huicochea, J. C. Clavel and J. Aburto, *Fuel*, 2013, **110**, 310.
44. N. Ferlin, L. Duchet, J. Kovensky and E. Grand, *Carbohydr. Res.*, 2008, **343**, 2819.
45. V. Neto, R. Granet and P. Krausz, *Tetrahedron*, 2010, **66**, 4633.
46. S. Liu, R. Sang, S. Hong, Y. Cai and H. Wang, *Langmuir*, 2013, **29**, 8511.
47. (a) M. Gosenca, J. Mravljak, M. Gašperlin and A. Obreza, *Acta Chim. Slovaca*, 2013, **60**, 310; (b) J. Mravljak and A. Obreza, *Tetrahedron Lett.*, 2012, **53**, 2234.
48. L. Chaveriat, I. Gosselin, C. Machut and P. Martin, *Eur. J. Med. Chem.*, 2013, **62**, 177.
49. F. T. Brahimi, M. Belkadi and A. A. Othman, *Arabian J. Chem.*, 2013, DOI: 10.1016/j.arabjc.2013.06.016.
50. V. Neto, A. Voisin, V. Héroguez, S. Grelier and V. Coma, *J. Agric. Food Chem.*, 2012, **60**, 10516.
51. M. El-S. Abdel-Raouf, A.-R. M. Abdul-Raheim and A.-A. A. Abdel-Azim, *J. Surfactants Deterg.*, 2011, **14**, 113.
52. D. Dax, P. Daniel, J. Eklund, Hemming, J. Sarfraz, P. Backman, C. Xu and S. Willfor, *BioResources* , 2013, **8**, 3771.
53. A. I. Mohammed, N. H. Mansour and L. S. Mahdi, *Arabian J. Chem.* 2014, http://dx.doi.org/10.1016/j.arabjc.2014.02.016.
54. A. I. Mohammed, *Asian J. Chem.*, 2012, **24**, 5585.
55. X. Zhou, G. Wang and T. Lei, *Phosphorus, Sulfur Silicon Relat. Elem.*, 2013, **188**, 701.
56. D. Zhang, L. Wang, X. Li and G. Wang, *J. Dispersion Sci. Technol.*, 2013, **34**, 1188.
57. X. Zeng, Z. Lu and Y. Liu, *J. Surfactants Deterg.*, 2013, **16**, 131.
58. (a) F. Han, Y.-d. Gao, Y.-l. Liu, Y.-l. Liang, Y.-w. Zhou and B.-c. Xu, *Tenside Surfact. Det.*, 2014, **51**, 6; (b) F. Han, Y.-l. Liu, Y.-d. Gao, Y.-l. Liang, Y.-w. Zhou and B.-c. Xu, *J. Surfactants Deterg.*, 2014, **17**, 733; (c) F. Han, Y.-y. Deng, P.-l. Wang, J. Song, Y.-w. Zhou and B.-c. Xu, *J. Surfactants Deterg.*, 2013, **16**, 155.
59. F. Han, Y.-y. Deng, Y.-w. Zhou and B.-c. Xu, *J. Surfactants Deterg.*, 2012, **15**, 123.
60. (a) C. Bize, J.-C. Garrigues, J.-P. Corbet, I. Rico-Lattes and M. Blanzat, *ChemPhysChem*, 2013, **14**, 1126; (b) C. Bize, M. Blanzat and I. Rico-Lattes, *J. Surfactants Deterg.*, 2010, **13**, 465.
61. (a) T. Yoshimura, S. Umezawa, A. Fujino, K. Torigoe, K. Sakai, H. Sakai, M. Abe and K. Esumi, *J. Oleo Sci.*, 2013, **62**, 353; (b) K. Sakai, S. Umezawa, M. Tamura, Y. Takamatsu, K. Tsuchiya, K. Torigoe, T. Ohkubo, T. Yoshimura, K. Esumi, H. Sakai and M. Abe, *J. Colloid Interface Sci.*, 2008, **318**, 440; (c) K. Sakai, M. Tamura, S. Umezawa, Y. Takamatsu, K. Torigoe, T. Yoshimura, K. Esumi, H. Sakai and M. Abe, *Colloids Surf. A*, 2008, **328**, 100.
62. K. Torigoe, A. Tasaki, T. Yoshimura, K. Sakai, K. Esumi, Y. Takamatsu, S. C. Sharma, H. Sakai and M. Abe, *Colloids Surf., A*, 2008, **326**, 184.

63. X. Zeng, H. Wang, Y. Chen and L. Wang, *Tenside Surfact. Det*, 2014, **51**, 427.

64. D. Zhang and Y. Qiao, *Res. Chem. Intermed.*, 2015, **41**, 3047.

65. G. Wang, W. Qu, Z. Du, Q. Cao and Q. Li, *J. Phys. Chem. B*, 2011, **115**, 3811.

66. , For recent articles see:(a) M. Durand, Y. Zhu, V. Molinier, T. Féron and J.-M. Aubry, *J. Surfactants Deterg.*, 2009, **12**, 371; (b) K. S. Arias, M. J. Climent, A. Corma and S. Iborra, *ChemSusChem*, 2014, **7**, 210.

67. A. Ranoux, L. Lemiegre, M. Benoit, J.-P. Guegan and T. Benvegnu, *Eur. J. Org. Chem.*, 2010, 1314.

68. A. Ranoux, L. Lemiegre, T. Benvegnu, *Sci. China: Chem.* 2010, **53**, 1957.

69. P. M. Foley, A. Phimphachanh, E. S. Beach, J. B. Zimmerman and P. T. Anastas, *Green Chem.*, 2011, **13**, 321.

70. A. Corma, S. B. A. Hamid, S. Iborra and A. Velty, *ChemSusChem*, 2008, **1**, 85.

71. A. I. Mohammed, Z. A. Abboud and A. H. O. Alghanimi, *Tetrahedron Lett.*, 2012, **53**, 5081.

72. N. A. Negm and A. S. Mohamed, *J. Surfactants Deterg.*, 2008, **11**, 215.

73. B. Lakhrissi, A. Benksim, M. Massoui, E. M. Essassi, V. Lequart, N. Joly, D. Beaupere, A. Wadouachi and P. Martin, *Carbohydr. Res.*, 2008, **343**, 421.

74. B. Lakhrissi, L. Lakhrissi, M. Massoui, E. M. Essassi, F. Comelles, J. Esquena, C. Solans and C. Rodriguez-Abreu, *J. Surfactants Deterg.*, 2010, **13**, 329.

75. (a) M. Abla, G. Durand, C. Breyton, S. Raynal, C. Ebel and B. Pucci, *J. Fluorine Chem.*, 2012, **134**, 63; (b) M. Abla, G. Durand and B. Pucci, *J. Org. Chem.*, 2011, **76**, 2084; (c) M. Abla, G. Durand and B. Pucci, *J. Org. Chem.*, 2008, **73**, 8142; (d) C. Breyton, F. Gabel, M. Abla, Y. Pierre, F. Lebaupain, G. Durand, J.-L. Popot, C. Ebel and B. Pucci, *Biophys. J.*, 2009, **97**, 1077.

76. T. Yoshimura, R. Ohori and K. Esumi, *J. Oleo Sci.*, 2013, **62**, 571.

77. G. Wang, Z. Du, Q. Li and W. Zhang, *J. Phys. Chem. B*, 2010, **114**, 6872.

78. S. M. Salman, T. Heidelberg and H. A. Bin Tajuddin, *Carbohydr. Res.*, 2013, **375**, 55.

79. C. Boyere, G. Broze, C. Blecker, C. Jerome and A. Debuigne, *Carbohydr. Res.*, 2013, **380**, 29.

80. C. Racles and V. Cozan, *Soft Mater.*, 2012, **10**, 413.

81. F. Han, P. Wang, J. Song, Y. Zhou and B. Xu, *J. Dispersion Sci. Technol.*, 2012, **33**, 1708.

82. (a) M. Deleu, S. Gatard, E. Payen, L. Lins, K. Nott, C. Flore, R. Thomas, M. Paquot and S. Bouquillon, *C. R. Chim.*, 2012, **15**, 68; (b) M. Deleu, C. Damez, S. Gatard, K. Nott, M. Paquot and S. Bouquillon, *New J. Chem.*, 2011, **35**, 2258.

83. V. Neto, R. Granet, G. Mackenzie and P. Krausz, *J. Carbohydr. Chem.*, 2008, **27**, 231.

84. (a) E. Soussan, C. Mille, M. Blanzat, P. Bordat and I. Rico-Lattes, *Langmuir*, 2008, **24**, 2326; (b) D. Vivares, E. Soussan, M. Blanzat and I. Rico-Lattes, *Langmuir*, 2008, **24**, 9260.
85. (a) A. Richel, F. Nicks, P. Laurent, B. Wathelet, J.-P. Wathelet and M. Paquot, *Green Chem. Lett. Rev.*, 2012, **5**, 179; (b) A. Richel, P. Laurent, B. Wathelet, J.-P. Wathelet and M. Paquot, *Tetrahedron Lett.*, 2010, **51**, 1356.
86. P. Laurent, H. Razafindralambo, B. Wathelet, C. Blecker, J.-P. Wathelet and M. Paquot, *J. Surfactants Deterg.*, 2011, **14**, 51.
87. N. Ferlin, D. Grassi, C. Ojeda, M. J. L. Castro, E. Grand, A. Fernandez Cirelli and J. Kovensky, *Carbohydr. Res.*, 2008, **343**, 839.
88. (a) N. Ferlin, D. Grassi, C. Ojeda, M. J. L. Castro, E. Grand, A. Fernandez Cirelli and J. Kovensky, *Carbohydr. Res.*, 2010, **345**, 598; (b) N. Ferlin, D. Grassi, C. Ojeda, M. J. L. Castro, A. Fernandez Cirelli, J. Kovensky and E. Grand, *J. Surfactants Deterg.*, 2012, **15**, 259; (c) N. Ferlin, D. Grassi, C. Ojeda, M. J. L. Castro, A. Fernandez Cirelli, J. Kovensky and E. Grand, *Colloids Surf. A: Physicochem. Eng. Asp.*, 2015, **480**, 439.
89. J. Mravljak and S. Pečar, *Tetrahedron Lett.*, 2009, **50**, 567.
90. H. S. Ewan, C. S. Muli, S. Touba, A. T. Bellinghiere, A. M. Veitschegger, T. B. Smith, W. L. Pistel II, W. T. Jewell, R. K. Rowe, J. P. Hagen and H. Palandoken, *Tetrahedron Lett.*, 2014, **55**, 4962.
91. S. M. Salman, T. Heidelberg, R. S. D. Hussen and H. A. Bin Tajuddin, *J. Surfactants Deterg.*, 2014, **17**, 1141.
92. A.-R. M. Abdul-Raheim, M. El-S. Abdel-Raouf, N.El-S. Maysour, A. F. El-Kafrawy, A. Z. Mehany and A.-A. A. Abdel-Azim, *J. Surfactants Deterg.*, 2013, **16**, 377.
93. (a) C. Boyere, A. Favrelle, A. F. Leonard, F. Boury, C. Jerome and A. Debuigne, *J. Mater. Chem. A*, 2013, **1**, 8479; (b) C. Boyere, A. Favrelle, G. Broze, P. Laurent, K. Nott, M. Paquot, C. Blecker, C. Jerome and A. Debuigne, *Carbohydr. Res.*, 2011, **346**, 2121.
94. E. D. Oldham, S. Seelam, C. Lema, R. J. Aguilera, J. Fiegel, S. E. Rankin, B. L. Knutson and H.-J. Lehmler, *Carbohydr. Res.*, 2013, **379**, 68.
95. F. Han, Y.-h. Chen, Y.-w. Zhou and B.-c. Xu, *J. Surfactants Deterg.*, 2011, **14**, 515.
96. C. Racles, A. Airinei, I. Stoica and A. Ioanid, *J. Nanopart. Res.*, 2010, **12**, 2163.
97. E. V. Shmendel', A. A. Timakova, M. A. Maslov, N. G. Morozova and V. V. Chupin, *Russ. Chem. Bull.*, 2012, **61**, 1497.
98. (a) E. V. Shmendel', M. A. Maslov, N. G. Morozova and G. A. Serebrennikova, *Russ. Chem. Bull.*, 2010, **59**, 2281; (b) M. A. Maslov, D. A. Medvedeva, D. A. Rapoport, R. N. Serikov, N. G. Morozova, G. A. Serebrennikova, V. V. Vlassov and M. A. Zenkova, *Bioorg. Med. Chem. Lett.*, 2011, **21**, 2937.
99. O. S. Maldonado, R. Lucas, F. Comelles, M. J. González, J. L. Parra, I. Medina and J. C. Morales, *Tetrahedron*, 2011, **67**, 7268.

100. A. Nakagawa, H. Kamitakahara and T. Takano, *Carbohydr. Res.*, 2011, **346**, 1671.

101. F. A. Sani, T. Heidelberg, R. Hashim and Farhanullah, *Colloids Surf. B*, 2012, **97**, 196.

102. C. Gauche, V. Soldi, S. Fort, R. Borsali and S. Halila, *Carbohydr. Polym.*, 2013, **98**, 1272.

103. A. Bhangale, S. Wadekar, S. Kale and A. Pratap, *Biotechnol. Bioprocess Eng.*, 2013, **18**, 679.

104. S. Miao, N. Callow, S. S. Dashtbozorg, J.-L. Salager and L.-K. Ju, *J. Surfactants Deterg.*, 2014, **17**, 1069.

105. S. Miao, N. V. Callow and L.-K. Ju, *Eur. J. Lipid Sci. Technol.*, 2015, **117**, 156.

CHAPTER 6

Oligosaccharides for Pharmaceutical Applications

JANI RAHKILA, TIINA SALORANTA AND REKO LEINO*

Johan Gadolin Process Chemistry Centre, c/o Laboratory of Organic Chemistry, Åbo Akademi University, FI-20500 Åbo, Finland
*Email: reko.leino@abo.fi

6.1   Introduction

Carbohydrates are the most abundant biomolecules on earth with functions ranging from the fundamental building block of RNA (ribose) to the most important energy store of human bodies (glucose). The latter is also the building block of cellulose, the most abundant biopolymer on earth. Compared to proteins, which consist of amino acids connected in a linear fashion, carbohydrate arrays are often branched with the monosaccharides connected to each other in various ways giving them great potential for storage of biological information. For example, a simple disaccharide consisting of only two hexopyranose moieties can be constructed in 11 different ways. Upon moving to larger oligosaccharides, the number of permutations increases exponentially. While not all possible combinations of monosaccharides exist in nature, the number of different oligosaccharides available by contemporary synthetic methods allows for a high degree of customization. This, in turn, results in highly specific binding achievable by fine-tuning of the three-dimensional structure of carbohydrates to match the binding sites of biological receptors. If the binding site is known, carbohydrates can be designed to match in an optimal, specific way. Transport of the carbohydrate systems to their active sites is, however, not trivial. Due to

RSC Green Chemistry No. 44
Biomass Sugars for Non-Fuel Applications
Edited by Dmitry Murzin and Olga Simakova

the usually high polarity of the carbohydrate molecules, they are unable to efficiently pass through the enterocyte layer of the small intestine. Consequently, carbohydrates are poorly suited for oral administration. Furthermore, the high polarity leads to short biological half-lives even in cases where the compounds are administered parenterally.[1,2] While the binding of carbohydrates is often highly specific, it may simultaneously be rather weak leading also to a weaker biological response. This may, however, be overcome by construction of multivalent carbohydrate assemblies where several identical saccharide moieties are connected to a common core molecule. Such multivalency typically enhances the binding compared to monovalent analogues.[3]

6.2 Synthesis

While many of the naturally occurring oligo- and polysaccharides have significant biological functions, their isolation is often not a straightforward task. Carbohydrates present in living organisms are typically heterogeneous by composition, which makes the isolation of pure compounds in sufficient quantities difficult. Pure, well-defined compounds, in turn, are a prerequisite for understanding their biological functions. Compound mixtures may induce biological effects different from those characteristic for their individual constituents. Thus, complementary synthetic methods are often required for construction of well-defined, pure oligosaccharides.

6.2.1 Traditional Synthesis of Oligosaccharides

The traditional way of constructing oligosaccharides from their building blocks is based on the use of carbohydrate donor and acceptor reagents. Donors typically contain a good leaving group which upon activation (usually by an acid, such as TMSOTf or $BF_3 \cdot OEt_2$) is cleaved off resulting in an oxocarbenium ion, which then reacts with a nucleophilic acceptor (Scheme 6.1). Common donors include trichloroacetimidates, thioglycosides, glycosyl halides, glycosyl acetates and glycosyl phosphates. The most common type of acceptor in oligosaccharide synthesis is an alcohol, yielding

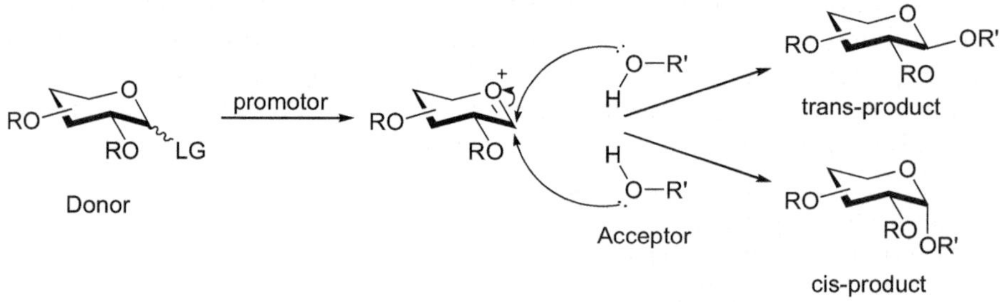

Scheme 6.1 General glycosylation reaction.

an *O*-glycoside, although amines and thiols can also be used resulting in *N*-glycosides and *S*-glycosides, respectively.

The reaction intermediate, *i.e.* the oxocarbenium ion, has a planar geometry. Consequently, the nucleophile can attack the ion from both sides, potentially forming two different diastereomeric reaction products. Typically, only one of the two possible diastereisomers is the desired product. Because of this, the coupling reaction needs to be directed towards the desired product using specific protecting group strategies, such as neighboring group participation, which favors the formation of *trans*-glycosides, or specific activation protocols for formation of the corresponding *cis*-glycosides. In general, without any directing groups, the α-configuration is favored due to an anomeric effect.[4]

Due to very similar reactivities of the saccharide hydroxyl groups, carbohydrate chemistry has traditionally heavily depended on laborious protecting group manipulations. Characteristically, several temporary protecting groups may need to be applied only to be able to introduce another protecting group at a later stage in the synthesis. The required multistep synthesis procedures then lead to low atom economy, low overall yields and a large amount of generated waste.

For synthesis of oligosaccharides, two main approaches are applied—linear and convergent (Scheme 6.2). In linear synthesis, the oligosaccharide is prepared by adding one monosaccharide moiety at a time. Convergent synthesis, in turn, relies on preparation of larger building blocks and assemblies which are then coupled together at a later stage. The latter approach reduces the number of synthetic steps, but also complicates the selection of donor molecules. In convergent synthesis, the leaving group used (Y in Scheme 6.2) should be sufficiently stable to survive the first coupling reaction. Thus, donors such as thioglycosides, which are stable under a wide range of conditions but can be converted *in situ* to more active species such as triflates, are often applied. Another benefit of using thioglycosides is that they also function as temporary protecting groups which can be removed selectively, leaving behind a free hydroxyl group that can then be converted into various leaving groups, such as a trichloroacetimidate or a glycosyl phosphate.

Of course, the syntheses can be expanded beyond the preparation of the tetrasaccharide described in Scheme 6.2. It is easy to observe how the convergent approach becomes more beneficial, at least in terms of the number of synthetic steps as the oligosaccharide becomes larger, especially in the preparation of linear oligosaccharides consisting of 2^n monosaccharide units. For example, the synthesis of an octasaccharide requires a total of 14 linear synthetic steps whereas a convergent synthesis would only require nine steps.

6.2.2 Automated Synthesis of Oligosaccharides

Automated syntheses of oligonucleotides and oligopeptides have been available for a long time already. For these biomacromolecules, there are no

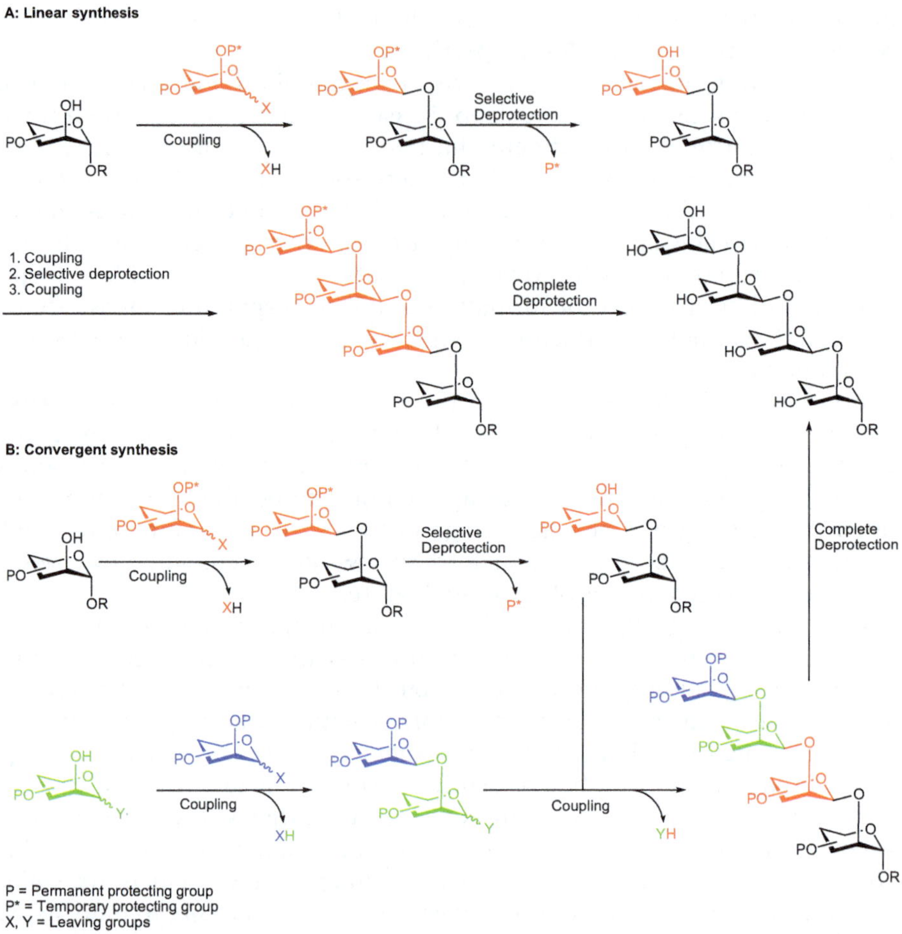

A: Linear synthesis

B: Convergent synthesis

P = Permanent protecting group
P* = Temporary protecting group
X, Y = Leaving groups

Scheme 6.2 Linear *vs.* convergent synthesis of a β-(1→2)-linked mannotetraose.

regio- or stereochemical issues involved in the coupling reactions as the backbones consist of smaller building blocks coupled together *via* phosphodiester or amide bonds in a linear fashion. Consequently, such syntheses can be carried out using a single type of protecting group in an iterative fashion. In addition, the synthesized nucleotide chains can be copied by utilizing the polymerase chain reaction. Proteins in turn, which are encoded by DNA, can be produced *via* recombinant DNA technology after the DNA sequence responsible for producing the protein is determined.[5]

The automated synthesis of oligosaccharides, however, is generally more challenging due to the aforementioned regio- and stereochemical issues. The large number of possible oligosaccharide structures also provides a great challenge, as the number of required building blocks is greater than the number of building blocks required for making oligopeptides or oligonucleotides of similar length. However, all of the possible oligosaccharide combinations are not present in living organisms. If the syntheses were carried out in a strictly linear fashion, similar to the synthesis of oligonucleotides and oligopeptides, it has been estimated that 36 monosaccharide building blocks would be required for creating 75% of all mammalian oligosaccharides. For the synthesis of 90% of such oligosaccharides, 65 building blocks, respectively, would be required.[6]

One of the key issues for automated synthesis in general is the purification of intermediates. In almost any reaction, some byproducts are formed and unreacted starting materials remain and need to be separated from the product. This problem has been solved in oligonucleotide and oligopeptide synthesis by attaching the growing chain to a solid support, such as polystyrene, polyethylene, PTFE or others,[6,7] which makes it possible to isolate the product by simple filtration and washing. The same approach is commonly used for automated carbohydrate synthesis as well. However, one of the problems encountered with solid supports is the inherently poor reaction kinetics of biphasic systems, which require large excesses of reagents to be used for driving the reactions to completion.

In order to overcome the aforementioned problems with solid supports, alternative approaches for automated oligosaccharide synthesis have evolved, for example based on affinity chromatography/solid-phase extraction using soluble (light) fluorous tags.[8] The fluorous tag is attached to the carbohydrate in a similar way as the solid support but with the benefit that the molecule as a whole will remain soluble, thus overcoming the problems with poor reaction kinetics of biphasic systems and reducing the required excess of reagents. After completion of the reaction, the target product can be isolated by fluorous solid-phase extraction.

When attaching the carbohydrate to either a fluorous tag or solid support, two alternatives emerge. The tag can be attached either to the nucleophile (glycosyl acceptor) or the electrophile (glycosyl donor). The preferred way is to attach the tag to the acceptor, which allows an excess of the more reactive donor to be used without concerns about potential side reaction products getting linked to the fluorous tag or the solid support.[6]

Automated synthesis does, however, introduce a new problem associated with stereoselectivity in the glycosylation reactions. Fundamentally, the problem is the same as in conventional oligosaccharide synthesis: it is often difficult to obtain complete stereoselectivity and stereochemical control in the glycosylation reaction. The new aspect of the problem arises from the fact that the diastereomers cannot be separated until the oligosaccharide has been released from the solid support or the fluorous tag has been removed. This results in an exponentially increasing number of products after each synthetic step. Already in the synthesis of a hexasaccharide, the theoretical number of final hexasaccharide products would be $2^5 = 32$. Only one of these products has the correct desired stereochemistry at each glycosidic linkage. In conventional carbohydrate synthesis, the byproducts with incorrect stereochemistry are typically removed after each coupling step by chromatographic methods, resulting finally in only two possible hexasaccharides. Because of this, new reaction conditions offering better stereoselectivities are still needed.

The coupling to the solid support resin or the fluorous tag is commonly achieved *via* a spacer unit. Some considerations then are needed for this spacer unit which should, if possible, incorporate functional groups for further modification of the oligosaccharide, *e.g.*, attachment to a carrier protein. One popular type of spacer unit contains a hydrocarbon chain with a double bond. This double bond can, for example, be cleaved by olefin metathesis to give a terminal double bond, or by ozonolysis to give an aldehyde that can then, for example, be conjugated to the free amine groups of lycine residues in a protein.[9]

In both solid and solution phase automated synthesis, the main synthetic cycles are rather similar. Iterative methods are used, which can be separated into five distinct steps (Scheme 6.3).[6,7]

1. (a) First, a single monosaccharide donor is attached to either a solid support or to a fluorous tag *via* a spacer containing a free hydroxyl group. (b) One or (for branched oligosaccharides) several protecting groups are selectively removed leaving free hydroxyl groups.
2. An excess of a donor molecule (usually 5–10 equivalents in solid phase synthesis and less than 2 equivalents in solution phase synthesis) and a promotor, *e.g.*, TMSOTf are added.
3. One or (for branched oligosaccharides) several protecting groups are again selectively removed leaving free hydroxyl groups.
4. The intermediate is purified by filtration (solid phase synthesis) or FSPE (solution phase synthesis).
   Steps 2–4 are then repeated until the synthesis is complete.
5. Upon completion of the iterative synthesis of an oligosaccharide, the product is cleaved from the support, all remaining protecting groups are removed and the final product is purified, usually by HPLC.

While automated solution-phase oligosaccharide synthesis strategies have been used to prepare a number of oligosaccharides, *e.g.* α-(1→2)-linked mannotetraoses and mannohexaoses,[7] their practical uses are still limited. Solid-phase synthesis, on the other hand, has been used to synthesize a large number of complex oligosaccharides, including blood group oligosaccharides, Lewis X pentasaccharide, Lewis Y hexasaccharide and the tumor associated antigen Le^x–Le^y nonasaccharide. The fully protected Le^x–Le^y nonasaccharide was synthesized using five monosaccharide building blocks in just 23 h with 6.5% (7 mg) overall yield (Scheme 6.4).[6,10]

6.2.3 Isolation and Fractionation of Polysaccharides from Natural Sources

While efficient synthetic strategies can be applied to produce well-defined oligosaccharides, many of the commercially available oligosaccharide

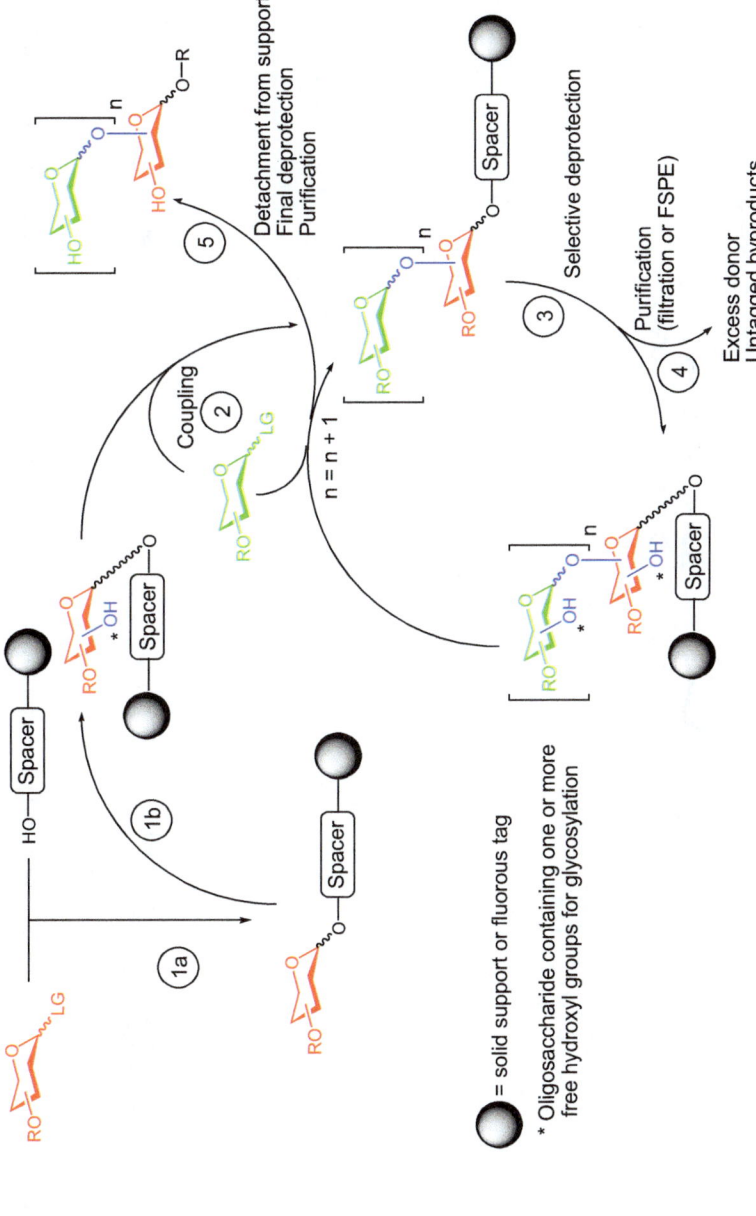

Scheme 6.3 Automated synthesis of oligosaccharides.

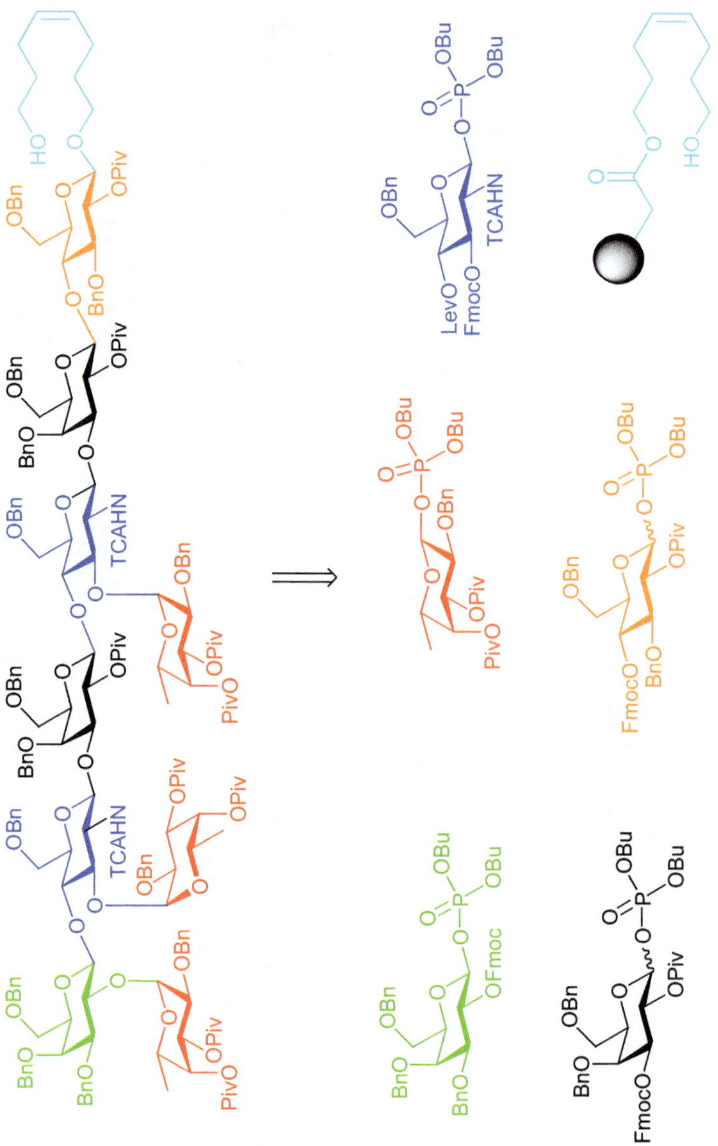

Scheme 6.4 Fully protected Le^x–Le^y nonasaccharide and the required building blocks for automated solid-phase synthesis.

pharmaceuticals still originate from natural sources. Organisms generally contain many different types of oligosaccharides with varying molecular weight distributions. Consequently, selective extractions of specific carbohydrates from different organisms in pure form are, in practice, not possible and the extracted compounds must undergo often tedious purification processes. Purification of oligo- and polysaccharides is commonly carried out by fractionation. Fractionation is not a single method but rather a combination of methods for separation of different molecules based on their physical or chemical properties. The process leads to better defined compounds in terms of molecular structure and molecular weight distribution.

The main methods for fractionation of polysaccharides are precipitation and different chromatographic methods.[11,12] Size exclusion chromatography separates compounds based on their actual size which is not necessarily directly proportional to the molecular weight. For the same type of polysaccharides, however, the physical sizes of the molecules are proportional to their molecular weight. This means that different types of polysaccharides might be difficult to separate by this technique alone, and combinations of different methods are often required.[13] One of the problems associated with chromatographic methods is that the polysaccharides need to be dissolved in water, which then is difficult to remove by energy efficient methods.[14]

Fractionation of polysaccharides by precipitation can be made more selective by choice of an appropriate precipitant, such as cetyltrimethylammonium bromide or copper salts. Commonly, the crude polysaccharide mixture is dissolved in water, followed by addition of the precipitant until precipitation occurs. The precipitate is then separated by filtration or centrifugation and the process is repeated until a determined amount of precipitant has been added or until no further precipitation occurs.[15,16]

6.3   Examples of Oligosaccharide Therapeutics on the Market

The oldest carbohydrate-based drug, heparin, one of the two oligosaccharide blockbuster drugs, is structurally a sulfated glycosaminoglycan, originally isolated as a heterogeneous mixture of polysaccharides from animal organs. Heparin has been used as an antithrombotic agent since the 1940s, already before its mechanism of action was known.[17] The main drawbacks associated with heparin are the half-life of only one hour and severe side effects such as thrombocytopenia and bleeding. The low molecular weight fractions of heparin that can be obtained after enzymatic or chemical fragmentation possess, in turn, longer half-lives and more predictable anticoagulant properties with fewer side effects. The pentasaccharide fragment, responsible for the anticoagulant properties of heparin, was identified in the early 1980s and synthetic procedures towards this active fragment were reported a few years later.[18] This pentasaccharide fragment (Figure 6.1) represents the

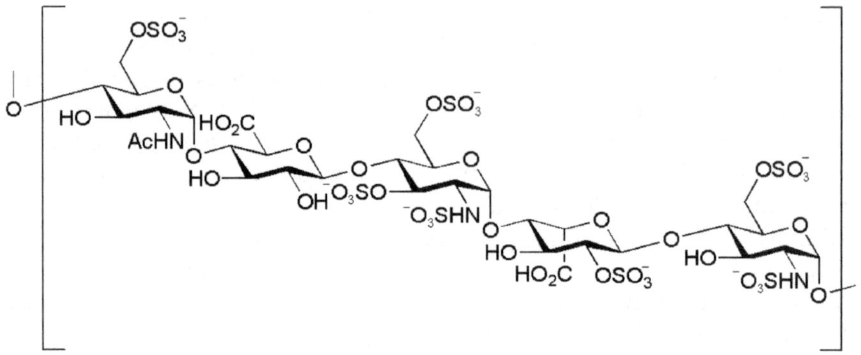

Figure 6.1 Pentasaccharide fragment responsible for the anticoagulant properties of heparin.

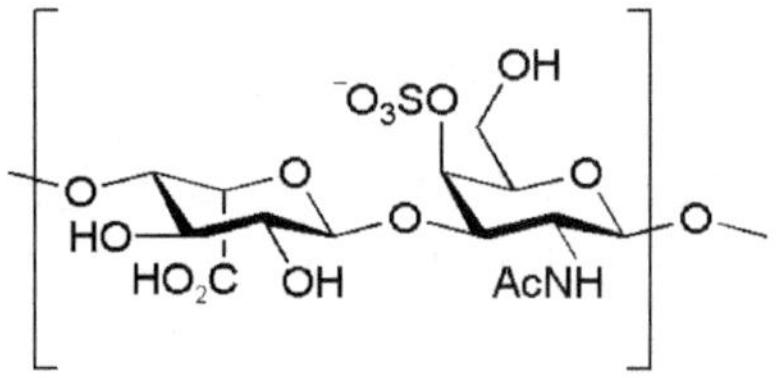

Figure 6.2 Dermatan sulfate.

essential heparin sequence that binds to antithrombin III (AT-III) and increases the AT-mediated inhibition of factor Xa.[19] The synthetic pentasaccharide corresponding to this fragment, fondaparinux, sold under the trade name *Arixtra* has been available since 2002 and a number of synthetic analogues have been explored since then.[20]

Sulodexide is a purified glycosaminoglycan extracted from porcine intestinal mucosa that is composed of a fast mobility heparin fraction (80%) and dermatan sulfate (20%). Dermatan sulfate, in turn, is a linear polysaccharide where the repeating disaccharide unit consists of N-acetylgalactosamine (GalNAc) and IdoA L-iduronic acid (IdoA) residues (Figure 6.2). Similar to heparin, sulodexide is used as an antithrombotic agent but characterized by a longer half-life and fewer side effects than unfractionated heparin. In addition, sulodexide has proved to be efficient in the treatment of other disorders, such as various cardiovascular and cerebrovascular diseases and diabetic nephropathy.[21]

Acarbose, the other oligosaccharide-based blockbuster drug, functions as an α-glucosidase inhibitor. It is structurally a pseudo-oligosaccharide consisting of a sequence of three monosaccharides attached to a valienamine residue (Figure 6.3). Acarbose is produced *via* a fermentation process and is marketed worldwide and used in the therapy of type 2 diabetes.[22]

Pentosan polysulfate, a semi-synthetic sulfated polysaccharide produced from xylan with xylose as the repeating unit is a new oligosaccharide-based drug on the market. Its production process comprises two basic steps: (1)

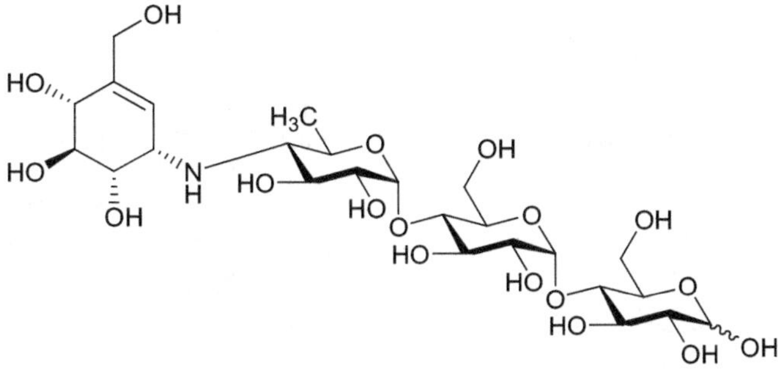

Figure 6.3 Acarbose.

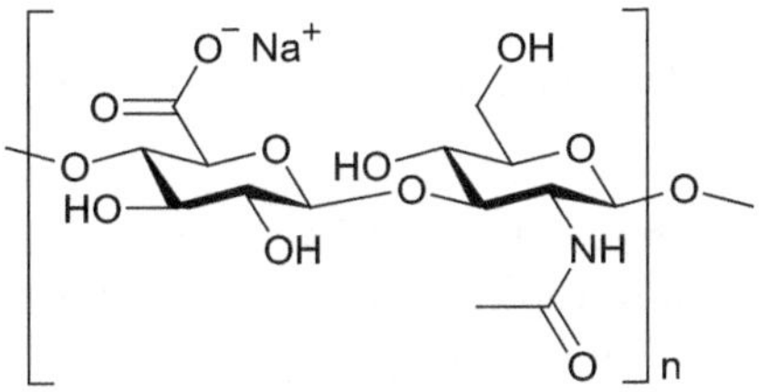

Figure 6.4 Sodium hyaluronate.

sulfation of xylan with chlorosulfonic acid in the presence of pyridine, and (2) oxidative depolymerisation in acidic or neutral aqueous media. The final product, a sulfated linear β-(1→4) linked polyxylose and its sodium salts with molecular weights ranging from 3 kDa to 10 kDa, is then obtained after dialysis and fractionation process.[23] Pentosan polysulfate is used for treating interstitial cystitis and is sold under name Elmiron.

Hyaluronic acid and more specifically its sodium salt (Figure 6.4), sodium hyaluronate, has been used for treating osteoarthritis of the knee since the 1980s. Its mechanism of action is based on the ability of this compound to increase the viscosity of the synovial fluid, which in turn increases the lubrication and cushioning properties of the fluid, resulting in reduction of pain in the joint.[24a] Water solutions of sodium hyaluronate are, due to their high viscosity, also used in ophthalmic surgery.[24b]

A final example of oligosaccharide-based drugs on the market is aminoglycosides, defined as naturally occurring pseudo-oligosaccharides which in turn consist of two to five monosaccharide units containing typically a one-to-one ratio between OH and NH_2 groups. These compounds, pioneered by streptomycin, have been utilized as antibiotics to treat diseases induced by Gram-negative bacteria. The antibiotic activities of aminoglycosides are due to their inhibition of protein synthesis resulting from binding to bacterial ribosomes.[25] The first discovered aminoglycoside was streptomycin, isolated from actinobacterium *Streptomyces griseus*, becoming the first effective treatment for tuberculosis.[26] Another aminoglycoside produced by bacteria belonging to the *Streptomyces* family is neomycin, which

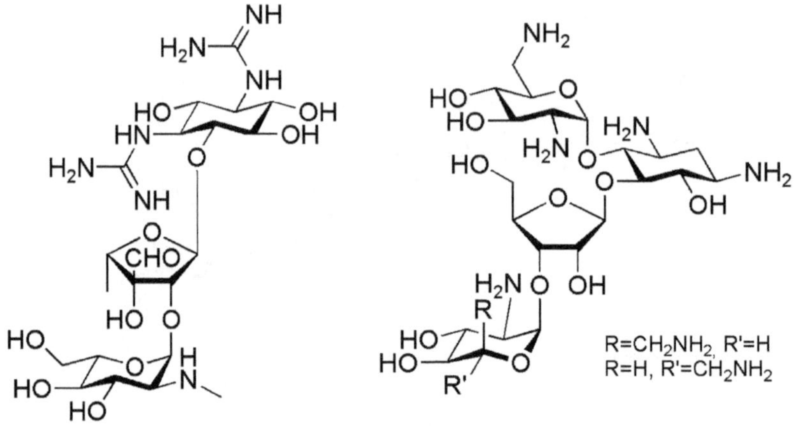

Figure 6.5 Streptomycin (left) and the two forms of neomycin (right).

can be isolated from *Streptomyces fradiae*. The two forms of neomycin are commonly found in many topical antibiotic preparations such as creams, ointments and eye drops (Figure 6.5).

6.4   Carbohydrate-based Vaccines and Adjuvants

Cells, both endogenous and exogenous, are covered by different types of molecules such as proteins, lipids and carbohydrates. When human bodies encounter microorganisms carrying unknown glycosides, proteins or other immunogenic molecules, an immune response is triggered and the immune system tries to kill the pathogen or neutralize its toxins. The first time our body encounters a certain pathogen, the immune response is highly complex. Consequently, the reaction times of the immune system towards infectious diseases are often much slower than the rates at which the pathogens multiply. In some cases the pathogens themselves, or the toxins they release, do not induce sufficiently strong immune responses for the body to be able to deal with the infection without aid. This is, for example, true for the anaerobic bacterium *Clostridium tetani* which releases tetanospasmin, an extremely potent neurotoxin that causes tetanus. This toxin is potent enough that even a lethal dose is insufficient to provoke an immune response.

Immune responses are complex cascades of different events, but in all simplicity the immune system can be classified into subsystems in different ways, such as adaptive and innate immunity, as well as humoral and cell-mediated immunity. Which one of these is activated depends on various factors and the detailed mechanisms are beyond the scope of this chapter.

In a simplified sense, when the human body encounters foreign antigens, the antigens are recognized by specific immunoglobulin (Ig) receptors on the surfaces of certain types of cells, B cells and other antigen-presenting cells

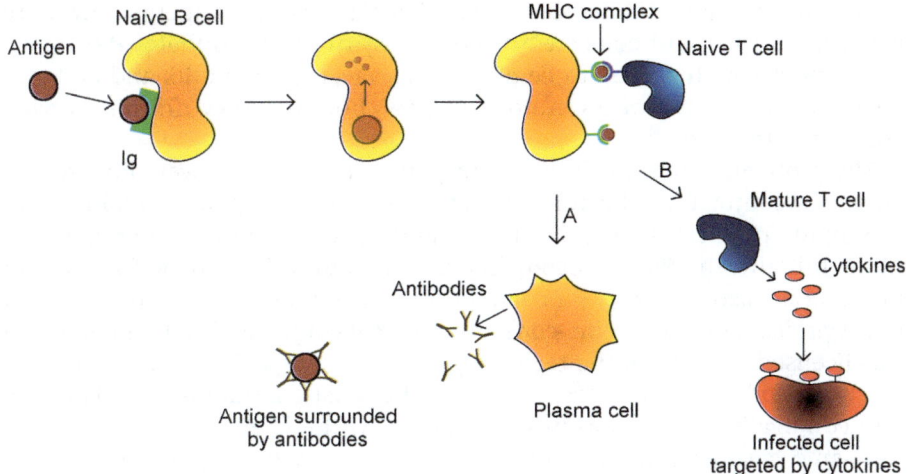

Figure 6.6 A simplified description of the activation of the immune system.

(APC). The antigens or pieces of them are later expressed on the surface of the cells as complexes (major histocompatibility complex, MHC, class I or II). These complexes are recognized by receptors on the surface of naïve T cells. During this recognition, the B and T cells are activated for different tasks depending on the MHC complex. B cells carrying MHC class II complexes mature into plasma cells (pathway A in Figure 6.6) which produce antibodies specific to the antigens that caused the immune response in the first place (humoral immunity)[27] and T cells that recognize MHC class I complexes mature into cytotoxic T cells (pathway B in Figure 6.6) which focus on killing cells either by direct interaction with the cells or by release of cytotoxins (cell-mediated immunity). Some of the B and T cells further mature into memory cells, which can survive in the body for a long time and react rapidly to prevent subsequent infections by the same pathogen.

As previously mentioned, the identification of cells, as well as many immunological responses, are often mediated by carbohydrates. Consequently, the development of new vaccines is often based on natural polysaccharides. While polysaccharides on cell surfaces are used for pathogen recognition, it has on several occasions been discovered that the polysaccharides alone do not usually produce sufficiently strong immune responses to be sufficiently effective as vaccines. Especially among young children without strong immune responses to T-cell-independent antigens such as polysaccharides, the immunization often fails. This is, for example, true for *Haemophilus influenza* type b (Hib), a Gram-negative coccobacillus responsible for several invasive bacterial infections especially among children. The bacterium is covered with a polysaccharide layer that cloaks underlying antigenic proteins which would in normal cases cause an immunogenic response. It is well known that bactericidal or opsonic antibodies that target this so-called capsular polysaccharide offer protection against invasive diseases caused by these types of encapsulated bacteria.[28] The first vaccines against Hib were licensed in 1985 and consisted of polyribosylribitol phosphate (PRP) poly-saccharides which proved to be effective as vaccines in adults and children above the age of two years. Among infants, however, they did not provide adequate protection.[29]

The problem with poor immune response can often be overcome, for example, by conjugating the polysaccharide to an immunogenic protein carrier or a lipid. This helps with the uptake and presentation of the antigens by APCs. Initially, the PRP was conjugated to a modified non-toxic fragment of diphtheria toxin, CRM_{197}, or an outer membrane protein of *Neisseria meningitidis*. Later, a vaccine consisting of PRP conjugated to tetanus toxoid was licensed, and since the technology used for preparing this vaccine was not protected by patent laws, it became the most commonly used Hib pro-tein conjugate vaccine worldwide.[29] A fully synthetic Hib vaccine based on the same carbohydrate structure (Figure 6.7) was developed and entered the market in Cuba in 2004 after more than a decade of experimentation (Quimi-Hib). When introduced, this was the first synthetic carbohydrate-based vaccine and it was shown to have a success rate of 99.7%.[30]

6.4.1 Pneumococcal Vaccines

Streptococcus pneumoniae is a bacterium covered in a layer of capsular polysaccharides. Besides pneumonia, it is also responsible for several other types of infections such as bronchitis, rhinitis, conjunctivitis, meningitis and many others.[31] Vaccine development against *S. pneumoniae* began al-ready in the early 20th century with the use of intact, heat killed bacteria. Later, in 1944, a four-type (containing capsular polysaccharides from four types of pneumococcus) vaccine was successfully used for immunizing humans, which led to a hexavalent capsular polysaccharide being licensed for general use after World War II.[32]

The development of new vaccines against pneumococcus ceased during the 1950s due to the increasing use of penicillin and other antimicrobial agents. It was not until two decades later, when the increasing antibiotic resistance was making the treatment of pneumococcal infections difficult, that the development started anew. Another issue with pneumococcus is that a number of different serotypes exist, and these carry different capsular polysaccharides, making vaccine development difficult. The currently used pneumococcal vaccine (PPV23) contains 23 purified capsular oligosacchar-ides from *S. pneumoniae* and that still only accounts for approximately 90% of the existing serotypes.[32–34]

6.4.2 β-Mannan Glycoconjugates

β-(1→2)-Linked mannosides are a group of polysaccharides, found on the surface of several fungi of the *Candida* species, known to be immunosti-mulatory. Out of these, *C. albicans* is the most common and considered to be

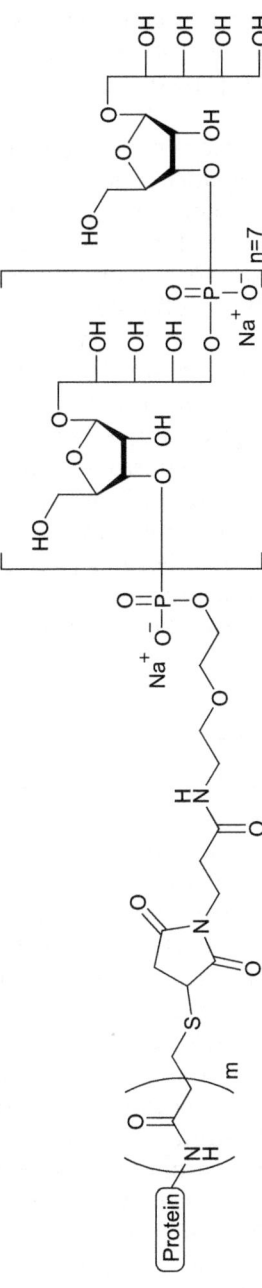

Figure 6.7 Structure of the synthetic Hib vaccine.

part of the human gut flora. *C. albicans* is, however, an opportunistic pathogen and the fourth most common cause for hospital-acquired bloodstream infections in the United States. For healthy people, *C. albicans* does not pose a significant threat, but it can prove especially dangerous for patients with suppressed immune systems, such as patients with acquired immunodeficiency syndrome (AIDS) or patients undergoing prolonged antibiotic treatments.[35]

The β-mannosides on the surface of the fungal cells show promising potential as vaccine candidates against candidemia. As previously mentioned, however, the oligosaccharides themselves are rather poorly immunogenic. Bundle and coworkers[35] have developed a vaccine consisting of a β-(1→2)-linked mannotriose conjugated to a tetanus toxoid. The vaccine has been tested by vaccinating rabbits followed by treatment with immunosuppressive drugs. Several days after the vaccinations, live *C. albicans* cells were administered directly into the bloodstream of the rabbits and several days after that the damage to inner organs was assessed. The vaccinated rabbits showed statistically significant reduction of fungal burden on the kidneys and liver compared to the control group of rabbits, vaccinated with the tetanus toxoid without the carbohydrate part. The direct administration of the pathogens to the bloodstream is, however, an extreme model, as in a normal case, the fungal cells would be slowly introduced to the bloodstream as a result of breakdown of mucosal barriers.[35] Despite several promising results from vaccinating rabbits, these vaccines are yet to enter clinical studies.

6.4.3 Anticancer Vaccines

Cancer cells are native cells that have mutated into cells growing and dividing uncontrollably. These cells usually display aberrant glycosylation patterns on the cell surface, which causes the immune system to destroy the cells. Problems arise when the immune system fails to recognize these cells as harmful.[36] Typically, cancer cells show overexpression of certain carbohydrates, *e.g.*, globo-H (GH) hexasaccharide (Fucα1→2Galβ1→3GalNAcβ1→3Galα1→4Galβ1-4Glcβ1) which can be found in excessive amounts on several epithelial tumors (including, but not limited to colon, ovarian, prostate and breast cancers).

As with most carbohydrates, the cancer-specific ones also need to be conjugated to an immunogenic carrier or be presented in a multivalent fashion to obtain significant immunological effects. Several vaccine candidates targeting the cell-surface carbohydrates expressed by cancerous cells have been developed and undergone clinical trials. Most of the vaccines have not proven satisfactory in advanced critical trials, mainly due to low reproducibility and poor selectivity of the immune response.[37,38] There are, however, recent breakthroughs in the field of carbohydrate-based anticancer vaccines. Danishefsky and coworkers have developed a vaccine consisting of five tumor-associated carbohydrate antigens (TACAs), GH, Le[y],

STn, TF and T_N conjugated to a keyhole limpet hemocyanin (KLH). This vaccine did, however, not manage to induce antibodies against Le^y in phase I clinical trials which was supposedly due to Le^y having a relatively high endogenic expression.[39] Consequently, the Le^y pentasaccharides were replaced by a GM2 tetrasaccharide, a tetrasaccharide associated with breast and prostate cancer. The new, improved vaccine (Figure 6.8) has shown great potential as a vaccine against prostate and breast cancer in phase I clinical trials and is currently in phase II/III clinical trials for breast cancer.[36,39,40]

Many vaccines targeting the same TACAs as the one developed by Danishefsky are currently being developed and are in, or preparing for, clinical trials.[37,41,42] Unfortunately, the immunogenic protein carriers used for these types of vaccines often trigger strong B-cell responses, which can suppress the antibody response against the target TACAs.[37,43] Boons and co-workers have developed an anticancer vaccine candidate where the antigen (T_N) is covalently linked to a T-cell epitope and a Toll-like receptor 2 ligand, Pam3CysSerK4. Activation of these TLR receptors is known to enhance local inflammation thus triggering the adaptive immune system. This compound has shown potential as an anticancer vaccine, eliciting high antibody responses, but has not yet entered clinical trials.[43]

6.4.4 Adjuvants

Due to safety reasons, vaccine development is moving away from using whole micro-organisms towards using fully synthetic antigens. This provides well-defined structures with fewer side-effects, and side-effects that can be studied in more detail. As previously mentioned, however, the problem with such molecules is that they are far less immunogenic than those used in traditional vaccines.[44]

Adjuvants are, by definition, compounds that stimulate the immune system without acting as a specific antigen. They enhance the immune response in one of three ways:[45]

- By providing a "depot" for the antigens so that they are released slowly and thus are not eliminated from the body before a sufficiently strong immune response is achieved
- By facilitating the phagocytosis of the antigens by APCs by helping the cells to target the antigens
- By enhancing and modulating the type of immune response caused by the antigen, for example by switching bias from cell-mediated to humoral immunity

Only a handful of adjuvants have been approved for human vaccine use. Of these alum, *i.e.*, aluminium salts such as aluminium oxohydroxide, aluminium hydroxyphosphate and aluminium hydroxide sulfate,[46] are the most widely used, and can be found in most of the human vaccine formulations

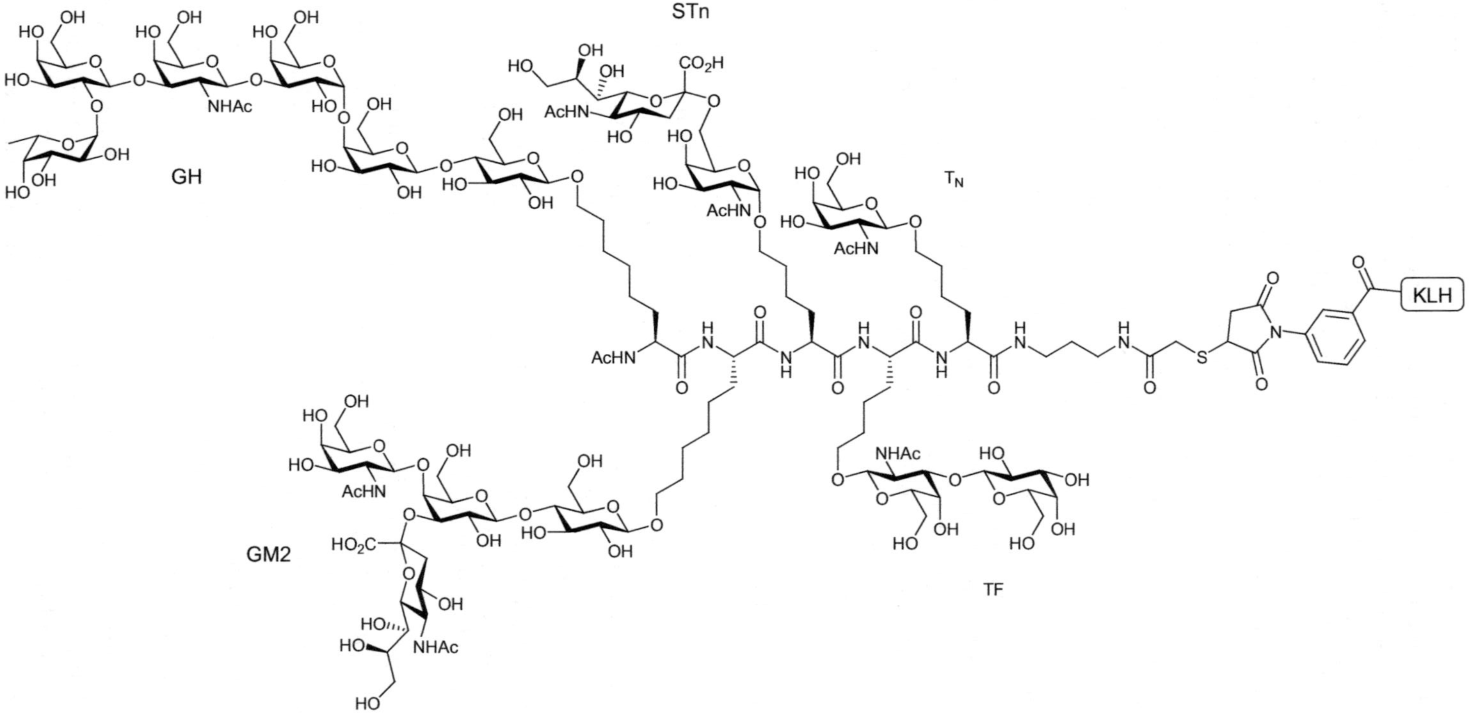

Figure 6.8 Structure of the anticancer vaccine developed by Danishefsky *et al.*

on the market.[44] Other types of adjuvants approved for human vaccines are, for example, oil–water emulsions, monophosphoryl lipid A and various protein-based structures. Two oligosaccharide-based vaccine adjuvants with great potential, inulin and chitosan, will be presented in the following paragraphs.

6.4.4.1 Inulin as a Vaccine Adjuvant

Inulins (Figure 6.9) are a group of polysaccharides naturally occurring in many types of plants *e.g.*, wheat, onions and garlic. Inulins consist of a linear chain of up to 100 β-(2→1)-linked D-fructofuranosyl moieties terminated by a single α-D-glucopyranosyl moiety. Inulins exist as three polymorphic forms: α-, β, and γ-inulin. These are all interconvertible, but only higher molecular weights can attain the γ-form, which is virtually insoluble in water at 37 °C (as opposed to the two other forms which are both water soluble).[47a]

Advax™ is an adjuvant derived from inulin which in mouse models has been shown to increase the humoral responses by an increase in the influenza-specific immunoglobulins. Furthermore, it has also been shown to stimulate the cell-mediated immunity, increasing T-cell responses, *i.e.*, cytokine secretion.[47b]

6.4.4.2 Chitosan as a Vaccine Adjuvant

Chitosan (Figure 6.10) is a linear β-(1→4)-linked D-glucosamine polysaccharide produced by deacetylation of chitin, the main component in the exoskeletons of arthropods. Chitosan has several attractive properties, such as biodegradability, biocompatibility, low toxicity and mucoadhesion. Being non-allergenic, it is also a good candidate for adjuvant development.[27,44]

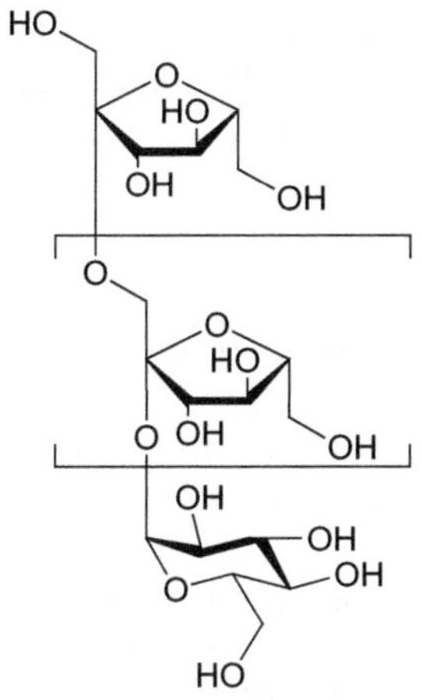

Figure 6.9 Structure of inulin.

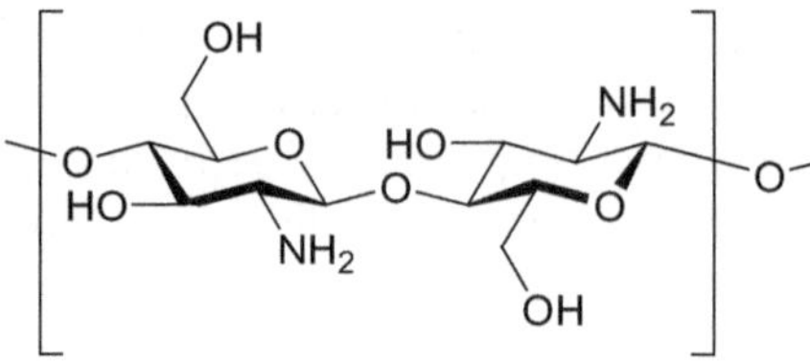

Figure 6.10   Structure of chitosan.

Chitosan has been shown to considerably enhance both the humoral and cell-mediated immunities, increasing both Ig and cytokine production when used as an adjuvant for nasally administered diphtheria vaccine.[48] This adjuvant effect is based on prevention of the enzymatic degradation of the antigen (mutated non-toxic diphtheria toxin), as well as on enhancing the uptake of the antigen through the mucosal barrier. In addition to these effects, chitosan also directly stimulates the immune system by activating macrophages and enhancing cytokine production.[44]

6.5   Conclusions

Due to their abundance, structural diversity and biological functions, oligosaccharides have found many uses in pharmaceutical development. Enormous efforts have been made towards developing methods for the isolation and synthesis of new oligosaccharides and many of the problems associated with the synthesis and isolation of this unique class of molecules have been overcome already. Nevertheless, relatively few generally applicable methods exist today and new targets often present whole new sets of problems that need to be tackled individually. Despite the amount of work put into this field of research over the last decades, relatively few oligosaccharide-based pharmaceuticals have entered the market to date.

In this chapter, selected existing and potential future applications of oligosaccharides as pharmaceuticals have been presented and discussed. Exciting research is ongoing in the field worldwide and exciting future breakthroughs are to be expected.

References

1. M. C. Galan, D. Benito-Alifonso and G. M. Watt, *Org. Biomol. Chem.*, 2011, **9**, 3598.
2. B. Ernst and J. L. Magnani, *Nat. Rev. Drug Discovery*, 2009, **8**, 661.
3. L. L. Kiessling, J. E. Gestwicki and L. E. Strong, *Curr. Opin. Chem. Biol.*, 2000, **4**, 696.
4. For a general overview see for example: (a) H. M. I. Osborn, *Carbohydrates* Elsevier Science Ltd, Oxford, 1st edn, 2003; (b) R. V. Stick, *Carbohydrates – The Sweet Molecules of Life*, Academic Press, London, 2001.

5. C. H. Hsu, S.-C. Hung, C.-Y. Wu and C.-H. Wong, *Angew. Chem., Int. Ed.*, 2011, **50**, 11872.

6. (a) P. H. Seeberger, *Chem. Soc. Rev.*, 2008, **37**, 19; (b) D. B. Werz and P. H. Seeberger, *Chem. – Eur. J.*, 2005, **11**, 3194; (c) P. H. Seeberger and D. B. Werz, *Nat. Rev. Drug. Discovery*, 2005, **4**, 751.

7. N. L. Pohl and G. Park, (Iowa State University Research Foundation, Inc., Ames, IA, US), Automated Solution-Phase Iterative Synthesis. *U. S. Pat.* Application Publication 11/767098, Jun. 24, 2010.

8. F. A. Jaipuri and N. L. Pohl, *Org. Biomol. Chem.*, 2008, **6**, 2686.

9. F. R. Carrel, K. Geyer, J. D. C. Codée and P. H. Seeberger, *Org. Lett.*, 2007, **9**, 2285.

10. K. R. Love and P. H. Seeberger, *Angew. Chem., Int. Ed.*, 2004, **43**, 602.

11. F. Yang and Y. Ito, *Anal. Chem.*, 2002, **74**, 440.

12. (a) A. Ziegler and J. Zaia, *J. Chromatogr. B: Anal. Technol. Biomed. Life Sci.*, 2006, **837**, 76; (b) D. Lecacheux and G. Brigand, *Carbohydr. Polym.*, 1988, **8**, 119.

13. E. R. Suárez, R. Syvitski, J. A. Kralovec, M. D. Noseda, C. J. Barrow, H. S. Ewart, M. D. Lumsden and T. B. Grindley, *Biomacromolecules*, 2006, **7**, 2368.

14. S. Pongasamart, (Chulalongkorn University, Bangkok, Thailand), Poly-saccharide Products From Durian: Process for Isolation and Purification and Their Application. *U. S. Pat.* Application Publication 10/085897, Sep. 4, 2003.

15. A. J. Erskine and J. K. N. Jones, *Can. J. Chem.*, 1956, **34**, 821.

16. N. Volpi, *Anal. Biochem.*, 1994, **218**, 382.

17. For a historical overview, see: D. Wardrop and D. Keeling, *Br. J. Haematol.*, 2008, **141**, 757.

18. (a) C. A. A. van Boeckel, T. Beetz, J. N. Vos, A. J. M. De Jong, S. F. Van Aelst, R. H. Van den Bosch, J. M. R. Mertens and F. A. Van den Vlugt, *J. Carbohydr. Chem.*, 1985, **4**, 293; (b) M. Petitou, P. Duchaussoy, I. Lederman and J. Choay, *Carbohydr. Res.*, 1986, **147**, 221.

19. (a) I. Capila and R. J. Linhardt, *Angew. Chem., Int. Ed.*, 2002, **41**, 390; (b) M. Petitou, B. Casu and U. Lindahl, *Biochimie*, 2003, **85**, 83.

20. M. Petitou and C. A. A. van Boeckel, *Angew. Chem., Int. Ed.*, 2004, **43**, 3118 and references therein.

21. For reviews see: (a) J. Harenberg, *Med. Res. Rev.*, 1998, **18**, 1; (b) D. A. Lauver and B. R. Lucchesi, *Cardiovasc. Drug Rev.*, 2006, **24**, 214.

22. U. F. Wehmeier and W. Piepersberg, *Appl. Microbiol. Biotechnol.*, 2004, **63**, 613.

23. P. B. Deshpande, P. Luthra, A. K. Pandey, D. J. Paghdar, P. S. Gowthamaiah and V. Govardhana, (Gujarat, India), Process for the Preparation of Pentosan Polysulfate or Salts Thereof. *U. S. Pat.* Application Publication 12/530320, Apr. 29, 2010.

24. (a) F. Tascioglu and C. Öner, *Clin. Rheumatol.*, 2003, **22**, 112; (b) T. Higashide and K. Sugiyama, *Clin. Ophthalmol.*, 2008, **2**, 21.

25. C.-H. Wong, *Carbohydrate-based Drug Discovery*, Wiley-VCH GmbH & Co. KgaA, Weinheim, 2003, vol. 2, pp. 661–684.
26. (a) H. E. Carter, R. K. Clark, S. R. Dickman, Y. H. Loo, O. S. Skell and W. A. Strong, *J. Biol. Chem.*, 1945, **160**, 337; (b) B. Singh and D. A. Mitchison, *Br. Med. J.*, 1954, **1**, 130.
27. R. P. McGeary, C. Olive and I. Toth, *J. Pept. Sci.*, 2003, **9**, 405.
28. A. H. Lucas, M. A. Apicella and C. E. Taylor, *Vaccines*, 2005, **41**, 705.
29. (a) S. K. Morris, W. J. Moss and N. Halsey, *Lancet Infect. Dis.*, 2008, **8**, 435; (b) R. E. O'Loughlin, K. Edmond, P. Mangtani, A. L. Cohen, S. Shetty, R. Hajjeh and K. Mulholland, *Vaccine*, 2010, **28**, 6128.
30. (a) V. Verez-Bencomo, V. Fernandez-Santana, E. Hardy, M. E. Toledo, M. C. Rodríguez, L. Heynngnezz, A. Rodriguez, A. Baly, L. Herrera, M. Izquidero, A. Villar, Y. Valdés, K. Cosme, M. L. Deler, M. Montane, E. Garcia, A. Ramos, A. Aguilar, E. Medina, G. Toraño, I. Sosa, I. Hernandez, R. Martinez, A. Muzachio, A. Carmenates, L. Costa, F. Cardoso, C. Campa, M. Diaz and R. Roy, *Science*, 2004, **305**, 522; (b) V. Fernández-Santana, F. Cardoso, A. Rodriguez, T. Carmenate, L. Peña, Y. Valdéz, E. Hardy, F. Mawas, L. Heynngnezz, M. C. Rodríguez, I. Figueroa, J. Chang, M. E. Toledo, A. Musacchio, I. Hernández, M. Izquierdo, K. Cosme, R. Roy and V. Verez-Bencomo, *Infect. Immun.*, 2004, **72**, 7115.
31. R. A. C. Siemieniuk, D. B. Gregson and M. J. Gill, *BMC Infect. Dis.*, 2011, **11**, 314.
32. M. A. Barocchi, S. Censini and R. Rappuoli, *Vaccine*, 2007, **25**, 2963.
33. D. C. Powers and E. L. Anderson, *J. Infect. Dis.*, 1996, **173**, 1014.
34. L. Morelli, L. Poletti and L. Lay, *Eur. J. Org. Chem.*, 2011, 5723.
35. (a) T. Lipinski, X. Wu, J. Sadowska, E. Kreiter, Y. Yasui, S. Cheriaparambil, R. Rennie and D. R. Bundle, *Vaccine*, 2012, **30**, 6263; (b) H. Xin, S. Dziadek, D. R. Bundle and J. E. Cutler, *Proc. Natl. Acad. Sci. U. S. A.*, 2008, **105**, 13526.
36. J. Zhu, J. D. Warren and S. J. Danishefsky, *Expert Rev. Vaccines*, 2009, **8**, 1399.
37. C. Nativi and O. Renaudet, *ACS Med. Chem. Lett.*, 2014, **5**, 1176.
38. D. Miles, H. Roché, M. Martin, T. J. Perren, D. A. Cameron, J. Glaspy, D. Dodwell, J. Parker, J. Mayordomo, A. Tres, J. L. Murray, N. K. Ibrahim and The Teratope Study Group, *Oncologist*, 2011, **16**, 1092.
39. J. Zhu, Q. Wan, D. Lee, G. Yang, M. K. Spassova, O. Ouerfelli, G. Ragupathi, P. Damani, P. O. Livingston and S. J. Danishefsky, *J. Am. Chem. Soc.*, 2009, **131**, 9298.
40. R. M. Wilson and S. J. Danishefsky, *J. Am. Chem. Soc.*, 2013, **135**, 14462.
41. Y.-L. Hang, J.-T. Hung, S. K. C. Cheung, H.-Y. Lee, K.-C. Chu, S.-T. Li, Y.-C. Lin, C.-T. Ren, T.-J. R. Cheng, T.-L. Hsu, A. L. Yu, C.-Y. Wu and C.-H. Wong, *Proc. Natl. Acad. Sci. U. S. A.*, 2013, **110**, 2517.
42. H. Cai, Z.-H. Huang, L. Shi, Z.-Y. Sun, Y.-F. Zhao, H. Kunz and Y.-M. Li, *Angew. Chem., Int. Ed.*, 2012, **51**, 1719.

43. S. Ingale, M. A. Wolfert, T. Buskas and G.-J. Boons, *ChemBioChem*, 2009, **10**, 455.
44. P. Riese, K. Schulze, T. Ebensen, B. Prochnow and C. A. Guzmán, *Curr. Top. Med. Chem.*, 2013, **13**, 2562.
45. J. H. Wilson-Welder, M. P. Torres, M. J. Kipper, S. K. Mallapragada, M. J. Wannemuehler and B. Narashiman, *J. Pharm. Sci.*, 2009, **98**, 1278.
46. C. J. Clements and E. Griffiths, *Vaccine*, 2002, **20**(Suppl. 3), S24.
47. (a) D. G. Silva, P. D. Cooper and N. Petrovsky, *Immunol. Cell. Biol.*, 2004, **82**, 611; (b) Y. Honda-Okubo, F. Saade and N. Petrovsky, *Vaccine*, 2012, **30**, 5373.
48. (a) E. A. McNeela, D. O'Connor, I. Jabbal-Gill, L. Illum, S. S. Davis, M. Pizza, S. Peppoloni, R. Rappuoli and K. H. G. Mills, *Vaccine*, 2001, **19**, 1188; (b) E. A. McNeela, I. Jabbal-Gill, L. Illum, M. Pizza, R. Rappuoli, A. Podda, D. J. M. Lewis and K. H. G. Mills, *Vaccine*, 2004, **22**, 909.

CHAPTER 7

Non-fuel Applications of Sugars in Brazil

PETER R. SEIDL,* ESTEVÃO FREIRE AND
SUZANA BORSCHIVER

Brazilian Green Chemistry School, School of Chemistry, Federal University
of Rio de Janeiro, Brazil
*Email: pseidl@eq.ufrj.br

7.1 Introduction

The use of biomass for the production of fuels and chemicals can contribute towards mitigating several of the problems involving the depletion of the world's non-renewable resources and the increasing impact of climate change caused by emissions resulting from human activity. In fact, biobased chemicals and polymers are making steady inroads into traditional markets for consumer products derived from fossil fuels.[1]

Today the concept of a biorefinery is widely applied to facilities that integrate biomass conversion processes and equipment for sustainable processing of biomass into a spectrum of value-added biobased products (food, animal feed, chemicals, materials) and bioenergy (biofuels, power and/or heat).[2] It has two strategic goals: the displacement of fossil fuels by renewable raw materials and the establishment of a sustainable biobased industry. But, despite its high volume, fuel is a low value product and the return on investment in biofuels-only operations presents a significant barrier to realizing the biorefinery's economic goal. Industrial adoption of renewable raw materials requires a financial incentive to justify the shift to new starting materials, the development of processes to convert them into useful

RSC Green Chemistry No. 44
Biomass Sugars for Non-Fuel Applications
Edited by Dmitry Murzin and Olga Simakova
© The Royal Society of Chemistry 2016
Published by the Royal Society of Chemistry, www.rsc.org

products and the capital investment necessary to assure their production on a commercial scale.[3]

High value, lower volume biobased chemicals provide this incentive and a number of evaluations have appeared that identify structures most easily obtained from a given biomass conversion process. The advantage of this approach is the tailoring of broad-based processes to the building blocks available from certain biorefinery operations.[3]

In 2004, the US Department of Energy (DOE) released the first of two reports outlining the research needs for biobased products.[4] By developing a list of specific structures, the report adopted product identification as a guide for research. Following economic and technical criteria, this list served as a guide for selecting structures that could serve as building blocks for a platform for the production of derivatives. In the years following the original DOE report, considerable progress in biobased product development has been made. Recently, Bozell and Petersen[3] revised that report and presented an update on the candidate structures. Several of these structures are available from chemical and biological transformations of sugar cane. Those that appear to be the most promising for the chemical industry, currently under investigation by Brazilian groups, are discussed in this chapter.

7.2   The Sugar Cane Industry in the Brazilian Economy

Sugar has been an important component of Brazil's economy since shortly after the country was discovered. In the five centuries since the first sugar mill was built, this small rural activity developed into a large-scale business that was responsible for the growth of villages that today are important cities like São Paulo and Santos. Besides producing different grades of sugar, the juice could be fermented and also became a source of an alcoholic beverage known as *cachaça* that soon became very popular when the country was still a colony. Ethanol has been produced in sugar mills ever since.[5]

As early as 1910, ethanol was also an important feedstock for chemicals and materials. Several companies, such as Rhodia, Usina Colombina, and Elekeiroz, produced ethyl chloride, acetic acid, acetic anhydride, cellulose acetate and ethyl ether from ethanol. In the 1940's, the Usina Victor Sence also produced acetone, butyl alcohol and butyl acetate from ethanol (and was, at the time, the sole manufacturer in South America) while in the 1950's and 1960's, it was also widely used in the production of ethylene, vinyl chloride and butene. Ethanol was first used as an automotive fuel in 1925 and, in the wake of a petroleum shortage, the Proalcohol program was established in 1975 to produce it on a large scale as a substitute for gasoline. It also gave the conversion of ethanol into chemicals a boost that lasted until subsidies for the production of ethanol and the severe drop in the price of crude oil rendered these processes uneconomical.[6]

The production of first-generation ethanol from sugar cane juice (essentially sucrose) has made the country a leading producer of biofuels. At present, approximately 90% of the automobiles made in Brazil are dual-fuel.[7] Production increased in 2012 when the government offered incentives such as raising the percentage of ethanol that is added to gasoline from 20 to 25%. The prospect of producing 2G ethanol led some companies to invest in new production technologies that significantly increased their yields.

7.2.1   The Sucrose Platform

Sucrose is an abundant, low-cost, highly pure (up to 99.96% on an anhydrous basis) raw material from a renewable source.[8] It can be hydrolysed to form glucose and fructose and is a starting material in the synthesis of non-ionic surfactants, polymers, sweeteners, emulsifiers, *etc.*,[9] as shown in Scheme 7.1.

In spite of all these advantages, sucrose also presents some difficulties as a raw material. The main one is its high functionality (eight hydroxyl groups with almost similar reactivity) making selective syntheses extremely difficult[10] and controlling this selectivity presents a serious problem.[11]

However, in Brazil today, most sucrose derivatives are imported and their main use is in food products and in formulations of drugs and cosmetics.[12]

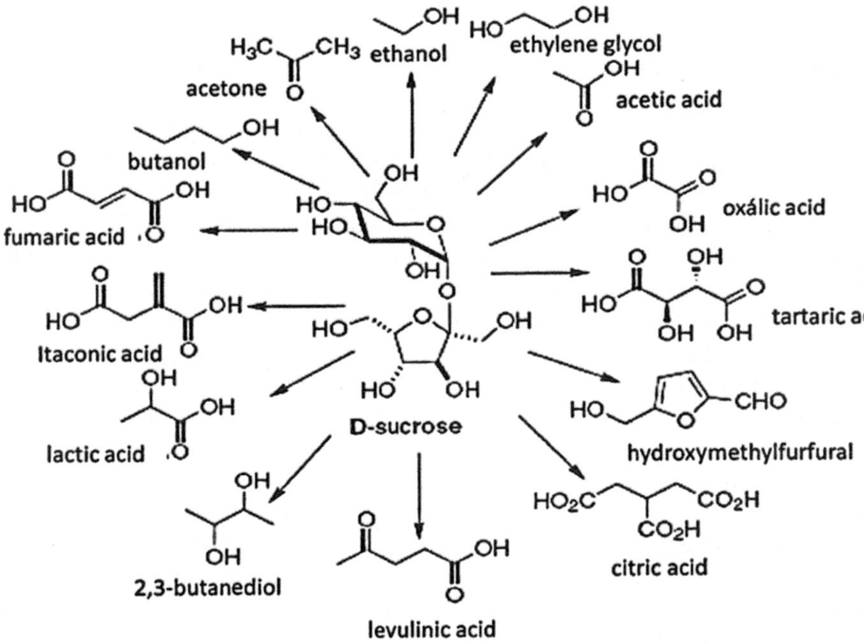

Scheme 7.1   Low molecular weight chemicals obtained from sucrose (reproduced from ref. 9 with permission of the Sociedade Brasileira de Química).

On the other hand, sucrose shows much promise as a starting material in bioprocesses. Two of the more promising products of current interest are:

- Succinic acid—one of the platform chemicals identified in the original DOE study.4 It is a driver in the production of polymers such as poly-butene succinate (PBS), a new biodegradable plastic and 1,4-butanediol (BDO) and other BDO derivatives such as polytetramethylene glycol ether (PTMEG). It has a high conversion coefficient on fermentation of sugar. Production of succinic acid, principally from the fermentation of sugars, is the object of much interest of several companies and some of their plants already operate on a commercial scale. In Brazil, DSM is studying the production of succinic acid based on sugar cane in a project financed by the BNDES, the national development bank, as part of a program for the support of the sugar and ethanol industry (see Section 7.5).
- Farnesene (and a set of closely related sesquiterpenes)—obtained from the fermentation of sugars using a genetically modified strain of *Saccharomyces cerevisiae*. This process was patented by Amyris, which operates a commercial scale plant in Brazil with an estimated capacity of 40 million litres per year.[13] Amyris also has been developing new applications of farnesene including biofuels, aromas and fragrances, and chemical specialties, among others.

7.2.2   Chemicals from Ethanol

Ethanol was specifically omitted from DOE's original list because its expected high production volume categorized it as a so-called super-commodity. Recent technology developments and strategic commercial partnerships have improved its platform potential, leading it to be included in the new list of top chemical opportunities from biorefinery carbohydrates.[3]

Ethanol dehydration was the source of most ethylene in the early part of the 20th century and can be carried out at high conversion and selectivity.[14] As has been pointed out in Section 7.2, ethanol was once an important raw material for the Brazilian chemical industry. Although the availability of cheap oil all but wiped out its commercial application, it is employed to produce "green" polyethylene by Braskem, Brazil's largest plastics producer, which has, in fact, been investing in process optimization and plant expansion. Ethylene produced from ethanol dehydration may still find niche applications in situations in which the sustainability of raw materials is as important a consideration as their unit price.

Challenges related to an increase in the use of ethanol from biomass as a chemical feedstock are based on the development of novel high performance catalysts and reductions in costs of replacing well-established processes and products with biomass-derived ones.[14] Bioethanol has an advantage compared to other biomass feedstocks, such as lignin, cellulose, hemicellulose and fatty acids: it can be directly converted, in one-pot processes, into "drop-in" chemicals. This means that ethanol can be used to obtain some of the same building block chemicals that are currently obtained from petroleum, such as ethylene, 1,3-butadiene, propylene, and higher hydrocarbons. In this case, changes in the chemical industry to incorporate ethanol as a feedstock could be minimized. Sousa et al.[15] have shown that conversion of ethanol into hydrocarbons occurs over zeolites with different acidic properties and porous structures (HZSM-5 and HMCM-22). The samples used were active for ethanol conversion and both their acidic properties and porous structure played an important role in product selectivity. HZSM-5 zeolite was the most promising catalyst for propene formation,[15] but catalysts for industrial use of these routes must still be developed and the costs for producing these drop-in chemicals from ethanol are still higher than the highly optimized petroleum-based processes.[14]

The production of chemicals that contain an oxygenated functional group has been an active area of research on ethanol conversion. Combinations of low cost catalysts can provide promising one-step synthetic routes to the production of chemicals such as n-butanol,[16] acetic acid,[17] acetone[18] and others such as ethyl acetate.[19]

At present, there is much interest in developing one-step processes to obtain n-butanol from ethanol. n-Butanol has a wide range of applications and may be obtained from mixed Mg-Al oxides. Adjacent acid and medium basic sites are needed in order to generate the intermediate compounds. Carvalho et al.[16] observed that the higher the concentration of Mg, the higher the hydrogenation of the catalyst. Strong basic sites and specific superficial atom arrangements, on the other hand, seem not to be essential for the reaction to take place.

Acetic acid, another important chemical raw material, may be obtained from ethanol in one step by direct oxidation. Reactions may be run in the liquid or gas phase. Several catalysts have been proposed for this reaction and PdO/ZrO_2 seems very promising.[17]

Acetaldehyde and ethyl acetate can be obtained from ethanol by two different routes.[20] In an inert atmosphere, ethanol can be dehydrogenated to acetaldehyde under aerobic conditions in the presence of a redox catalyst.[19] The acetaldehyde that is formed may undergo a second oxidation to acetic acid. If the catalyst also has an acidic functionality, acetic acid readily esterifies with ethoxide species adsorbed on the catalyst forming ethyl acetate directly, without going through acetic acid oxidation. On the other hand, the reaction under an aerobic atmosphere can be carried out under milder conditions in the presence of a copper catalyst, forming hydrogen. However, this reaction also forms butanone, a hazardous substance that is difficult to eliminate. Different routes have been proposed for the direct synthesis of ethyl acetate from ethanol. Dehydrogenation and oxidative processes and a variety of heterogeneous catalytic systems with gold-based catalysts are able to oxidize primary alcohols. Copper-based catalysts have been successfully employed for the selective conversion of ethanol to ethyl acetate or acetaldehyde while the best results for selective conversion to ethyl acetate have been achieved with ZrO_2-supported copper catalysts.[20]

7.3 The Bioethanol Plant as an Integrated Biorefinery

The biorefinery takes advantage of the various components in biomass and their intermediates therefore maximizing the value derived from the biomass feedstock. It employs a multidisciplinary approach that integrates physical and mechanical methods, chemical and biological conversion, catalysis and biocatalysis to attain high efficiency, low cost, and low energy consumption. Biorefineries are continuously evolving as new advances in biomass feedstock research, related processes and products become available for sustainable energy production—a challenge and an opportunity that are currently faced in endeavors to transition to a biobased economy and society.[2]

An integrated forest biorefinery (IFBR)[2] is a biorefinery that can process forest-based biomass such as wood and forestry residues to bioenergy and bioproducts including cellulosic fibers for pulp and paper production (Figure 7.1). As lignocellulose consists of four major components—cellulose, hemicelluloses, lignin and extractives—the IFBR has four production platforms that can be used in an integrated manner for production of biofuels and high-value bioproducts.[2]

Second-generation biofuels which use, for example, biomass feedstock, agricultural wastes and wood residue, represent an efficient and complementary approach to increase liquid biofuel production. The adoption of second-generation ethanol production from sugar cane lignocellulosic (LC) biomass is attractive from a number of perspectives. By making use of all

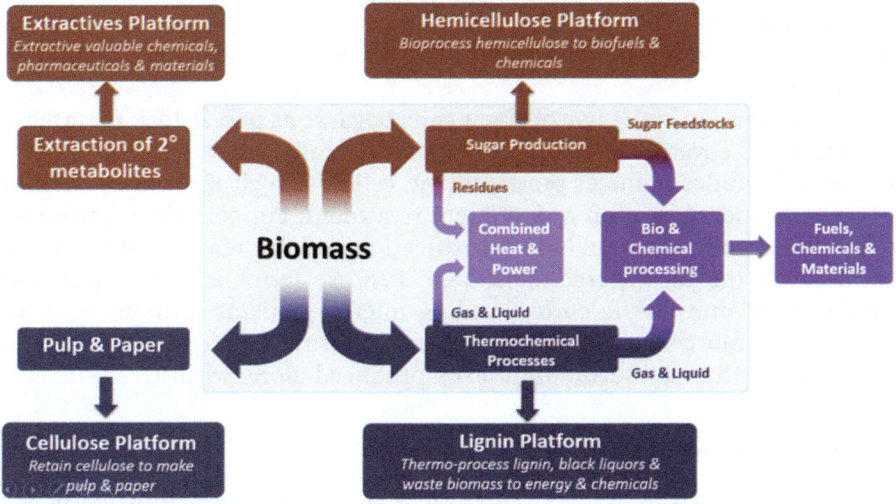

Figure 7.1   The IFBR concept (reproduced from ref. 2).

available biomass, such approaches can improve the carbon footprint of biofuels further, as well as increase the yield of ethanol per hectare and provide a means to sustain the operation of these plants throughout the year, instead of their current seasonal operation.[21–23]

Ethanol from biomass plants is responsible for LC residues that correspond to approximately two thirds of the total weight of the sugar cane crop. They also generate large quantities of vinasse and carbon dioxide. The potential use of carbon dioxide for the production of chemicals and fuels has been recently reviewed[24] and will not be covered here. Vinasse has been used in the production of biogas and treatment of the soil[25] but not as a source of chemicals.

The concept of a LC biorefinery is central to the production of chemicals and fuels from sugar cane. It can be applied to a facility that converts sugar cane into sugar and/or ethanol since bagasse and straw are generated in large quantities and chemical transformations may be employed in order to add value to these potential raw materials.[24] This type of biorefinery is based on a selective separation of residual biomass fractions and on processes that transform them in accordance with their chemical characteristics and desired target products. In this context, LC materials, especially agroindustrial residues, have been the subject of intense research since they are renewable sources of carbon and energy that are available in large quantities. The integrated and rational utilization of this abundant feedstock can revolutionize a variety of industries and bring immeasurable benefits to countries with extensive regions of high biological productivity such as Brazil.[26]

However, LC materials do not contain saccharides that can be directly converted using known technology. For the integral utilization of LC feedstocks, it is necessary to develop pretreatment processes for the selective and efficient fractionation of their main polysaccharide components, which should undergo hydrolysis to generate a high concentration of sugar monomers. These are the building-blocks for biotechnological and chemical processes.[27]

A recent review on the use of sugar cane bagasse as a feedstock for ethanol production[28] covers the main issues of the transformations that are involved. It includes biomass pretreatment, cellulose hydrolysis, fermentation of hexoses and separation and effluent treatment. For complete conversion to ethanol, it must also include detoxification and fermentation of pentoses released during the pretreatment step. Pentose fermentation requires an independent unit as the corresponding microorganisms ferment pentoses and hexoses more slowly than those that only ferment hexoses and are more sensitive to ethanol and the inhibitors produced along with it.

7.3.1 Treatment of LC Biomass

LC materials, such as sugar cane bagasse and straw, are mainly composed of two polysaccharidic fractions (cellulose and hemicelluloses) and a polyphenolic macromolecule (lignin). The cellulose platform in an IFBR is reserved mainly for production of cellulosic fibers for pulp and paper, which is the core business of the pulp and paper industry. This is in contrast to a lignocellulosic feedstock biorefinery, such as an ethanol biorefinery, where cellulose is hydrolyzed and fermented to ethanol and chemicals.

Presently, research on cellulose for chemicals and polymeric materials is mainly focused on the following aspects:[2]

1. The catalytic conversion of cellulose to fuels and chemicals;
2. The development of environmentally new solvent systems to dissolve cellulose and following applications;
3. The modification of functional material with cellulose derivatives or cellulose graft copolymers.

7.3.1.1 Structures and Properties of Cellulose

A cellulose molecule has the generic chemical formula $(C_6H_{12}O_5)_n$ and is a polydisperse linear 1,4-β-glucan. It is part of a renewable resource which is the most abundant natural polymer on earth. It has been estimated that the global production of cellulose is 1.5 trillion tons each year and is considered an almost inexhaustible source of raw materials for environmentally friendly and biocompatible products.[2]

The structure and properties of cellulose have been the subject of a large volume of work. It consists of a skeletal linear polysaccharide, and is connected by β-1,4-glycosidic linkages. The glucose units are further tightly bound by numerous extensive inter- and intramolecular hydrogen bonds as shown in Figure 7.2.[29]

The glucopyranose units of cellulose chains range from approximately 100 to 14 000. Accordingly, cellulose has an average molecular weight in the range of 300 000–500 000. One of the most interesting characteristics is that cellulose consists of several crystal polymorphs, with the possibility of conversion from one form to another. It is these highly crystalline regions that increase resistance to decomposition processes.

7.3.1.2 Structures and Properties of Hemicellulose

Hemicellulose, the second most abundant polysaccharide after cellulose, is an amorphous heterogeneous polymer comprising 15–35% of lignocellulosic biomass with a degree of polymerization (DP) of 80–200.[2] Hemicellulose forms an interface in the cell wall matrix with binding properties mediated by covalent and noncovalent interactions with lignin, cellulose and other polysaccharides. The close association between the biopolymers in plant biomass is realized *via* chemical bonds, predominantly between lignin and hemicelluloses, in lignin-carbohydrate complexes that include benzyl-ether, benzyl-ester and phenyl-glycoside types of linkages. The composition and structure of hemicelluloses (a heteropolymer) are more complicated than

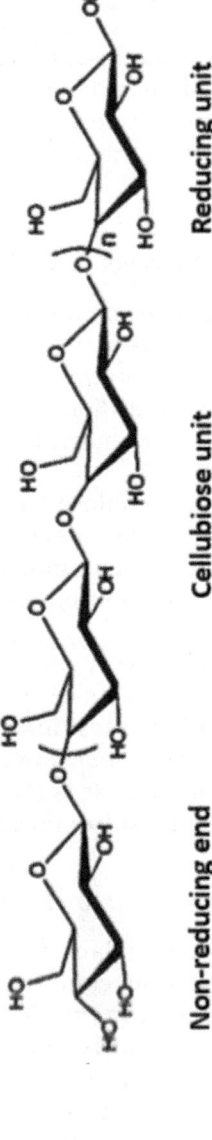

Figure 7.2 The central part of a cellulose molecular chain (reproduced from ref. 29).

those of cellulose (a homopolymer) and can vary quantitatively and qualitatively in various LC species.[2] Details of these interactions are discussed in Section 7.3.3.

The building blocks of hemicelluloses (polyoses) include pentoses (D-xylose and L-arabinose) and hexoses (D-glucose, D-galactose and D-mannose). Sugar acids (acetic, 4-*O*-methyl glucuronic acid, ferulic/coumaric acids) make up the remainder of hemicellulose structures. Hardwood xylans as complex heteropolysaccharides, comprising β-4-linked D-xylopyranose units, are highly substituted (Figure 7.3). The xylopyranose unit of the xylan main chain can be substituted at the C2 and/or C3 positions with acetic acid (at both the C2 and C3 positions in hardwoods), 4-*O*-methylglucuronic acid (at the C2 position in both hardwoods and softwoods), and arabinose (at the C3 position in softwoods). Arabinose may be further esterified by phenolic acids which crosslink xylan and lignin in the cell wall matrix. Sugar cane bagasse hemicelluloses are composed of heteroxylans, with a predominance of xylose, which is part of a chain that can be chemically hydrolyzed more easily than cellulose. There are many possible processes and products that would have to be considered in exploring all the possible alternatives and it would be impractical and probably impossible to present all pathways of the hemicellulose platform.[2]

7.3.1.3 Structures and Properties of Lignin

Lignin, the second most abundant biopolymer on earth, is the only large-scale renewable feedstock composed of aromatics[30,31] and represents the polyphenolic fraction. Lignins are complex, three-dimension biopolymers consisting of phenylpropanoid units containing both aromatic and aliphatic groups.[2] The phenylpropane units (C9 or C6C3), known as monolignols or lignin precursors, are linked together through C–C and C–O–C bonds and have different amounts of methoxy groups (Me). The dominant bond is the β-O-4 linkage. Three types of monolignols have been identified: *p*-coumaroyl alcohol, coniferyl alcohol and synapyl alcohol (Figure 7.4).

The lignin macromolecule also contains a variety of functional groups that have an impact on its reactivity such as methoxyl groups, phenolic hydroxyl groups, and a few terminal aldehyde groups. Only a small proportion of the phenolic hydroxyl groups are free since most are occupied in linkages to neighboring phenylpropane linkages. Carbonyl and alcoholic hydroxyl groups are incorporated into the lignin structure during the enzymatic dehydrogenation. Lignin is more concentrated in the middle lamella and primary cell wall. Lignin surrounds the cellulose–hemicellulose matrix to provide stiffness to the cell walls and glue the cells together. Lignin as a hydrophobic polymer serves as a barrier against water penetration and is resistant toward degradation by most microorganisms except white-rot fungi and some bacteria.[2] Lignin interacts with hemicellulose covering the cellulose matrix and conferring resistance to enzymatic and chemical degradation.[32,33] Research on the production of fuels and chemicals from LC

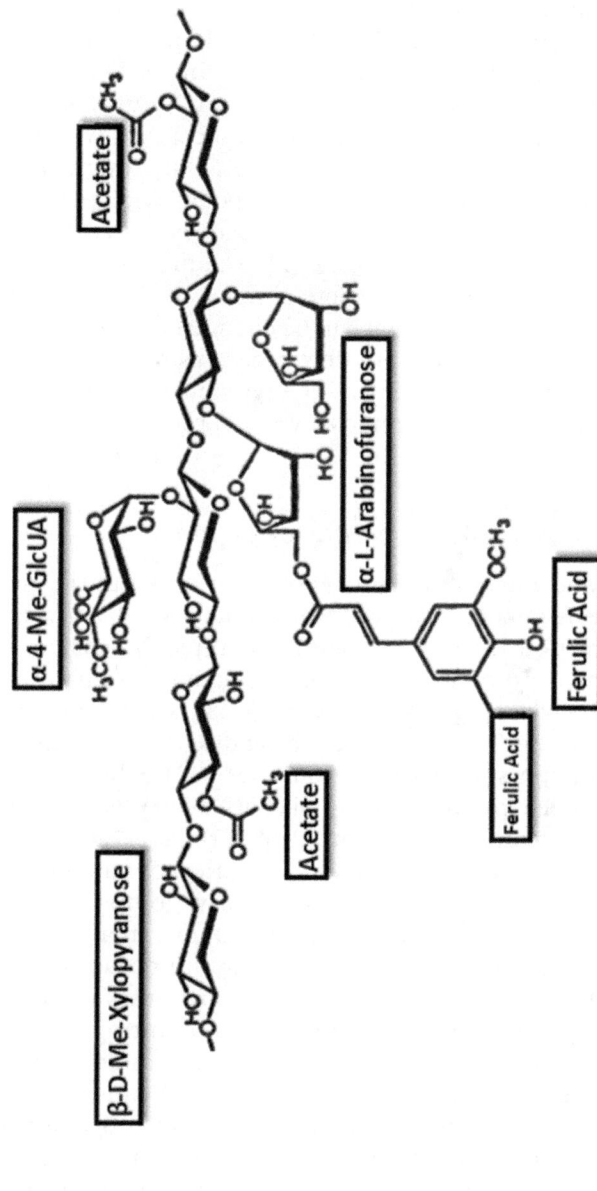

Figure 7.3 The chemical structure of xylan (reproduced from ref. 2).

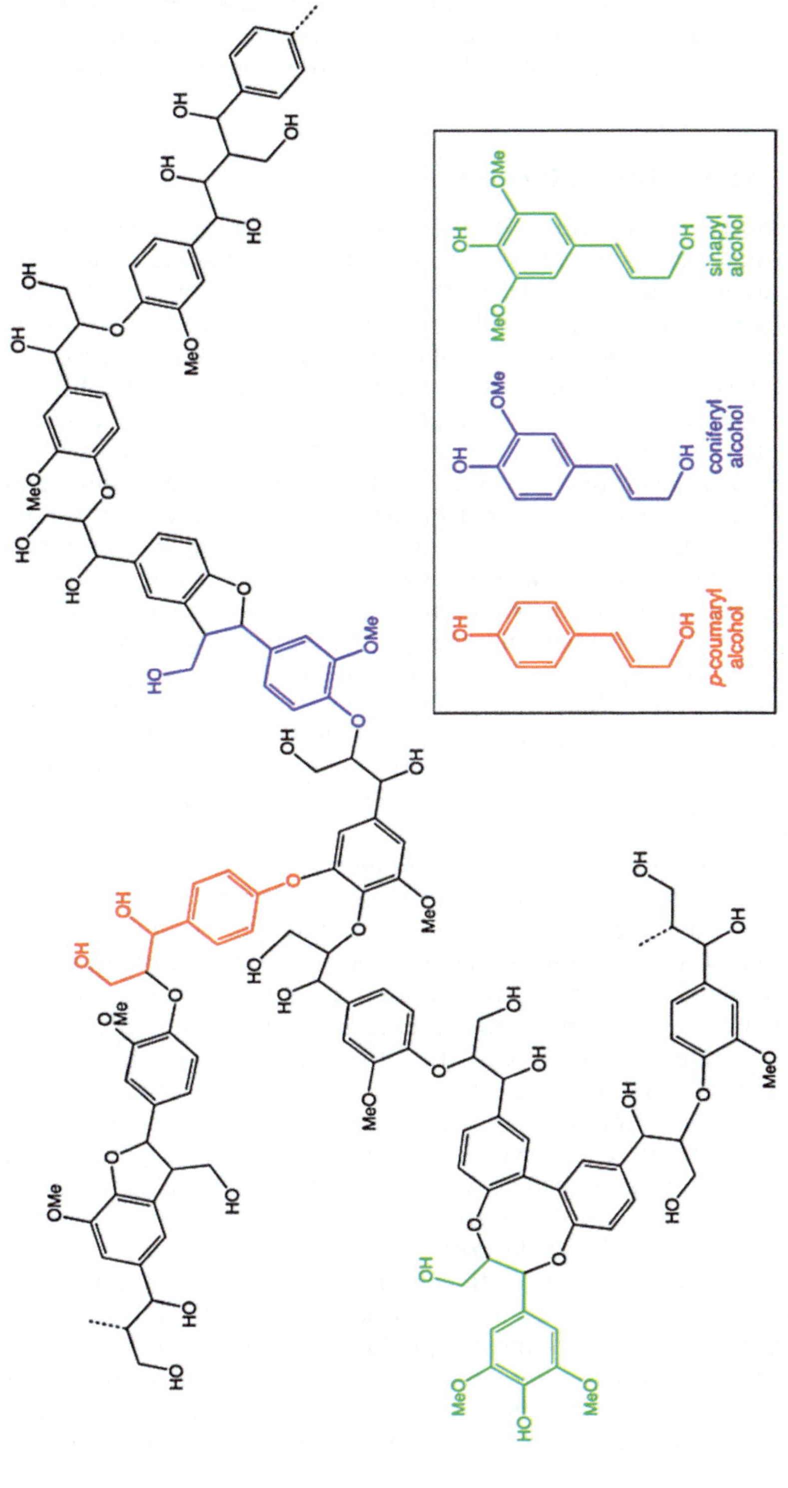

Figure 7.4 The structural model of lignin (reproduced from ref. 2).

residues is thus aimed at processes that increase access to the polysaccharidic components by overcoming the recalcitrance of plant cell walls, arguably the most fundamental unsolved problem of plant-based green technologies.[34]

7.3.2 Pretreatment Processes

Pretreatment of lignocellulosic feedstock is probably the single most crucial step in sugar cane biorefinery processes since it has a large impact on the yield and efficiency of subsequent treatments.[35] It is envisaged as the first step of sequential refining of lignocellulosics and extensive research has been performed in the evaluation of pretreatments.[36] Ideally, a pretreatment should preserve the hemicellulose and lignin fractions, limit inhibitor formation, minimize the energy input, be cost-effective, warrant the recovery of high value-added co-products and minimize the production of toxic waste.[37] There is a consensus regarding the need for a pretreatment to remove and/or modify the matrix of lignin and hemicellulose surrounding the cellulose fraction, to enable efficient enzymatic saccharification of cellulose.[37] Pretreatment should also reduce cellulose crystallinity, and increase porosity and the accessible surface area of particles to facilitate processability of the cellulose-enriched residue. Different pretreatment methods have been suggested[36] including physical, such as milling and irradiation (γ-ray, electron beam, microwave), physico–chemical (steam explosion, ammonia fiber explosion, CO_2 explosion, SO_2 explosion), hydrothermal (autohydrolysis), chemical (alkali, dilute acid, gas, oxidizing agents, organic solvents, ionic liquids), biological (white-rot fungi), and electrical methods or combinations of these.

Pretreatment technology must be selected in accordance with the specific requirements of the LC biomass used and, at present, there is no single feasible method optimal for pretreatment of all types of lignocellulosics.[38] Pretreatment may result in compounds known for their inhibitory effect on the steps that follow (enzymatic hydrolysis and fermentation). Production of these compounds ranging from carbohydrate constituents (*e.g.*, acetic acid) and their degradation products (*e.g.*, furfural and 5-hydroxymethyl furfural) to solubilized extractives (*e.g.*, phenolic compounds, terpenes, sterols) and lignin degradation products (phenolic compounds) must be avoided or diminished; a careful optimization of pretreatment to reduce the production of inhibitors increases the efficiency and yields of hydrolysis/fermentation. At present, different methods are under research and development to improve efficiency and lower the cost.

Overall, a successful pretreatment should be scalable to industrial size, minimizing the use of energy, chemicals, and capital investment, minimizing the loss of sugars and the production of chemicals toxic to the enzymes or fermenting micro-organisms, and maximizing the enzymatic convertibility and the production of valuable byproducts such as lignin.[37] Certain aspects of selected pretreatment processes that are of importance to the transformation of sugar cane bagasse and straw into chemicals will be described in more detail.

For example, the selective separation of hemicelluloses can be performed using various pretreatments. Each pretreatment confers particular characteristics to the obtained phases, with the main purpose of increasing the susceptibility of the solid phase to enzymatic hydrolysis, and in some cases to make available the monomeric sugars of the hemicelluloses. Among these treatments, acid hydrolysis stands out when compared with non-catalytic pretreatment because it allows for the generation of a liquid phase (hemicellulose hydrolysate) rich in xylose and with minor amounts of lignin derivatives that can inhibit cell metabolism relative to the levels observed with alkaline and organic pretreatment. Additionally, xylose, the main hemicellulose-derived pentose, and hexose may be degraded during acid hydrolysis into furfural and 5-(hydroxymethyl) furfural (HMF), respectively, which are also reported to be inhibitors of cell metabolism.[26] The complexity and heterogeneity found in the lignocellulosic biomass of different species makes it is advisable to optimize a pretreatment for each feedstock.

7.3.2.1 Effects of Feedstock Composition on Pretreatment Methods

A recent study by Moutta *et al.*[38] illustrates the effect of LC feedstock composition on the outcome of pretreatments. The generation of monosaccharides (C5 and C6) from sugarcane biomass was studied *via* processing bagasse or straw and mixtures of both materials in the following proportions: bagasse–straw 3 : 1, 1 : 1 and 1 : 3. Samples were pretreated with sulfuric acid which resulted in approximately 90% of hemicellulose solubilization, corresponding to mostly xylose. Pretreated straw showed greater susceptibility to enzymatic hydrolysis in comparison to bagasse, as shown by glucose yields of 76% and 65%, respectively, whereas the mixtures showed intermediate yields. Thus, one strategy to balance sugar cane biomass availability and possibly increase 2G ethanol production would be to use bagasse–straw mixtures in appropriate ratios according to market fluctuations.

The composition of different biomass affects the efficiency of processing, influencing thus the choice of pretreatments required to maximize the recovery of sugars. For example, in an extensive investigation of these effects, Lima *et al.*[37] pretreated six LC feedstocks such as sugar cane, grasses and bark residues, under acid, alkaline, sulfite and hot water conditions over a range of temperatures. Average cellulose, hemicellulose and lignin content was determined by several different methods. Hot water pretreatment showed a similar effect on the chemical composition of the different biomass, removing mainly the hemicellulose fraction. The lignin content remained fairly constant (varying between 27% and 23%), while the average cellulose content increased from around 40% to 60% as the temperature increased to 130 °C. This enrichment in cellulose is a direct consequence of the removal of hemicellulose. However, at 180 °C, the cellulose content was lower, possibly due to the production of degrading compounds such as furaldehydes, rather than a reduction in hemicellulose removal. On average, pretreatment at 180 °C resulted in a reduction in the hemicellulose fraction from approximately 25% (untreated feedstock) to 13% (pretreatment at 180 °C). The acid pretreatment was highly efficient for hemicellulose removal, and an increase in temperature (up to 130 °C) had a further positive effect when compared to hot water treatment. However, at 180 °C, the degrading hemicellulose product, furfural (2-furfuraldehyde), was detected for all three grasses. At the highest temperature (180 °C), higher cellulose losses were also observed, and the average cellulose content decreased to around 60% after acid pretreatment at 180 °C, compared to 70% at 130 °C. However, even with the increase in temperature, acid pretreatment was not sufficient for lignin removal. At the highest temperature applied in the study (180 °C), approximately 20% of remaining dry matter was lignin. The highest cellulose enrichment was observed in samples subjected to the alkaline pretreatment using sodium hydroxide, which removed higher quantities of both lignin and hemicellulose fractions. The average lignin content for all feedstock was reduced from around 27% to 9% at 180 °C. However, at this temperature, some cellulose losses were observed, particularly in sugar cane bagasse. The chemical composition of biomass submitted to treatment with sodium bisulfite at increasing temperatures was observed to be similar to hot water pretreatment. In all feedstock, an increase in cellulose enrichment was observed up to 130 °C, reaching around 60%. At 180 °C, a slight decrease in cellulose content was observed for sugar cane bagasse while cellulose fractions from grasses remained constant. The discrete cellulose enrichment observed after sulfite pretreatment is associated with a low removal of both hemicellulose and lignin. A clear increase in the crystalline portion of the cellulosic fraction up to 130 °C was observed for all species and all pretreatments used. At 180 °C, however, some losses in the crystalline fraction could be observed, particularly after hot water and acid pretreatments for the grasses. A clear correlation between pretreatment conditions and the amorphous cellulose fraction was observed. However, considering the glucose content in the soluble fraction from pretreatment, it is possible that at lower temperatures this fraction was mainly removed, while at higher temperatures there was also a degree of biomass amorphization. Hemicellulose fractions were analyzed after pretreatment to evaluate the changes in monosaccharide composition. Sugar cane bagasse showed that the hemicellulose fraction was composed mainly of xylose, arabinose and glucose, followed by lower amounts of galactose and fucose.[37]

Solid state NMR spectra of the solid fractions indicated that, among the considered pretreatments, sulfuric acid was the most effective in the removal of hemicellulose but sodium hydroxide was most efficient in the removal of hemicellulose together with a reduction in lignin content in both grasses and eucalyptus bark biomasses. However, the pretreatment temperature was also an important parameter and the use of higher temperatures promoted the removal of amorphous cellulose. In this sense, the results point to the intrinsic advantages of grass samples, which require lower pretreatment temperatures than eucalyptus barks.[37]

To evaluate the generation of inhibitors and potential valuable products in the soluble phase of the protocol, a profile of compounds moved by the pretreatment solution was determined. The monosaccharide composition of the soluble fraction from hot water, sulfuric acid, sodium hydroxide and sodium bisulfite pretreatments at increasing temperatures, ranging from 50 °C to 180 °C, was studied. The potential formation of 2-furaldehyde and 5-hydroxymethylfuraldehyde as a result of sulfuric acid pretreatment was also investigated. For pretreatments conducted at 50 °C, glucose was the main monosaccharide in the soluble fraction from most of the biomass types and was detected together with xylose and other hemicellulose sugars. It can be related to an easier solubilization of glucose from hemicellulose, as well as the removal of the amorphous cellulose fraction. This enrichment in glucose was particularly evident in hot water, acid and sulfite pretreatments. In the soluble fraction from sodium hydroxide pretreatment, the xylose amount was higher than glucose for all grasses even at 50 °C, while for the bark samples the opposite was observed. This difference is associated with the efficient removal of the hemicellulose fraction by alkaline pretreatment, even at lower temperature, and the different composition of hemicelluloses in eucalyptus bark, which has a lower content of xylans. With increasing temperatures, a gradual increase in xylose, arabinose, galactose and other monosaccharides was also observed for all pretreatments, indicating an efficient removal of the hemicellulose fraction. However, acid and alkaline pretreatments indicated a higher content of monosaccharides in the soluble fraction for all types of biomass.[37]

At a higher temperature (180 °C), a decrease in glucose content for all samples became evident, in spite of a xylose increase, most notably with the acid pretreatment. The decrease in glucose observed at higher temperatures can be explained by the formation of inhibitors. The highest 5-hydroxymethylfurfural content was found for all samples pretreated at 180 °C using sulfuric acid. However, lower amounts could be observed at 90 °C or higher. Acidic conditions led to a rapid decay of glucose into 5-hydroxymethylfurfural by dehydration. Sugarcane bagasse and bark were more susceptible to cellulose dehydration, whereas the grasses showed low levels of hydroxymethylfurfural. C5 conversion into 2-furaldehyde was found mainly in the soluble fraction from the grasses.[37]

To evaluate the effect of pretreatments on the morphology of different biomass to improve enzymatic digestibility, scanning electron microscopy was used for investigation of the morphological changes produced by sodium hydroxide pretreatment at 130 °C. This pretreatment resulted in significant lignin and hemicellulose removal and, consequently, in a higher cellulose enrichment, without the production of high levels of the inhibitor. The effects of different pretreatments on sugar cane bagasse, compared to other raw materials were also investigated. A sample obtained from hot water pretreatment showed a surface similar to that obtained for raw bagasse, where there was a continuous covering layer (possibly formed by lignin and hemicellulose). After acid pretreatment, cellulose bundles were more evident, with less cohesion between them. This can be associated with the high level of hemicellulose removal, thereby enabling enzyme access to the cellulose fiber. A continuous layer over the cellulose bundles' surface was also observed after sodium bisulfite treatment, but in this case some parts of the bundles were already evident. Furthermore, it was possible to observe some residues over the surface, which could be associated with lignin modification and precipitation. Among the pretreatments described, the largest morphological changes were produced by sodium hydroxide, reflecting the removal of the covering layer, mainly lignin and a consequent loss of biomass structure, with separation of fiber bundles. Lignin precipitation was also observed on the surface of fibers. At higher magnification, the presence of microfibrillated cellulose on the surface of the samples could be observed.[37]

Cellulose nanofillers are mainly native cellulose, extracted by traditional bleaching treatments of lignocellulosic fibers. However, the extraction conditions (time, temperature, chemical concentration) are fundamental to the efficient extraction of cellulose nanoparticles with the required characteristics. These microfibrillated celluloses were not observed for sugar cane bagasse in a previous pretreatment condition using the same 1% sodium hydroxide concentration when preceded by a sulfuric acid pretreatment step at 120 °C and a residence time for the alkaline step of 1 h, however, the residence time was 40 min and treatment temperature was 130 °C, without the acid step.[37]

The effects of sodium hydroxide pretreatment at 130 °C were also studied in the other samples. Scanning electron microscopy images of the grasses also revealed longer and isolated fibers of crystalline cellulose, compared to those of sugar cane bagasse. This indicates their potential for the generation of natural fillers after efficient enzymatic hydrolysis, when the crystalline cellulose can be easily accessed and cleaved by cellulases.[37]

Saccharification screening was performed to verify the effect of pretreatments on the saccharification potential of different biomasses. Results of this analysis indicated that sulfuric acid and sodium hydroxide greatly improved the sugar release from sugarcane bagasse and the grasses, whilst for the eucalyptus bark samples, sodium hydroxide pretreatment was significantly better. This differential effect could be related to the different hemicelluloses and different composition of lignin in eucalyptus bark. The amounts of sugar released by the feedstock pretreated with hot water and sodium bisulfite were very similar to that of the control, and for all biomasses there was only a discrete effect of increased temperature. However, the increase of pretreatment temperature significantly affected the enzymatic digestibility of sugar cane bagasse and grasses submitted to acid and sodium hydroxide. A gradual increase of sugar release was observed up to 130 °C, followed by a decrease at 180 °C.[37]

7.3.3 Recalcitrance of Lignocellulosic Materials

The complexity of biomass has led to multidisciplinary efforts of both an analytical and theoretical nature dedicated to understanding LC recalcitrance at the molecular level. High resolution microscopic analysis has shown that small lignin aggregates formed within the bulk dilute-acid-pretreated biomass, coalesce into larger droplets, and redistribute to the cell wall surfaces. Their analysis may reveal details of the architecture of pretreated biomass.[34]

The three major components of plant biomass—cellulose, hemicelluloses and lignin—form highly organized entities in plant cell walls. Recalcitrance arises mainly from the amorphous cell-wall matrix containing lignin and hemicellulose assembled into a complex supramolecular network that coats the cellulose fibrils. The non-cellulosic components of plant biomass, lignin and hemicelluloses, naturally assemble into a supramolecular structure that protects cellulose microfibrils in secondary plant cell walls and is considered a major reason for biomass recalcitrance. Lignin consists of phenylpropanoid units bonded together to form a complex 3D supramolecular network. Together with hemicelluloses, that are amorphous heteropolysaccharides, lignin fills the voids between cellulose microfibrils in the secondary plant cell wall and creates a highly resistant barrier that impairs cellulose digestibility by reducing its accessibility to carbohydrate modifying enzymes.[34]

The overall performance of biomass saccharification may be attributed to the synergistic action of many complementary enzymes—including a variety of cellulases, hemicellulases, and accessory enzymes—which makes it difficult to study one factor at a time.[39] Traditional solution methods have suffered from the classical ensemble average limitation presented by analysis of these mixtures of complex biomass, therefore, the data gathered are sometimes inconclusive and, in part, contradictory. To overcome these problems, Ding *et al.*[39] visualized the action of these enzyme systems on untreated and delignified plant cell walls under controlled digestion conditions in real time with the use of a multimodal microscopy suite. They examined two naturally existing enzyme systems of commercial relevance for saccharification of LC biomass: (i) the secretome of the anaerobe *Clostridium thermocellum*, which is representative of multi-enzyme bacterial cellulosomes and (ii) a commercially available blended enzyme mixture derived from the fungus *Trichoderma reesei* (*Hypocrea jecorina*), which is representative of the fungal or free cellulases. The authors used tagged carbohydrate-binding modules to identify exposed cellulose surfaces and green dye labeled enzymes to examine overall cell wall accessibility to cellulases and verified that certain cellulases specifically recognize the planar face of crystalline cellulose and play a critical role in the hydrolysis of crystalline cellulose. They found that treatment effectively removes lignins, thereby exposing microfibrils to enzyme access and results in near-complete digestion of all cell walls. Enhanced digestibility by fungal cellulases due to their penetration into the pore structure of microfibril networks was also observed. In contrast, the larger cellulosome complexes could only penetrate the larger wall lamella gaps, resulting in fragmentation of walls. The advantageous degradation properties exhibited by the fungal cellulases on cell walls may be compromised when digesting purified forms of crystalline cellulose, in which the porous architecture of the native cellulose microfibril network has been completely destroyed during its preparation process.[39]

Despite the different mechanisms of fungal cellulases and cellulosomes, cell wall materials are completely digestible when lignins are effectively removed. Thermochemical pretreatment strategies to enhance biomass digestibility by partial removal or re-distribution of lignin result in sugar degradation and loss at high severities[40] and the challenge now is to effectively and economically modify lignins *via* strategies that maintain the integrity of fermentable sugars. Among these strategies, researchers have recently focused on genetically engineering plants for desirable lignin contents or compositions that are more amenable to classical pretreatment at low severity.[41]

To better understand how lignins may be efficiently modified, spatial distributions of lignin and cellulose in model samples have been assessed in 3D to investigate structural aspects underlying their interactions. Silveira *et al.*[34] employed a statistical–mechanical approach to reveal the supramolecular interactions in this network and provide molecular-level insight into the effective lignin–lignin and lignin–hemicellulose thermodynamic interactions. They found that such interactions are hydrophobic and entropy-driven, and arise from the expelling of water from the mutual interaction surfaces. The molecular origin of these interactions is carbohydrate–π and π–π stacking forces, their strength depending on the lignin chemical composition. Methoxy substituents on the phenyl groups of lignin promote substantial entropic stabilization of the ligno-hemicellulosic matrix. Although these results provide a detailed molecular view of the fundamental interactions within the secondary plant cell walls that lead to recalcitrance, the underlying molecular mechanisms by which lignin chemical composition affects cell wall recalcitrance remain elusive.[34]

7.3.4 Enzymatic Treatment of LC Biomass

As has been pointed out, sugar cane bagasse may serve as an excellent raw material for second generation (2G) ethanol production since it contains a high amount of carbohydrates such as glucose and xylose.[42] It is equally important to consider the hemicellulosic fraction along with the cellulosic part of the cell wall since it is rich in pentose residues, mainly xylose, which are not fermented by native *S. cerevisiae*. For the production of hemicellulosic ethanol, *Scheffersomyces shehatae* (Syn *Candida shehatae*) has been considered a promising microorganism which provides high ethanol productivities. However, a balanced nutrient supplementation is required for optimal growth for the production of ethanol with desired yields and productivities. Ethanol production from sugar cane bagasse hemicellulosic hydrolysate as the main carbon source was evaluated by Antunes *et al.*[42]

using the yeast *Shehatae UFMG-HM 52.2* in four different fermentation media formulations. Three of these fermentation media showed similar ethanol yield and productivity while a fourth medium demonstrated lower ethanol production compared to the others. The authors note that the medium with a simple composition showed good ethanol yields and productivity.

The relative concentrations of glucose and xylose in lignocellulosic hydrolysates will depend on the type of biomass, pretreatment and hydrolysis technology that is used. Gonçalves *et al.*[43] investigated the role of the major *Saccharomyces cerevisiae* genes on xylose uptake, which is believed to be one of the rate-limiting steps of xylose-to-ethanol fermentation. Anaerobic xylose or xylose plus glucose co-fermentations by recombinant yeast strains with overexpression of individual genes revealed that none of these transporters has the ideal properties for all possible industrial processes. The transporter encoded by the gene that was investigated is considered a low affinity (but high capacity) sugar permease. A strain overexpressing this permease as a unique hexose transporter led to the highest consumption of sugars and ethanol production rates during xylose plus glucose co-fermentation, but did not allow significant xylose consumption when this sugar was the only carbon source. A strain overexpressing the high affinity transporter allowed efficient xylose consumption and fermentation, but, during xylose plus glucose co-fermentation, it showed a clear preference for glucose with slow and delayed xylose consumption. While the data indicate that both sugars compete for the permease, some mutant permeases show reduced inhibition of glucose transport in the presence of xylose. This structural information can be used to engineer more suitable permeases for the fermentation of sugars present in lignocellulosic hydrolysates, an opportunity only recently being addressed. This work indicates that approaches to engineer certain transporters to increase their affinity towards pentoses or to avoid their sugar-induced degradation are promising strategies to improve second generation bioethanol production by xylose-fermenting yeasts. Furthermore, approaches to engineer *S. cerevisiae* transporters to improve second generation bioethanol production need to also consider the composition of the biomass sugar syrup whereby the transporter seems more suitable for hydrolysates containing xylose/glucose blends whereas other permeases would be a better choice for xylose-enriched sugar streams such as those obtained from acid stream explosion and/or hydrothermal biomass pretreatment processes.[43]

A recent report indicates that sugar cane recalcitrance varies significantly with internode regions and cultivar type.[44] Sugar cane internodes can be divided diagonally into four fractions, of which the two innermost ones are the least recalcitrant pith and the moderately accessible pith–rind interface. These fractions differ in enzymatic hydrolysability due to structural differences. In general, cellulose hydrolysis in plants is hindered by its physical interaction with hemicellulose and lignin. Lignin is believed to be linked covalently to hemicelluloses through hydroxycinnamic acids, forming a compact matrix around the polysaccharides. Acetyl xylan esterase and three feruloyl esterases were evaluated for their potential to fragment the LC network in sugar cane and to indirectly increase the accessibility of cellulose.[44] Although this region represents only 24 to 26% of the sugar cane dry mass, it has interesting fractions suitable for direct (without pretreatment) enzymatic hydrolysis, giving moderate to high glucose yields. These fractions include most of the naturally occurring chemical linkages present in the lignocellulose, which is not the case with severely pretreated materials.[44] Hence, natural sugar cane fractions are attractive substrates for studying the role of accessory enzymes in experimental hydrolysis cocktails. They were characterized and evaluated for their overall chemical composition. From those clones, a hybrid with significantly lower lignin and significantly higher hemicellulose content was chosen and its susceptibility to hydrolysis was compared to that of a reference cultivar since understanding the fundamental structure/anatomy of plant tissues is key in predicting whether the enzymatic hydrolysis of plant tissues into monomeric sugars is viable.[44]

Hydrolysability of the two innermost (pith and pith–rind interface) fractions of two sugar cane clones with different lignin content was studied by supplementing cellulases with xylanases and esterases. Acetyl xylan esterase enhanced accessibility and hydrolysis of cellulose and xylan by cellulases and xylanases in all fractions. However, the effect of feruloyl esterases was less clear. The three feruloyl esterases had distinct product profiles on non-pretreated sugar cane substrate, indicating that sugar cane could function as a possible natural substrate for activity measurements. Of the three feruloyl esterases tested, only one type released *p*-coumaric acid, while the other types released ferulic acid from both pith and interface fractions. Ferulic acid release was higher from the less recalcitrant clone/fraction (pith), whereas more *p*-coumaric acid was released from the clone/fraction (interface) with higher lignin content. In addition, compositional analysis of the four fractions revealed that *p*-coumaroyl content correlated with lignin, while feruloyl content correlated with arabinose content, suggesting differences in the esterification patterns of these two hydroxycinnamic acids. In sugar cane, feruloyl groups may also be more likely to decorate xylan, while *p*-coumaroyl groups decorate lignin, as in other grass species. The hydrolysis data suggest that only one esterase was able to release *p*-coumaroyl groups associated with the lignin. Despite the extensive release of phenolic acids, feruloyl esterases only moderately promoted enzyme access to cellulose or xylan.[44]

7.3.5 Complementary Sources of LC Biomass

The development of second-generation biofuels requires a diverse set of feedstock that can be grown sustainably and processed cost effectively. In particular, many biofuel production plants operate seasonally and stand idle for several months of the year. This is an unsatisfactory situation that denotes an inefficient use of capital and provides only intermittent employment for workers. One way to avoid discontinuous biofuel production is to use a wider range of biomass sources that may be available during the current idle periods. The diversification of feedstock for LC derived fuels requires an innovative approach that extends beyond agricultural wastes. Perennial grasses have been proposed as important sources of biomass in Europe and the US, based on their low input and marginal land requirements. Biomass grasses could also make a substantial contribution to the Brazilian energy matrix, serving as an alternative to the sugar cane interseason, when there is no bagasse or straw production. There are a number of candidate grasses that are already established and characterized from an agronomical point of view. The tropical climate in Brazil supports the efficient growth of a range of grasses with high productivity and preliminary studies have shown promising average productivity yields (dry mass) for different perennial grass species as compared to sugar cane. The paper on novel sources of Brazilian biomass for biorenewables production,[37] commented on in Section 7.3.2.1, illustrates potential complementary sources of carbohydrates for bioethanol production. Feedstock such as grasses and bark residues from commercial wood products can be sustainably grown and applied to local production during the inter-season, when no sugar cane bagasse is produced.

7.4 Platform Chemicals from Lignocellulosic Materials

7.4.1 Furans

As has been pointed out throughout this chapter, the production of chemicals and fuels from lignocellulosic biomass requires effective pretreatment and hydrolysis of cellulose and hemicelluloses polymers of the biomass to the corresponding pentose and hexose sugar units. This step is followed by catalytic dehydration of sugars to obtain the corresponding furfural and HMF products.[45] The subsequent transformation of HMF into other value added chemicals, such as promising next generation polyester building block monomers 2,5-furandicarboxylic acid (FDCA), 2,5-bis(hydroxymethyl)furan (BHMF), and 2,5-bis(hydroxymethyl)tetrahydrofuran (BHMTF) and potential biofuel candidates 2,5-dimethylfuran (DMF), 5-ethoxymethylfurfural (EMF), ethyl levulinase (EL) and γ-valerolactone (gVL) (Figure 7.5), has also been explored by different researchers using HMF as a starting substrate or directly from biomass in a one-pot process.[45]

The dehydration of 5- and 6-carbon sugars is a well-known transformation for the preparation of furfural and hydroxymethylfurfural (HMF). These chemicals were omitted from the original DOE list because of market and conversion process considerations. Technology development has improved the dehydration of sugars to furans improving their potential as platform chemicals.[3]

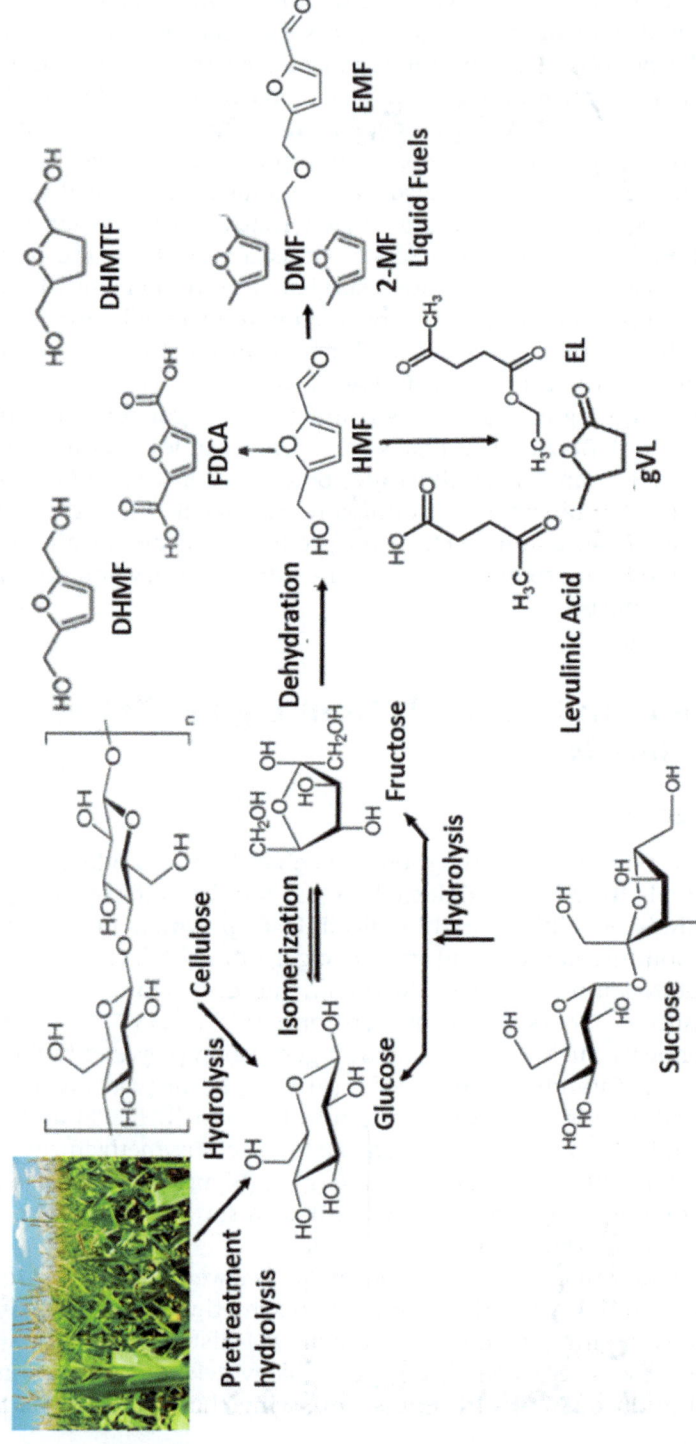

Figure 7.5   HMF production and its utilization routes for chemicals and liquid fuels (reproduced from ref. 45).

7.4.1.1 5-Hydroxymethylfurfural (5-HMF)

The hexose dehydration product, 5-hydroxymethylfurfural (HMF), is an important intermediate in the chemical transformation in biomass. This bifunctional, six-carbon molecule can be easily converted into a variety of useful derivatives, incorporated into a variety of polymers, or upgraded into fuels. Although HMF has long been synthesized in high yield from fructose, Binder *et al.*[46] and others demonstrated recently that HMF can be produced in high yield from glucose and in moderate yield from cellulose and lignocellulosic biomass using chromium salts. In fact, HMF and its derivatives levulinic acid, 5-bis(hydroxymethyl)furan (2,5-BHF), 2,5-diformylfuran (2,5-DFF) and 2,5-furandicarboxylic acid (2,5-FDCA) were identified early as very promising chemical intermediates obtained by the catalytic conversion of carbohydrates based on C_6 units. HMF was obtained by dehydration of fructose in the presence of soluble or solid acid catalysts or from glucose or even polysaccharides by more complex catalytic systems and reaction media. The key issue was to prepare 5-HMF with economically acceptable processes that could be scaled up to the industrial level. This is not yet achieved in spite of intensive research efforts conducted over the past few years. A recent review compares the different catalytic systems and solvents employed for the production of 5-HMF and examines their relevance to green chemistry.[47]

7.4.2 Organic Acids

Organic acids constitute a significant fraction of chemicals available from LC biorefineries' carbohydrate streams and thus have received considerable attention as platform chemicals.[2] The main applications of these organic acids are given in Table 7.1. Among these acids, lactic, succinic and levulinic acids are under consideration as promising platform chemicals for Brazilian companies.

7.4.3 Sugar Alcohols

Sugar alcohols are promising intermediates for the production of hydrocarbons as drop-in products. Among these, xylitol and sorbitol are specifically included in recent reviews on biobased products (Figure 7.6).[2,47] Xylitol can be produced by biological routes at low temperatures. These routes as well as applications of xylitol are discussed in ref. 2.

7.5 Prospects for Chemicals from Brazilian Sugar Cane

Research on sugar cane for the production of biofuels and chemicals has received much support, both at the federal and state levels. Brazilian researchers take part in several international studies related to this topic and the State of São Paulo, where most of the sugar cane is grown, has a very

Table 7.1 Applications of organic acids.[2]

Organic acid	Applications
Citric acid	70% of total production used in confectionary and beverage products, 30% in pharmaceuticals (anticoagulant blood preservative, antioxidant) and metal cleaning; selling price decreased with marked shift from pharmaceuticals to food applications (879 000 t produced in 2002).
Lactic acid	Acidulant, flavor enhancer, food preservative, feedstock for calcium stearoyl-2-lactylates (baking), ethyl lactate (biodegradable solvent) and polylactic acid plastics (100% biodegradable) for packaging, consumer goods, biopolymers (approved by FDA); estimated U.S. consumptions of 30 million lb with 6% growth pa; potential demand of 5.5 billion lb as a very large volume-commodity chemical.
Itaconic acid	Feedstock for syntheses of polymers for use in carpet backing; paper coating N-substituted pyrrolidinones for use in detergents and shampoos; cements comprising copolymers of acrylic and itaconic acid.
Aspartic acid	For synthesis of aspartame; monomer for manufacture of polyesters and polyamides; polyaspartic acid as substitute for EDTA with potential market of $450 million per year.
Fumaric acid	For manufacture of synthetic resins, biodegradable polymers; intermediate in chemical and biological synthesis.
Malic acid	Acidulant in food products; citric acid replacement; raw material for manufacture of biodegradable polymers; for treatment of hyperammonemia and liver dysfunction; component for amino acid infusions.
Succinic acid	Used as acidulant, pH modifier, flavoring and antimicrobial agent, ion chelator in electroplating to prevent metal corrosion, surfactant, detergent, foaming agent; for production of antibiotics, amino acids and pharmaceuticals; 270 000 ton in 2004; U.S. domestic market estimated at $1.3 billion per year with 6–10% annual growth.
Levulinic acid	For synthesis of methyl tetrahydrofuran (gasoline extender), diphenolic acid (for epoxy resins), tetrahydrofuran (solvent), 1,4-butanediol (polymer intermediate), succinic acid (specialty chemical), delta-aminolevulinic acid (active chemical in herbicides and pesticides), sodium levulinate (antifreeze ingredient), ethyl levulinate (diesel oxygenate).

powerful and active agency for the promotion of R&D. The Foundation for the Support of Research of the State of São Paulo, better known by its abbreviation FAPESP, established the Bioenergy Research Program (BIOEN) in 2008 to bring together public and private R&D efforts to advance and apply knowledge in fields related to ethanol production. It includes five divisions, among them are Biomass Research, focused on sugar cane and including plant improvement and farming; Ethanol Industrial Technologies; Biorefineries Technologies and Alcohol Chemistry; and Research on Sustainability and Impacts. The Program also supports academic exploratory research and training of scientists and technical staff in technologies related to ethanol. It also promotes partnerships with industry for cooperative R&D

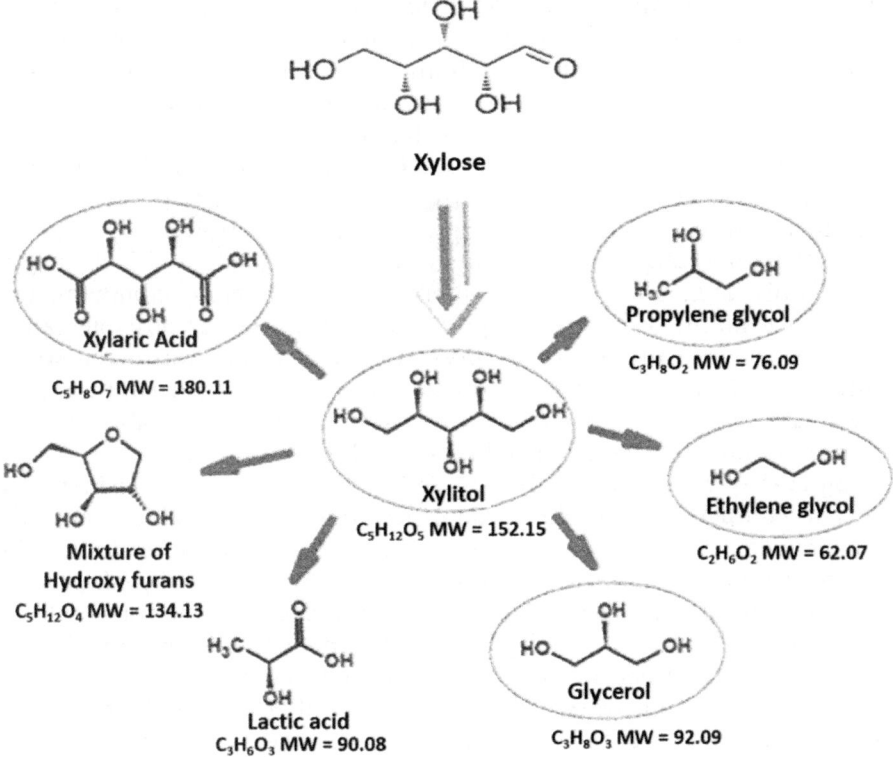

Figure 7.6   Xylitol as a building block chemical (reproduced from ref. 2).

activities. It has a specific program on new products from sugar cane that includes development of biotechnological processes and scale-up and integration of these processes.

More recently, in September 2012, the National Bank for Economic and Social Development (BNDES) in collaboration with the Federal Finance Agency for Innovation (FINEP) established the Program for the Support of Industrial Technological Innovation in the Sucroenergetic and Sucrochemical Sector (PAISS),[48] which supports projects on development, production and commercialization of new industrial technologies on processing biomass from sugar cane. At the beginning of 2013, federal support had reached approximately one hundred million dollars, almost 80% of which was invested in genetic improvement of sugar cane and most of the rest in the development of machinery and equipment. Recently, BNDES also invested about ten million dollars on the conversion of sugar cane biomass into 2G ethanol.

Last year BNDES promoted a study on the diversification of the chemical industry and specifically analyzed the potential for production of chemicals from renewable raw materials. It concluded that the most promising substrates in the carbohydrate chains are mono- and disaccharides that can be readily fermented. Among the first generation products are those that are already obtained mainly from sugar fermentation such as ethanol, isoamyl alcohol and amino acids such as lysine, threonine and tryptophan. Obtaining second generation sugars that are competitive compared to those of a first generation still represents a technological challenge. On the other hand, there are strong incentives for obtaining large volumes of fermentable sugars that do not compete with those produced as part of the food chain and there may be synergies in integrating the use of residues from sugar production to obtain renewable chemicals.[49] Compounds such as acrylic acid, adipic acid, butadiene, 1,4-butanodiol , isobutanol, *n*-butanol, isoprene, methionine and propyleneglycol were identified as possible candidates for the substitution of traditional routes by innovative renewable routes. World markets for each compound and its main derivatives were investigated and the prospects for their substitution by renewable raw materials were qualitatively evaluated based on known processes.

Opinions of specialists in the respective fields were also considered in order to complete a scenario of the expectations of renewable chemicals' competitiveness. For other selected compounds, such as levulinic acid, succinic acid and farnesene, the prospects for attaining market-share were based on the expectations of developing these compounds as building blocks and on the time that is estimated for them and their derivatives as well as their new applications to be commercially confirmed.[49]

Another study, that is a part of the "Plano Brasil Maior" (or "Bigger Brazil Plan,") conducted in 2013 by the Brazilian government, mentioned that the chemical sector based on renewable raw materials is in the process of being structured. Its competition dynamics depend on innovation in an environment of a high degree of uncertainty. Brazilian comparative advantages are significant but they depend on technological and industrial initiatives that would assure the country a relevant position in a future chemical industry. This study is a basis for the establishment of government policies and financial support.

7.6   A Strategy for Biobased Products

A considerable investment has been made in R&D on sugar cane and its use as a source of fuels and chemicals. However, in order to establish a comprehensive strategy for the production of biobased chemicals, there are other considerations that must be taken into account, for example aspects regarding how this area is related to trends such as fusions and acquisitions in the sector, formation of partnerships, venture capital startups, technology transfer, *etc.*, as well as how the regulatory aspects of this environment may be influencing a company's strategic position in the face of the new challenges related to the use of renewable raw materials. This aspect must include the legislation involving questions relative to carbon dioxide emissions, as well as the specific legislation related to the use of biopolymers and bioproducts as, for example BioPreferred (as defined by the USDA), compostability, biodegradability, *etc.* The answer to these questions may serve as an input for proposals of national policies that promote competitiveness at the national level and provide incentives to firms that face the challenges of the new bioindustry that is taking shape.

Finally, there are also important considerations related to industrial property rights. The development of efficient bioprocesses depends on selection of microorganisms that are specifically tailored for certain functions. Particularly in the case of biotechnologically important yeasts that can ferment xylose and other sugars present in biomass hydrolysates, collections from certain ecosystems are protected by laws that regulate the use of genetic information. This is a serious barrier for engineering certain traits that are important for industrial strains.

Acknowledgements

We thank the Graduate Program in Technology of Chemical and Biochemical Processes for support of the Brazilian Green Chemistry School. N. Pereira Jr., L. Appel, J. A. R. Rodrigues, and V. F. Ferreira made useful suggestions and pointed out material that should be included. S. L. Meirelles collected and organized a large part of this information, F. G. de Andrade provided a list of researchers working on relevant topics, P. R. S. has a research fellowship from the CNPq.

References

1. M. M. Bomgardner, *Chem. Eng. News*, 2014, **92**, 10–14.
2. L. P. Christopher, *Integrated Forest Biorefineries*, ed. L. P. Christopher, The Royal Society of Chemistry, RSC Publishing, Cambridge, 2013, ch. 1, pp. 1–66.
3. J. J. Bozell and G. R. Petersen, *Green Chem.*, 2010, **12**, 539–554.
4. T. Werpy and G. Petersen, *Top Value Added Chemicals from Biomass Volume I — Results of Screening for Potential Candidates from Sugars and Synthesis Gas Top Value Added Chemicals From Biomass Volume I : Results of Screening for Potential Candidates*, US DOE, Oak Ridge, 2004.
5. J. A. R. Rodrigues, *Quim. Nova*, 2011, **34**, 1242–1254.
6. O. V. Perrone and A. Industria, *Petroquimica no Brasil*, IBP, Rio de Janeiro, 2010, pp. 139–142 (in Portuguese).
7. E. S. Lora and R. V. Andrade, *Renewable Sustainable Energy Rev.*, 2009, **13**, 777–778.
8. F. S. S. Cavalcante, E. Freire and P. R. Seidl, *Rev. Quim. Ind.*, 2014, **744**, 28–34.
9. V. F. Ferreira, D. R. da Rocha and F. C. da Silva, *Quim. Nova*, 2009, **32**, 623–638.
10. K. Buchholz, J. Anders, R. Buczys, E. Lampe, M. Walter and E. Yaacoub, *Carbohydr. Res.*, 2006, **341**, 322–331.
11. M. Boscolo, *Quim. Nova*, 2003, **26**, 906–912.

12. F. S. S. Cavalcante, MSc Dissertation, Federal University of Rio de Janeiro, 2011 (in Portuguese).
13. https://amyris.com/pt.br, Dec. 11, 2014.
14. J. M. R. Gallo, J. M. C. Bueno and U. Schuchardt, *J. Braz. Chem. Soc.*, 2014, **25**, 2229–2243.
15. Z. S. B. Sousa, D. V. Cesar, C. A. Henriques and V. T. da Silva, *Catal. Today*, 2014, **234**, 182–191.
16. D. L. Carvalho, R. R. de Avillez, M. T. Rodrigues, L. E. P. Borges and L. G. Appel, *Appl. Catal., A*, 2012, **415-416**, 96–100.
17. C. R. K. Rabelo, L. G. Appel, A. B. Gaspar, S. Letichevsky and P. C. Zonetti, WO2013053032 (2013), to Petróleo Brasileiro S.A.
18. C. P. Rodrigues, P. C. Zonetti, C. G. Silva, A. B. Gaspar and L. G. Appel, *Appl. Catal., A*, 2013, **458**, 111–118.
19. P. C. Zonetti, J. Celnik, S. Letichevsky, A. B. Gaspar and L. G. Appel, *J. Mol. Catal. A: Chem.*, 2011, **334**, 29–34.
20. I. C. Freitas, S. Damyanova, D. C. Oliveira, C. M. P. Marques and J. M. C. Bueno, *J. Mol. Catal. A: Chem.*, 2004, **281**, 26–37.
21. P. Havlík, U. A. Schneider, E. Schmid, H. Böttcher, S. Fritz, R. Skalský, K. Aoki, S. De Cara, G. Kindermann, F. Kraxner, S. Leduc, I. McCallum, A. Mosnier, T. Sauer and M. Obersteiner, *Energy Policy*, 2011, **39**, 5690–5702.
22. E. Lichtfouse, M. Navarrete, P. Debaeke, S. Véronique and C. Alberola, *Agron. Sustainable Dev.*, 2009, **29**, 1–6.
23. C. B. Granda, L. Zhu and M. T. Holtzapple, *Environ. Prog.*, 2007, **26**, 233–250.
24. E. F. S. Aguiar, L. G. Appel, P. C. Zonetti, A. C. Fraga, A. A. Bicudo and I. Fonseca, *Catal. Today*, 2014, **234**, 13–23.
25. S. V. Junior, *Strategies for the Use of Biomass in Renewable Chemistry*, ISSN 2177-4439, Embrapa Agroenergia, Brasilia, 2012.
26. G. J. V. Betancur and N. Pereira Jr, *Electron. J. Biotechnol.*, 2010, **13**(5), 6–9.
27. S. I. Mussato and I. C. Roberto, *Biotecnolog. Cienc.Desenvolv.*, 2002, **5**, 35–39.
28. C. A. Cardona, J. A. Quintero and I. C. Paz, *Bioresour. Technol.*, 2010, **101**, 4754.
29. J. Chang, *Renewable Resources for Biorefineries*, ed. C. S. K. Lin and R. Luque, The Royal Society of Chemistry, RSC Publishing, Cambridge, 2014, pp. 146–175.
30. A. J. Ragauskas, G. T. Beckham, M. J. Biddy, R. Chandra, F. Chen, M. F. Davis, B. H. Davison, R. A. Dixon, P. Gilna, M. Keller, P. Langan, A. K. Naskar, J. N. Saddler, T. J. Tschaplinski, G. A. Tuskan and C. E. Wyman, *Science*, 2014, **344**, DOI: 10.1126/science.1246843.
31. C. O. Tuck, E. Pérez, I. T. Horváth, R. A. Sheldon and M. Poliakoff, *Science*, 2012, **37**, 695–699.
32. Y. Sun and J. Cheng, *Bioresour. Technol.*, 2002, **83**, 1–11.
33. C. Ververis, K. Georghiou, D. Danielidis, D. G. Hatzinikolaou, P. Santas, R. Santas and V. Corleti, *Bioresour. Technol.*, 2007, **98**, 296–301.

34. R. L. Silveira, S. R. Stoyanov, S. Gusarov, M. Skaf and A. Kovalenko, *J. Phys. Chem. Lett.*, 2015, **6**, 206–211.
35. A. García, M. G. Alriols and J. Labidi, *Ind. Crops Prod.*, 2014, **53**, 102–110.
36. T. E. Amidon, B. Bujanovic, S. Liu, A. Hasan and J. R. Howard, *Integrated Forest Biorefineries*, ed. P. Christopher, The Royal Society of Chemistry, RSC Publishing, Cambridge, 2013, pp. 151–179.
37. M. A. Lima, L. D. Gomez, C. G. Steele-King, R. Simister, O. D. Bernardinelli, M. A. Carvalho, C. A. Rezende, C. A. Labate, E. R. de Azevedo, S. J. McQueen-Mason and I. Polikarpov, *Biotechnol. Biofuels*, 2014, **7**, 10–29.
38. R. O. Moutta, V. S. Ferreira Leitão and E. P. S. Bon, *Biocatal. Biotransform.*, 2014, **32**, 93–100.
39. S.-Y. Ding, Y.-S. Liu, Y. Zeng, M. E. Himmel, J. O. Baker and E. A. Bayer, *Science*, 2012, **338**, 1055–1060.
40. B. Kumar and C. B. Wyman, *Biotechnol. Bioeng.*, 2009, **102**, 1544–1557.
41. C. X. Fu, J. R. Mielenz, X. Xiao, Y. Ge, C. Y. Hamilton, M. Rodriguez Jr., F. Chen, M. Foston, A. Ragauskas, J. Bouton, R. A. Dixon and Z.-Y. Wang, *Proc. Natl. Acad. Sci. U. S. A.*, 2011, **108**, 3803–3808.
42. F. A. F. Antunes, A. K. Chandel, T. S. S. Milessi, J. C. Santos, C. A. Rosa and S. S. da Silva, *Int. J. Chem. Eng.*, 2014, **2014**, 180681, 8 pages.
43. D. L. Gonçalves, A. Matsushika, B. B. Sales, T. Goshima, E. P. Bon and B. Stambuk, *Enzyme Microb. Technol.*, 2014, **63**, 13–20.
44. A. Várnai, T. H. F. Costa, C. B. Faulds, A. M. F. Milagres, M. Siika-aho and A. Ferraz, *Biotechnol. Biofuels*, 2014, **7**, 153.
45. B. Saha and M. M. Abu-Omar, *Green Chem.*, 2014, **16**, 24–38.
46. J. C. Binder, A. V. Cefali, J. J. Blank and R. T. Rines, *Energy Environ. Sci.*, 2010, 765–771.
47. P. Gallezot, *Chem. Soc. Rev.*, 2012, **41**, 1538–1558.
48. www.finep.gov.br/pagina . asp?pag = programas_paiss, 11 dec 2014.
49. Bain & Company and Gas Energy, *Potencial de diversificação da indústria química brasileira, Relatório 4 – Químicos com base em fontes renováveis*, Nov. 2014. (in Portuguese) www.bndes.gov.br/siteBNDES/e3xport/sites/default/bndes_pt/galerias/arquivos/produtos/download/aep_fep/chamada-publica_FEPpro_pec0311_químicos_cana_de_açucar.pdf.

Subject Index

Page references to **figures, tables and reaction sequences** are shown in **bold**.